Lecture Notes in Computer Science   14289

Founding Editors

Gerhard Goos
Juris Hartmanis

Editorial Board Members

The series Lecture Notes in Computer Science (LNCS), including its subseries Lecture Notes in Artificial Intelligence (LNAI) and Lecture Notes in Bioinformatics (LNBI), has established itself as a medium for the publication of new developments in computer science and information technology research, teaching, and education.

LNCS enjoys close cooperation with the computer science R & D community, the series counts many renowned academics among its volume editors and paper authors, and collaborates with prestigious societies. Its mission is to serve this international community by providing an invaluable service, mainly focused on the publication of conference and workshop proceedings and postproceedings. LNCS commenced publication in 1973.

Oscar Pedreira · Vladimir Estivill-Castro
Editors

Similarity Search and Applications

16th International Conference, SISAP 2023
A Coruña, Spain, October 9–11, 2023
Proceedings

Editors
Oscar Pedreira ⓘ
University of A Coruña
Coruña, Spain

Vladimir Estivill-Castro ⓘ
Pompeu Fabra University
Barcelona, Spain

ISSN 0302-9743 ISSN 1611-3349 (electronic)
Lecture Notes in Computer Science
ISBN 978-3-031-46993-0 ISBN 978-3-031-46994-7 (eBook)
https://doi.org/10.1007/978-3-031-46994-7

This Springer imprint is published by the registered company Springer Nature Switzerland AG
The registered company address is: Gewerbestrasse 11, 6330 Cham, Switzerland

Paper in this product is recyclable.

Preface

This volume contains the papers presented at the 16th International Conference on Similarity Search and Applications (SISAP 2023), held in A Coruña from October 9–11, 2023. The *Universidade da Coruña* hosted the SISAP 2023 conference, continuing the tradition of offering an annual forum for researchers and application developers in the area of similarity data management.

More extensive and more diverse collections of information are assembled digitally, and only some allow formulation of a retrieval query with exact information. Moreover, one naturally also wants to explore nearby, and thus similar, items. Simultaneously, many methods in machine learning use massive data sets and require efficient data management during the training process. We are always in search of even faster, more effective approaches to similarity search. SISAP continues to be the international meeting point for scholars and practitioners whose research aims at continuous improvement and whose applications pose challenges to the state of the art. While there is vast literature and technological progress, the increased size of datasets, the new modes of data and the emergent applications continue to motivate the community for further advances.

Therefore, this edition of the conference invited participation in a technical challenge. This new activity represents a milestone in the evolution of the technology and the dispersion across the community of best methods and practices. It offers a benchmark and a point of reflection for reproducible results out of a common open data set. These proceedings include the description of the challenge by its organisers as well as the reports of participating teams. The program had a special session held on the first day of the conference for the presentation by the challenge organisers and the exposition of their approach and result by each team.

The conference also brought three invited keynote speakers, creating a wide view of theoretical, practical and applied insights by remarkable speakers. Each day of the conference started with a keynote presentation, and an abstract is included in this volume for the lectures delivered by Michael Houle, Yury Malkov and Julio Gonzalo.

For this edition of the SISAP conference, we received 33 submissions from authors from 20 different countries. The Program Committee (PC) was composed of 50 members from 17 countries. Each submission received three single-blind reviews, and the chairs and PC members thoroughly discussed the papers and reviews. Based on the reviews and discussions, the PC chairs accepted 20 papers, 16 of which were accepted as full papers and four as short/demo papers, resulting in an acceptance rate of 48% for the full articles and a cumulative acceptance rate of 60% for full and short essays. SISAP 2023 also called for papers for the Doctoral Symposium. However, all submissions in this edition were considered too preliminary for publication. Nevertheless, those PhD students were given the opportunity to present orally and receive feedback from discussion with the Doctoral Symposium chair and committee.

Springer has published the proceedings of SISAP 2023 in this volume in the Lecture Notes in Computer Science (LNCS) series. We also acknowledge their support and generous contribution to the Best Student Paper Award. We also awarded, as is traditional, a Best Paper Award (as judged by the PC chairs and the Steering Committee). Moreover, the authors of selected excellent papers (based on reviews and presentations) were invited to submit detailed, updated and extended versions for publication in a special issue of the Information Systems (Elsevier) journal.

We are extremely thankful to all members of the Program Committee for their comprehensive and timely reviews, which ensured the rigorous scientific and academic standard of the papers, as well as constructive suggestions that have assisted authors in polishing their final version and improving even further the quality of these proceedings.

October 2023   Oscar Pedreira
                                                                Vladimir Estivill-Castro

Organization

General Chair

Oscar Pedreira   University of A Coruña, Spain

Program Committee Chairs

Vladimir Estivill-Castro   Universitat Pompeu Fabra, Spain
Oscar Pedreira   University of A Coruña, Spain

Steering Committee

Giuseppe Amato   ISTI-CNR, Italy
Edgar Chávez   CICESE, Mexico
Stéphane Marchand-Maillet   University of Geneva, Switzerland
Pavel Zezula   Masaryk University, Czech Republic

Doctoral Symposium Chair

Amalia Duch Brown   Polytechnic University of Catalonia, Spain

Publicity Chairs

Conrado Martínez   Polytechnic University of Catalonia, Spain
Cole Foster   Brown University, USA

Doctoral Symposium Program Committee

Giuseppe Amato	ISTI-CNR, Italy
Richard Connor	University of St Andrews, UK
Eric S. Tellez	CICESE-INFOTEC-CONACYT, Mexico

Main Track Program Committee

Giuseppe Amato	ISTI-CNR, Italy
Laurent Amsaleg	CNRS-IRISA, France
Fabrizio Angiulli	University of Calabria, Italy
Diego Arroyuelo	Universidad Técnica Federico Santa María, Chile
Ilaria Bartolini	University of Bologna, Italy
Michal Batko	Masaryk University, Czech Republic
Panagiotis Bouros	Johannes Gutenberg University Mainz, Germany
Benjamin Bustos	University of Chile, Chile
K. Selcuk Candan	Arizona State University, USA
Edgar Chavez	CICESE, Mexico
Mario Giovanni C. A. Cimino	University of Pisa, Italy
Richard Connor	University of St Andrews, UK
Alan Dearle	University of St Andrews, UK
Vlastislav Dohnal	Masaryk University, Czech Republic
Amalia Duch Brown	Polytechnic University of Catalonia, Spain
Vladimir Estivill-Castro	Universitat Pompeu Fabra, Spain
Rolf Fagerberg	University of Southern Denmark, Denmark
Fabrizio Falchi	ISTI-CNR, Italy
Karina Figueroa	Universidad Michoacana de San Nicolás Hidalgo, Mexico
Qiang Gao	Southwestern University of Finance and Economics, China
Yunjun Gao	Zhejiang University, China
Claudio Gennaro	ISTI-CNR, Italy
Thi Thao Nguyen Ho	Aalborg University, Denmark
Akihiro Inokuchi	Kwansei Gakuin University, Japan
Björn Þór Jónsson	Reykjavik University, Iceland
Benjamin Kimia	Brown University, USA
Peer Kröger	Christian-Albrechts-Universität zu Kiel, Germany
Stéphane Marchand-Maillet	University of Geneva, Switzerland
Conrado Martínez	Polytechnic University of Catalonia, Spain
Vladimir Mic	Masaryk University, Czech Republic
Lia Morra	Politecnico di Torino, Italy

Vincent Oria	New Jersey Institute of Technology, USA
Marco Patella	University of Bologna, Italy
Oscar Pedreira	Universidade da Coruna, Spain
Miloš Radovanović	University of Novi Sad, Serbia
Marcela Ribeiro	Federal University of São Carlos, Brazil
Eric S. Tellez	CONACyT - INFOTEC, Mexico
Kunihiko Sadakane	University of Tokyo, Japan
Erich Schubert	TU Dortmund University, Germany
Jan Sedmidubsky	Masaryk University, Czech Republic
Tetsuo Shibuya	University of Tokyo, Japan
Yasin Silva	Loyola University Chicago, USA
Tomas Skopal	Charles University, Czech Republic
Caetano Traina-Junior	USP, Brazil
Goce Trajcevski	Iowa State University, USA
Lucia Vadicamo	ISTI-CNR, Italy
Takashi Washio	Osaka University, Japan
Pavel Zezula	Masaryk University, Czech Republic
Kaiping Zheng	National University of Singapore, Singapore
Arthur Zimek	University of Southern Denmark, Denmark

Additional Reviewer

Þór Guðmundsson, Gylfi	Reykjavik University, Iceland

Keynotes

From Intrinsic Dimensionality to Chaos and Control: Towards a Unified Theoretical View

Michael Houle⬤

New Jersey Institute of Technology, Newark, NJ 07102, USA
michael.houle@njit.edu

Abstract. Researchers have long considered the analysis of similarity applications in terms of the intrinsic dimensionality (ID) of the data. Although traditionally ID has been viewed as a characterization of the complexity of discrete datasets, more recently a local model of intrinsic dimensionality (LID) has been extended to the case of smooth growth functions in general, and distance distributions in particular, from its first principles in terms of similarity, features, and probability. Since then, LID has found applications—practical as well as theoretical—in such areas as similarity search, data mining, and deep learning. LID has also been shown to be equivalent under transformation to the well-established statistical framework of extreme value theory (EVT). In this presentation, we will survey some of the wider connections between ID and other forms of complexity analysis, including EVT, power-law distributions, chaos theory, and control theory, and show how LID can serve as a unifying framework for the understanding of these theories. Finally, we will reinterpret recent empirical findings in the area of deep learning in light of these connections.

The Rise of HNSW: Understanding Key Factors Driving the Adoption of Search Libraries in Machine Learning

Yury Malkov ⓘ

yurymalkov@mail.ru

Abstract. As representation learning and large language models continue to evolve, the need for efficient similarity search techniques has grown exponentially in the last few years. HNSW has emerged as a leading algorithm for nearest neighbor search, finding applications in a diverse range of products such as Weavite, Qdrant, Vespa, Milvus, Zilliz, Faiss, Elasticsearch, Redis and others. In this talk, we will explore the core principles and development of HNSW, as well as the key design decisions and factors that have contributed to its widespread adoption beyond its high performance. Through these insights, we aim to guide developers in creating innovative libraries and solutions to address the ever-increasing demand for efficient search libraries and machine learning tools in general.

Towards a Universal Similarity Function: The Information Contrast Model and Its Application as Evaluation Metric in Artificial Intelligence Tasks

Julio Gonzalo 🆔

Universidad Nacional de Educación a Distancia
julio@lsi.uned.es

Abstract. Computing similarity implies, at least, two aspects: how to represent items, and how to compare item representations (similarity functions). Item representation is a task-dependent problem, but what about similarity functions? Is it possible to study the design of optimal similarity functions from a universal, application-free perspective? In the talk, we start by proposing a set of formal constraints on the space of permissible similarity functions for Information Access problems and comparing it with other related axiomatic formulations of similarity in other fields (cognitive science and algebra). Then, we propose a new parameterized similarity function, ICM, which satisfies all constraints for a given range of values of its parameters. We discuss the usefulness of ICM in two very different application domains: first, to compute textual similarity under different application scenarios and representation paradigms, which was the original task for which ICM was designed. But ICM can be successfully applied outside its intended original scope: in the talk, we show how it can be used as an evaluation measure in Artificial Intelligence that computes the similarity between system outputs and gold standards, and how it may bring formal and empirical advantages in this area.

Contents

Indexing Challenge

Research Track

Finding HSP Neighbors via an Exact, Hierarchical Approach

Cole Foster[1]([✉]), Edgar Chávez[2], and Benjamin Kimia[1]

[1] Brown University, Providence, RI 02912, USA
{cole_foster,benjamin_kimia}@brown.edu
[2] Centro de Investigación Científica y de Educación Superior de Ensenada,
Ensenada, Mexico
elchavez@cicese.edu.mx

Abstract. The Half Space Proximal (HSP) graph is a low out-degree monotonic graph with wide applications in various domains, including combinatorial optimization in strings, enhancing kNN classification, simplifying chemical networks, estimating local intrinsic dimensionality, and generating uniform samples from skewed distributions, among others. However, the linear complexity of finding HSP neighbors of a query limit its scalability, except when sacrificing accuracy by restricting the test to a small local neighborhood estimated through approximate indexing. This compromise leads to the loss of crucial long-range connections, introducing false positives and excluding false negatives, compromising the essential properties of the HSP. To overcome these limitations, we propose a fast and exact HSP Test showing sublinear complexity in extensive experimentation. Our hierarchical approach leverages pivots and the triangle inequality to enable efficient HSP search in general metric spaces. A key component of our approach is the concept of the *shifted generalized hyperplane* between two points, which allows for the invalidation of entire point groups. Our approach ensures the desired properties of the HSP Test with exactness even for datasets containing hundreds of millions of points.

Keywords: Similarity Search · Half-Space Proximal · Exact

1 Introduction

The antidote to complexity in search is organization. In similarity search in a metric space, the local topology of a dataset element allows for efficient query search. Traditional approaches to capturing this local topology in the form of a graph have often focused on pairwise distances, such as the kNN graph, where each element is connected to its k nearest neighbors. Proximity graphs such as the Relative Neighborhood Graph (RNG), the Gabriel Graph, *etc.*, use triplets of elements at a time to determine the occupancy of a neighborhood region, *e.g.*, a lune defined by two elements, by a third element. While kNN graphs and

O. Pedreira and V. Estivill-Castro (Eds.): SISAP 2023, LNCS 14289, pp. 3–18, 2023.
https://doi.org/10.1007/978-3-031-46994-7_1

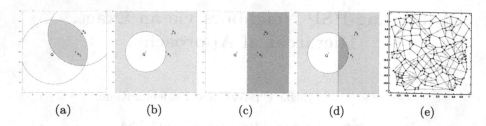

Fig. 1. (a) A link between Q and x_2 is disrupted by the presence of x_1 in the RNG lune between Q and x_2. The region of x_2 whose link with Q is invalidated is the intersection of the region corresponding to Eq. 1(a) shown in (b) and the region region corresponding to Eq. 1(b) shown in (c), with the intersection shown in (d). The RNG and HSP Graph on a 2D dataset of uniformly distributed points with the RNG links are shown in blue and the additional, directed HSP links are shown in red. (Color figure online)

other pairwise-distance graphs can unevenly distribute neighbors, clustered in one direction, the proximity graphs embed a sense of direction in their neighborhood definition, albeit requiring additional cost. Similarly, the Yao graph [20] and the Theta graph [4] construct a sense of direction by defining equal angle cones around each point through which links can be made between two points, but this is restricted to a Euclidean space.

The Half-Space Proximal (HSP) graph [6] aims to capture a sense of direction, but without the computational requirements of the RNG and without the Yao/Theta graphs' restrictions to a Euclidean space. Like an RNG, it connects two elements if the lune between them is empty, but unlike the RNG which checks occupancy by *all* other elements, HSP checks occupancy by those elements already established as HSP neighbors, Fig. 1(a).

This recursive definition requires a treatment of elements rank-ordered by distance to the query. Consider a dataset S of elements $\{x_1, x_2, \ldots, x_N\}$ and a query point Q. The only element whose lune with Q is unconditionally empty is the nearest neighbor, which is denoted as x_1. Then, the set of elements x_2 whose RNG lune between Q and x_2 is occupied with x_1 is the intersection of two regions,

$$\begin{cases} d(Q, x_1) < d(Q, x_2) & \text{(1a)} \\ d(x_1, x_2) < d(Q, x_2) & \text{(1b)} \end{cases}$$

with the first inequality defining the outside-disc yellow area in Fig. 1(b) and the second inequality defining the right half-space red area in Fig. 1(c), with their intersection denoted by the orange region in Fig. 1(d). This test is effective in discarding nearly half the space. The nearest neighbor among the surviving elements (later called the active list) is now also an undisputed HSP neighbor which in turn discards members in its own half-space, Fig. 2. The process is repeated until all members are considered. It should be clear that the HSP is a superset of the RNG, Fig. 1(e).

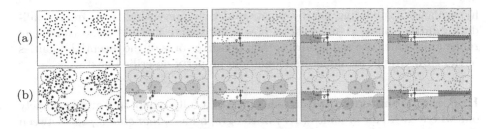

Fig. 2. (a) A visualization of finding the HSP neighbors of Q among the points shown on the left. Each of the nearest neighbors rules out a half-space (shaded area). (b) A visualization of the proposed Hierarchical HSP algorithm, where pivots are used to invalidate entire groups of points without having to calculate distances to query or testing inequalities (Color figure online)

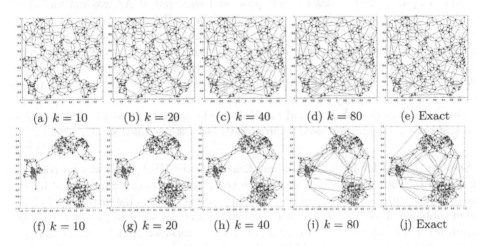

(a) $k = 10$ (b) $k = 20$ (c) $k = 40$ (d) $k = 80$ (e) Exact

(f) $k = 10$ (g) $k = 20$ (h) $k = 40$ (i) $k = 80$ (j) Exact

Fig. 3. A dataset of $N = 200$ 2D points is used to compare the exact HSP Graph to the approximate HSP Graph where the HSP neighbor candidates are restricted to the k closest points. The top and bottom rows correspond to uniformly distributed and clustered data, respectively (Color figure online)

Efficient Approaches to Computing the HSP: Despite the numerous benefits and applications of the HSP graph (see below), there has been surprisingly few attempts at reducing the $O(N)$ complexity in finding HSP neighbors of an element. All existing approaches restrict the HSP neighbors to a local area around Q: in the original paper [6], the HSP Test is performed on the Unit Disc Graph, thus constraining the test to a small radius around Q. More recently, an approximate kNN search (k = 300) by HNSW [11] was used to provide a similar constraint in the application of instance-based classification [19].

However, any method of restricting the HSP Test to some local area will inevitably result in an approximation. Consider Fig. 2, where the fourth HSP neighbor lies very far away from Q. These types of links, called *shooters*, are often those occupied-lune neighbors that are a result of near-parallel generalized hyperplanes invalidating nearby points and thus preventing them from invalidating further ones. Figure 3 shows how approximate methods miss these long range links which are significant as they preserve some of the essential qualities of the HSP Graph, including its monotonic property described next.

HSP is a Monotonic Graph: This idea of a fully monotonic graph, namely one where each pair of nodes (u, v) in the graph has a *monotonic path* between them, *i.e.*, a path $u = u_0, u_1, u_2, \ldots, u_{\ell+1} = v$ where $d(u_{i+1}, v) \leq d(u_i, v)$ for all i on the path, has been an elusive character in the history of similarity search. The notion emerged in 1985 when Dearholt [9] proposed the Monotonic Search Network (MSNET) as the ideal graph for Computer Vision databases to enable "search without backtracking", constructed by adding edges to the RNG Graph. In 2002, Navarro [17] proposed the Spatial Approximation Tree (SA-Tree) which constructs a monotonic path from the root node and to every point in the dataset. The conditions used to construct these monotonic paths in the SA-Tree are exactly the same as used for the HSP Test [6], which was introduced just a few years later. Chavez *et. al.* [18] later proved the HSP Graph to be a monotonic graph [18]. Although a monotonic graph does not guarantee exact search for a query outside of the dataset, the conditions used to provide monotonicity in the SA-Tree and the HSP are leveraged to provide diverse links in state-of-the-art graph-based approximate nearest neighbor search [10, 11].

Application of HSP: The initial application of the HSP graph was for routing between nodes in *ad-hoc networks* where the challenge is using only local information without central control. Another application is the challenging optimization problems in strings [15], where the HSP is used for selecting a central string from a set. Firstly the median string of a set is obtained and from the median string, which with high probability will not be in the set, the HSP neighbors are computed. The HSP test provides sufficient diversity within the members of the subset while at the same time fulfilling the centrality criterion.

The HSP neighbor finding has been used to enhance the majority-rule neighborhood classifiers where the kNN neighbors are replaced with the HSP neighbors, eliminating the need to set the parameter k. The candidate neighbors come from a probabilistic index such as the HNSW [11].

Another application of the HSP graph is in representing chemical networks [1, 2]. The typically used complete graph or an α-similarity graph are significantly reduced in size by retaining only HSP neighbors, reducing quadratic memory requirements to linear ones.

The HSP neighbors have also been used in local intrinsic dimensionality estimation [14]. The connection between the maximum degree of the HSP graph and the kissing or sphere packing number (the maximum number of mutually touching spheres in a Euclidean space) is used to define the indexability of a set.

In [13], the authors define the *hubness* HSP (HubHSP) graph. The geometric structure of the HSP neighbor definition remains intact, but instead of nearest neighbors in each step they use the node with the highest measure of "hubness", which they define. This tilts the process in favor of hubs, important to sampling skewed distributions, for example.

Exact Computation of HSP is Significant: Finding exact neighbors is key to local intrinsic dimension as well as local density estimation. Also, a standing conjecture about the HSP is about it being t-spanner of finite stretch t. This has a potential application as a distance oracle. Storing the entire distance matrix of a metric database implies $O(N^2)$ space of precomputed distances, while storing the HSP graph uses space proportional to $O(N \times \delta(HSP(S)))$, with $\delta(.)$ the degree of the graph. Using the monotonic property of the HSP the oracle can be consulted in time proportional to the diameter of the graph. For the above oracle to work, it is mandatory to compute the HSP exactly to obtain a proper bound.

Hierarchies in Exact Metric-Space Similarity Search: Brute force exact metric-space search is avoided by using indices which leverage *pivots*, select points in the dataset, and the triangle inequality to bound the distance between the query point and other members of the dataset, removing a large portion from consideration. For example, the List of Clusters (LOC) [7] organizes the dataset into an ordered list of pivots, where each pivot is responsible for a group of points within a radius of that pivot. The query can traverse the list, only considering the clusters that may contain the nearest neighbor. By the observation that an increase an N also increases the number of points in each cluster, the Recursive List of Clusters [12] brings greater efficiency by organizing each cluster into its own LOC. In fact, this recursive organization of the dataset into smaller and smaller groups is the basic concept of tree structures, which achieve logarithmic search complexity. Some metric-space tree structures of interest include the M-Tree [8], the Cover Tree [3], and several others [5,16,21].

Overview: The proposed approach constructs a multilayer hierarchy of coarse-scale pivots "owning" finer-scale pivots in their pivot domains defined by a radius. The hierarchy is then used to (i) find the nearest neighbors to a query without computing distance to each element by relying on conditions on pivots and (ii) use pivot conditions to discard entire domains from being HSP neighbors, preserve entire domains as active, or declare domains as indeterminant which require examination of its members. This results in significant savings and leads to scalable HSP computation on large-scale, high-dimensional datasets featuring clustered data. Our implementation is publicly available at https://github.com/cole-foster/HHSP.git.

2 Finding the Exact, Hierarchical HSP Neighbors (HHSP)

The main bottleneck in finding the HSP neighbors of a query is (i) the computation of the distance to *all* dataset members, and (ii) to a lesser extent, the validation of the inequalities in Eq. 1. The key intuition is to use a hierarchy where pivots representing a group of dataset points can be examined and used to either discard or retain from consideration the entire elements in the pivot domain, thus avoiding explicit computation of distances or validation of inequalities.

Specifically, recall that in the HSP algorithm the distance between Q and all dataset members are calculated to return x_1 as the closest point to Q. Since x_1 is the nearest neighbor of Q, it becomes the first HSP neighbor and is used to discard a half-space from being HSP neighbors of Q by checking Eq. 1 on the remaining dataset members, Fig. 2(a). Next, the closest point among the active list $\mathcal{A}$ which is the set of surviving dataset members becomes the second HSP neighbor and is used to check and discard the remainder of the active list that satisfy Eq. 1. This process is repeated until all members are labeled either as HSP neighbors or discarded, with overall complexity of $O(N)$.

A first key savings can be achieved through the computation of the nearest neighbor of Q by using pivots in a hierarchy. Consider, at first, a two-layer hierarchy where the "bottom" layer is the dataset and the "top" layer is a select group of elements of the dataset called pivots p_i such that each data point x_j is in the domain of one, and only one, p_i, namely, $d(x_j, p_i) \leq r$, where r is a fixed parameter of the hierarchy, Fig. 4(a). There are numerous ways to construct this hierarchy. The approach used here is to randomly consider the dataset members one by one and assign them to either belonging to an existing pivot or if no pivot can be found, assign it as a pivot. The parameter r determines the size of the pivot domain.

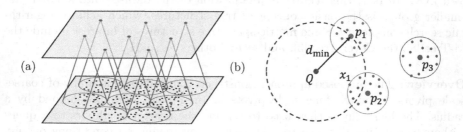

Fig. 4. (a) A two-layer hierarchy where the bottom layer contains all of $\mathcal{S}$ and the top layer has a select few pivots where the distance of dataset elements to its pivot parent is less than r. (b) Given an upper-bound distance $d_{\min}$, a pivot p may contain the nearest neighbor x_1 if $d(Q, p) \leq d_{\min} + r$ (Color figure online)

The pivot structure can be used to significantly reduce the computational effort in finding the nearest neighbor of Q. Let $d_{\min}$ denote the minimum distance of the query Q to the data elements already considered. Then, the distance from Q to any other element x in the pivot domain of p satisfies the triangle inequality

$$d(Q, p) - r \leq d(Q, p) - d(p, x) \leq d(Q, x) \leq d(Q, p) + d(p, x) \leq d(Q, p) + r. \quad (2)$$

Thus, if for a pivot p, $d(Q,p) > d_{min} + r$,

$$d_{min} < d(Q,p) - r \le d(Q,x), \tag{3}$$

and the pivot domain cannot contain the nearest element, see the blue pivot domain in Fig. 4(b). On the other hand, if $d(Q,p) \le d_{min} + r$, the green pivot domain in Fig. 4(b), the elements in the domain of p must be explicitly considered and if one has a lower distance than d_{min}, it updates d_{min}. This prevents the computation of distances to a vast majority of dataset elements. The process is repeated until all pivots have been considered in this way. A better performance can be achieved by using tighter bounds if each pivot would maintain the distance to its most distant member, r^*.

A second key savings can be achieved through wholesale examination of Eq. 1 for all members of a pivot domain *without* calculating distance to query or validation of inequalities by examining $d(Q,p)$. The following proposition prevents member-wise validation of the second inequality in Eq. 1 if the pivot satisfies certain conditions:

Proposition 1. *Let Q be a query, x_1 the furthest HSP neighbor of Q thus far, and p_2 a pivot with domain radius r satisfying the following:*

$$\begin{cases} d^2(Q,p_2) - d^2(x_1,p_2) > 2r\, d(Q,x_1) & (4a) \\ d(Q,p_2) \ge r. & (4b) \end{cases}$$

Then, all points $x_2 \in \mathcal{D}(p_2, r)$, i.e., where $d(x_2, p_2) \le r$, satisfy $d(x_1, x_2) < d(Q, x_2)$.

Proof. By the triangle inequality and $d(x_2, p_2) \le r$,

$$d^2(x_1, x_2) \le [d(x_1, p_2) + d(p_2, x_2)]^2 \le [d(x_1, p_2) + r]^2 = d^2(x_1, p_2) + 2r\, d(x_1, p_2) + r^2. \tag{5}$$

Applying the given Eq. 4,

$$\begin{aligned} d^2(x_1, p_2) + 2r\, d(x_1, p_2) + r^2 &< d^2(Q, p_2) - 2r\, d(Q, x_1) + 2r\, d(x_1, p_2) + r^2 \\ &= d^2(Q, p_2) - 2r\, [d(Q, x_1) - d(x_1, p_2)] + r^2 \\ &\le d^2(Q, p_2) - 2r\, d(Q, p_2) + r^2 \\ &= [d(Q, p_2) - r]^2 \\ &\le [d(Q, x_2) + d(x_2, p_2) - r]^2 \\ &\le [d(Q, x_2) + r - r]^2 \\ &= d^2(Q, x_2). \end{aligned} \tag{6}$$

∎

It is intriguing that the region where p_2 satisfies Eq. 4(a) is the half-space to the right of a shifted generalized hyperplane in a Euclidean space, because the quadratic terms involving coordinates of p_2 cancel out leaving a linear equation which represents a hyperplane. Returning now the first inequality in Eq. 1, the following proposition identifies the condition on a pivot so that the entire pivot domain can be discarded.

Proposition 2. *Let Q be a query and x_1 the furthest HSP neighbor of Q thus far, then a pivot p_2 with radius r satisfying both of the following inequalities*

$$\begin{cases} d^2(Q,p_2) - d^2(x_1,p_2) > 2r \ d(Q,x_1) & \text{(7a)} \\ d(Q,x_1) < d(Q,p_2) - r, & \text{(7b)} \end{cases}$$

invalidates all points $x_2 \in \mathcal{D}(p_2,r)$, i.e., $d(p_2,x_2) \le r$, as a HSP neighbors of Q.

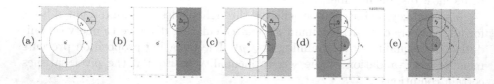

Fig. 5. (a) The point x_2 will satisfy Eq. 1(a) when its parent pivot p_2 is outside of the circle of radius $d(Q,x_1) + r$ centered at Q. (b) The point x_2 satisfies Eq. 1(b) when p_2 is to the right of the generalized hyperplane shifted by r. (c) The intersection of the two regions (orange) which defines locations for p_2 where its pivot domain members cannot be an HSP neighbor of Q. (d) The point x_2 does not satisfy both inequalities of Eq. 1 when p_2 falls into the purple region. (e) When p_2 falls into the blue region, it is undetermined if x_2 satisfies both inequalities of Eq. 1 (Color figure online)

Proof. First, let's show that $d(Q,x_1) < d(Q,x_2)$:

$$d(Q,x_1) < d(Q,p_2) - r \le d(Q,x_2) + d(x_2,p_2) - r \le d(Q,x_2) + r - r = d(Q,x_2). \tag{8}$$

Second, since $0 \le d(Q,x_1)$, Eq. 7(b) shows that $d(Q,p_2) > r$ which together with Eq. 7(a) satisfy Proposition 1 which states that $d(x_1,x_2) < d(Q,x_2)$, the second inequality of Eq. 1 holds. ∎

The regions corresponding to Eqs. 7(b) and Eqs. 7(a) are shown in Fig. 5(a) and 5(b), respectively, leading to their common intersection in Fig. 5(c).

In addition to determining which pivot domains are entirely ruled out, it is also possible to determine which pivot domains cannot get ruled out in their entirety because their members do not satisfy either the first or the second inequalities in Eq. 1 and can therefore remain on the active list in their entirety.

Proposition 3. *Let Q be a query, x_1 be the furthest HSP neighbor of Q thus far, and p_2 a pivot satisfying either of the following:*

$$\begin{cases} d^2(x_1,p_2) - d^2(Q,p_2) \ge 2r \ d(Q,x_1) \quad and \quad d(x_1,p_2) \ge r & \text{(9a)} \\ d(Q,p_2) \le d(Q,x_1) - r. & \text{(9b)} \end{cases}$$

Then, all points $x_2 \in \mathcal{D}(p_2,r)$, i.e., $d(p_2,x_2) \le r$, violate one of the inequalities of Eq. 1, i.e., either $d(Q,x_1) \ge d(Q,x_2)$ or $d(x_1,x_2) \ge d(Q,x_2)$.

Proof. First, Eq. 9(b) implies that

$$d(Q, x_2) \le d(Q, p_2) + d(p_2, x_2) \le d(Q, p_2) + r \le d(Q, x_1). \qquad (10)$$

Second, by the triangle inequality, Eq. 9(a), and $d(p_2, x_2) \le r, r \ge 0$,

$$
\begin{aligned}
d^2(Q, x_2) \le [d(Q, p_2) + d(p_2, x_2)]^2 &\le [d(Q, p_2) + r]^2 \\
&= d^2(Q, p_2) + 2r\ d(Q, p_2) + r^2 \\
&\le d^2(x_1, p_2) - 2r\ d(Q, x_1) + 2r\ d(Q, p_2) + r^2 \\
&= d^2(x_1, p_2) - 2r\ [d(x_1, Q) - d(Q, p_2)] + r^2 \\
&\le d^2(x_1, p_2) - 2r\ d(x_1, p_2) + r^2 \\
&= [d(x_1, p_2) - r]^2 \\
&\le [d(x_1, x_2) + d(x_2, p_2) - r]^2 \\
&\le d^2(x_1, x_2).
\end{aligned}
$$

$$(11)$$

∎

Figure 5(d) visualizes the regions defined by the inequalities in Eqs. 9(a) and 9(b), which is the union of the shifted half-space and a reduced radius disc. A pivot p_2 in the purple region is retained in the active list without detailed examination of its elements.

The pivots that are neither fully discarded (orange area) nor fully accepted as surviving in their entirety (purple area) can potentially contain elements which can be discarded and elements that survive (cyan area), Fig. 5(e). The elements in these pivot domains must be individually tested with the inequalities of Eq. 1. However, this determination can be delayed until the point where their elements need to be examined. It is entirely possible that this entire pivot domain would be discarded in the next steps.

The details of the procedure are in Algorithm 1. Basically, the hierarchy is used to efficiently find the nearest neighbor, Proposition 2 and 3 are used to discard entire pivot domains and retain an active pivot list $\mathcal{A}_1$ (purple area) of pivots, and an indeterminant list $\mathcal{I}$ (cyan area). The procedure is then repeated by finding the next nearest element by exploring pivots in $\mathcal{A}_1 \bigcup \mathcal{I}$. In the process, some of the pivots in $\mathcal{I}$ may have to be explicitly examined. The pivots are removed from $\mathcal{I}$, and added to an active point list $\mathcal{A}_2$ of elements. The process is repeated by finding the nearest element in $\mathcal{A}_1 \bigcup \mathcal{I} \bigcup \mathcal{A}_2$ until they are all exhausted.

Multi-layer Hierarchies: This two-layer hierarchical approach achieves efficiency by using pivots to avoid the consideration of a vast number of points. As N increases, the number of pivots and the number of points in each pivot domain must both increase. This motivates the use of additional layers, similar to other hierarchical indices [3,5,8,16,21]. Just as pivots are able to discard or

---

Algorithm 1: Hierarchical HSP Search on a 2-Layer Hierarchy

Input: Q as the query point, $\mathcal{P}$ as the set of top layer pivots in the 2-Layer Hierarchy constructed on $\mathcal{S}$ with top layer radius r.
Output: HSP(Q) as the exact HSP neighbors of Q in $\mathcal{S}$.

1 **begin**
2   Initialize HSP(Q) = $\emptyset$, $\mathcal{A}_1 = \mathcal{P}$, $\mathcal{I} = \emptyset$, $\mathcal{A}_2 = \emptyset$.
3
4   **while** $|\mathcal{A}_1| > 0$ *or* $|\mathcal{I}| > 0$ *or* $|\mathcal{A}_2| > 0$ **do**
5
           /* Finding the Next HSP Neighbor   */
6   Find x_i as the closest active point:
7   · Initialize d_{min} with the distance to the closest $p \in \mathcal{A}_1$ or $x \in \mathcal{A}_2$.
8   · Update d_{min} with $d(Q, p)$ for $p \in \mathcal{I}$ if p does not satisfy Eq. 1 for any $x_j \in$ HSP(Q).
9   · Iterate through $p \in \mathcal{A}_1$; search the domains of pivots where $d(Q, p) \leq d_{min} + r$, updating d_{min}.
10   · Iterate through $p \in \mathcal{I}$; if p satisfies $d(Q, p) \leq d_{min} + r$, then remove p from $\mathcal{I}$ and validate each member of the pivot domain against Eq. 1 for all $x_j \in$ HSP(Q). Those points that are retained are added to $\mathcal{A}_2$ and may update d_{min}.
11   · The closest active point x_i becomes the next HSP neighbor, x_i is added to HSP(Q).
12
           /* Validation of the Active Points   */
13   For each $p \in \mathcal{I}$:
14   1. If Prop. 2 satisfied for Q and x_i, remove p from $\mathcal{I}$.
15   2. Otherwise, continue.
16   For each $p \in \mathcal{A}_1$:
17   1. If Prop. 2 satisfied for Q and x_i, remove p from $\mathcal{A}_1$.
18   2. If Prop. 3 satisfied for Q and x_i, continue.
19   3. Otherwise, remove p from $\mathcal{A}_1$ and add p to $\mathcal{I}$.
20   For each $x \in \mathcal{A}_2$:
21   1. If Eq. 1 satisfied for Q and x_i, remove x from $\mathcal{A}_2$.
22   2. Otherwise, continue.
23   **end**
24 **end**

---

retain an entire pivot domain of elements, coarse-scale (large radius) pivots discard or retain pivot domains of finer-scale pivots, *e.g.*, Fig. 6. Additional layers can achieve enhanced efficiency when N is increased further, Fig. 7.

3   Experiments

Exact HSP Complexity on Uniformly Distributed Data: The HHSP's complexity is examined on uniformly distributed data of varying dimension, Fig. 7. Observe that the indexing construction time, number of distances for search, and search time all depict an approximately linear profile against N across dimension for varying numbers of layers in a log-log domain. A straight line in log-log is $\log(y) = \alpha \log(N) + \gamma$ or $y = \beta N^\alpha$. While theoretical complexity bounds have not yet been defined, the above experiments suggest an approach to characterizing complexity of the HHSP, Table 1.

Comparison to Brute Force: The traditional concern in using exact query search is the curse of dimensionality, where as the intrinsic dimension of the dataset grows the index becomes less effective, eventually being no more effective than brute-force search. In such arguments, the size of the dataset is kept

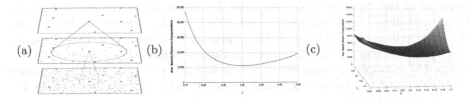

Fig. 6. (a) Depiction of a three-layer hierarchy. (b) The radius for a 2-layer hierarchy may be chosen to minimize the average number of distance computations on search. (c) Similarly, choosing radii for a three-layer hierarchy can be posed as a 2D optimization.

Table 1. The experimental complexity of the HHSP based on uniformly distributed data is captured by the value α for the experimental complexity $O(N^{\alpha})$ using a least-squares fit of the function $y = \beta N^{\alpha}$.

Complexity	2D	4D	6D	8D	10D
Index Construction Distances	1.060	1.113	1.189	1.534	1.542
Index Construction Time	1.131	1.150	1.221	1.605	1.687
Index Memory Usage	0.960	0.966	0.968	0.970	0.983
Search Distances	0.108	0.203	0.311	0.447	0.553
Search Time	0.979	0.968	0.917	0.840	0.810

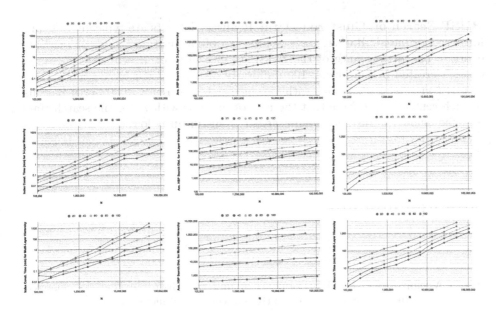

Fig. 7. Index construction time, number of distances, and search time for HHSP on 2D-10D data using 2, 3, and multilayer indices. Observe that these plots are approximately linear in the log-log domain

constant as the "volume" over which the dataset elements are spread becomes larger and thus density approaches zero. We posit that the key to determining the effectiveness of an index is to keep "density" constant. Our analysis shows that the index remains effective over a variety of dimensions and dataset sizes, Fig. 8. For example, in datasets of $N = 1,638,400$ points, the HHSP achieves time savings of 686.7x in 10D, enabling exact search on datasets of even 100 million points, Fig. 7.

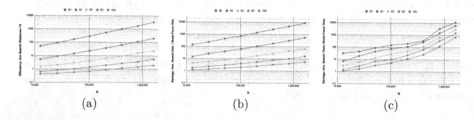

(a)   (b)   (c)

Fig. 8. The savings of the HHSP over the brute force HSP algorithm in ratios of (a) the average number of search distances per N, (b) the ratio of average distance computations in comparison over those required for the brute force HSP test, and (c) the ratio of the average search time and that required for brute force search. It is evident that the index remains effective and its efficiency increases with N

HHSP Search on Clustered Data: Realistic datasets are typically not uniformly distributed. Rather, data points are often clustered, *e.g.*, object categories. The performance of the HHSP is measured on a synthetic clustered dataset created by initializing 100 uniformly distributed clustered centers from $[-1, 1]^D$ and using Gaussians with variance=0.05 to create 1,000 perturbations of each center, Fig. 9. Note that the rate of increase with D for the search time and the number of distance computations is significantly lower.

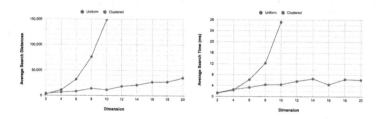

Fig. 9. A comparison of HHSP performance on uniformly distributed data vs. clustered data for $N = 100,000$ shows a significantly reduced rate of increase.

Real World Datasets: The performance of HHSP on real world datasets, Table 2. (1) *LA*, a 2D dataset of the geographic locations in Los Angeles with $N = 1,073,827$ points; (2) *Forest*, a 6D dataset of quantitative variables used for classifying forest cover types; and (3) *Corel*, a $N = 67,840$ collection of 32D color histograms of images. The HHSP significantly outperforms the brute force HSP Test in all cases, leading to 5,218x, 2,598x, and 8.64x savings over brute force time for LA, Forest, and Corel, respectively.

Table 2. Comparing the HHSP to the original HSP Test on Real-World Datasets.

Dataset	N	Ave. HSP Neigh	BF Time (ms)	BF Distances	Index Const. Time (s)	HHSP Time (ms)	HHSP Distances	Ratio of Times
LA (2D)	1,073,727	3.32	55,315.248	2,757,279.67	2.744	10.627	1,124.15	5,205.16
Forest (6D)	580,812	5.34	10,913.616	1,630,106.17	4.498	4.207	3,194.70	2,594.16
Corel (32D)	67,840	9.71	67.669	192,751.53	7.161	7.832	64,420.57	8.64

Comparison to Approximate HSP Search: The only existing, scalable approach at reducing the complexity of the HSP algorithm relies on performing an approximate kNN search to retrieve a large neighborhood around the query, and then apply the HSP Test on that neighborhood [19]. This approach leverages a state-of-the-art graph-based approximate search index, the Hierarchical Navigable Small World (HNSW) Graph [11]. This approach to HSP Search involves an inherent trade-off between search time and accuracy, requiring a large neighborhood to obtain exactness at the cost of longer search time, Fig. 10.

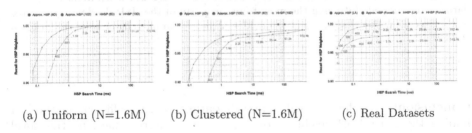

(a) Uniform (N=1.6M)   (b) Clustered (N=1.6M)   (c) Real Datasets

Fig. 10. Comparing the approximate HSP Search by HNSW to the exact HSP search by the HHSP Test. The point labels on the curves correspond to the size of the neighborhood returned by HNSW for the approximate HSP.

The results show that, while there is no advantage to the HHSP on uniformly distributed data, the approximate method saturates performance while HHSP is able to reach 100% recall at a reasonable time. A similar result is shown for the high-dimensional Forest data, but not for the low-dimensional LA data where there is negligible difference.

Improving Graph-Based Nearest Neighbor Search: Since the HNSW achieves state-of-the-art results as an approximate monotonic graph, there remains a question as to whether its performance may be improved by being a fully, exact monotonic graph. To showcase the impact of the exact HSP neighbors, we replace the HNSW links on each layer with the exact HSP links. Figure 11 shows the comparison between the HNSW and the HNSW with HSP links on a 6D uniform and clustered datasets with $N = 1,000,000$ points. The construction of the full HSP Graphs by the HHSP took just over 10 h, while it would take the brute force approach an estimated 200 d!

First, as guaranteed by the monotonic property of the HSP, Fig. 11(a,b) shows the HSP Graph ensure exact search for any member of the dataset, which is not the case for the original HNSW, especially in clustered data. Secondly, Fig. 11(c,d) shows that the HSP links provide a slight, yet modest improvement over the original HNSW links for queries that are not members of the dataset.

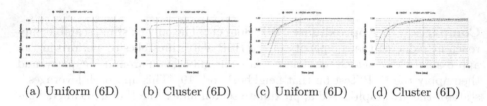

(a) Uniform (6D)   (b) Cluster (6D)   (c) Uniform (6D)   (d) Cluster (6D)

Fig. 11. Comparing the HNSW to HNSW with links are replaced with the exact HSP links. (a,b) Recall for the nearest neighbor when dataset items are queried. The monotonic property of the HSP graph guarantees a perfect recall, but this is not always the case for the original HNSW. (c,d) Recall for the nearest neighbor when items not in the dataset are queried. Although the monotonic property of the HSP does not guarantee perfect recall in this case, we see it provides a modest improvement over the original HNSW.

Conclusion: This hierarchical approach outlines a fast, efficient method of finding the exact HSP neighbors of a query in a metric space: by the novel definition of the shifted generalized hyperplanes between two points, pivots are able to discard or retain entire groups of points as consideration for being an HSP neighbors. While approximate methods to the HSP are able to achieve good recall with fast search times, they miss the vital, long-range links essential to the monotonic property of the HSP. By constructing the exact HSP Graph on a dataset of one-million points, which is a feat in itself, we show that the monotonic property can improve the performance of graph-based approximate nearest neighbor search.

Acknowledgement. We gratefully acknowledge the support of NIH Award 1S10OD025181 and NSF award 1910530.

References

1. Aguilera-Mendoza, L., et al.: Automatic construction of molecular similarity networks for visual graph mining in chemical space of bioactive peptides: an unsupervised learning approach. Sci. Rep. **10**(1), 18074 (2020)
2. Ayala-Ruano, S., et al.: Network science and group fusion similarity-based searching to explore the chemical space of antiparasitic peptides. ACS Omega **7**(50), 46012–46036 (2022)
3. Beygelzimer, A., Kakade, S., Langford, J.: Cover trees for nearest neighbor. In: Proceedings of the 23rd ICML, pp. 97–104 (2006)
4. Bose, P., Morin, P., van Renssen, A., Verdonschot, S.: The $\theta5$-graph is a spanner. Comput. Geom. **48**(2), 108–119 (2015)
5. Brin, S.: Near neighbor search in large metric spaces. In: VLDB, vol. 95, pp. 574–584. Citeseer (1995)
6. Chavez, E., et al.: Half-space proximal: a new local test for extracting a bounded dilation spanner of a unit disk graph. In: Anderson, J.H., Prencipe, G., Wattenhofer, R. (eds.) OPODIS 2005. LNCS, vol. 3974, pp. 235–245. Springer, Heidelberg (2006). https://doi.org/10.1007/11795490_19
7. Chávez, E., Navarro, G.: A compact space decomposition for effective metric indexing. Pattern Recogn. Lett. **26**(9), 1363–1376 (2005)
8. Ciaccia, P., Patella, M., Zezula, P.: M-tree: an efficient access method for similarity search in metric spaces. In: VLDB, vol. 97, pp. 426–435 (1997)
9. Dearholt, D.W., Schvaneveldt, R.W., Durso, F.T.: Properties of Networks Derived from Proximities. New Mexico State University, Computing Research Lab. (1985)
10. Fu, C., Xiang, C., Wang, C., Cai, D.: Fast approximate nearest neighbor search with the navigating spreading-out graph. Proc. VLDB Endowment **12**(5), 461–474 (2019)
11. Malkov, Y.A., Yashunin, D.A.: Efficient and robust approximate nearest neighbor search using hierarchical navigable small world graphs. IEEE Trans. Pattern Anal. Mach. Intell. **42**(4), 824–836 (2018)
12. Mamede, M.: Recursive lists of clusters: a dynamic data structure for range queries in metric spaces. In: Yolum, I., Güngör, T., Gürgen, F., Özturan, C. (eds.) ISCIS 2005. LNCS, vol. 3733, pp. 843–853. Springer, Heidelberg (2005). https://doi.org/10.1007/11569596_86
13. Marchand-Maillet, S., Chávez, E.: HubHSP Graph: effective data sampling for pivot-based representation strategies. In: Skopal, T., Falchi, F., Lokoč, J., Sapino, M.L., Bartolini, I., Patella, M. (eds.) Similarity Search and Applications: 15th International Conference, SISAP 2022, Bologna, Italy, October 5–7, 2022, Proceedings, pp. 164–177. Springer, Cham (2022). https://doi.org/10.1007/978-3-031-17849-8_13
14. Marchand-Maillet, S., Pedreira, O., Chávez, E.: Structural intrinsic dimensionality. In: Reyes, N., et al. (eds.) SISAP 2021. LNCS, vol. 13058, pp. 173–185. Springer, Cham (2021). https://doi.org/10.1007/978-3-030-89657-7_14
15. Mirabal, P., Abreu, J., Seco, D., Pedreira, Ó., Chávez, E.: Boosting perturbation-based iterative algorithms to compute the median string. IEEE Access **9**, 169299–169308 (2021)
16. Moore, A., Gray, A., et al.: New algorithms for efficient high dimensional nonparametric classification. In: Advances in NIPS 16 (2003)
17. Navarro, G.: Searching in metric spaces by spatial approximation. VLDB J. **11**(1), 28–46 (2002)

18. Ruiz, G., Chávez, E.: Proximal navigation graphs and t-spanners. arXiv preprint arXiv:1404.1646 (2014)
19. Talamantes, A., Chavez, E.: Instance-based learning using the half-space proximal graph. Pattern Recogn. Lett. **156**, 88–95 (2022)
20. Yao, A.C.C.: On constructing minimum spanning trees in k-dimensional spaces and related problems. SIAM J. Comput. **11**(4), 721–736 (1982)
21. Yianilos, P.N.: Data structures and algorithms for nearest neighbor. In: Proceedings of the ACM-SIAM Symposium on Discrete Algorithms, vol. 66, p. 311 (1993)

Approximate Similarity Search for Time Series Data Enhanced by Section Min-Hash

Ryota Tomoda$^{(\boxtimes)}$ and Hisashi Koga

University of Electro-Communications, Tokyo 182-8585, Japan
{tomoda,koga}@sd.is.uec.ac.jp

Abstract. Dynamic Time Warping (DTW) is a well-known similarity measure between time series data. Although DTW can calculate the similarity between time series with different lengths, it is computationally expensive. Therefore, fast algorithms that approximate the DTW have been desired. SSH (Sketch, Shingle & Hash) is a representative hash-based approximation algorithm. It extracts a set of quantized subsequences from a given time series and finds similar time series by means of Min-Hash, a hash-based set similarity search. However, Min-Hash does not care about the location of set elements (i.e., quantized subsequences) in the time series, so that hash collisions have a rather weak correlation with DTW. In this paper, to strengthen the correlation between hash collisions and DTW, we propose a new method termed Section Min-Hash that also allows position shifts required by DTW. After quantizing subsequences in a time series based on Euclidean distance, Section Min-Hash explicitly specifies multiple sections within the time series and generates the hash values from all the sections.

Keywords: Time series · Similarity search · Dynamic Time Warping

1 Introduction

In recent years, time series data have been utilized in various fields. As a result, similarity search for time series data has attracted much attention. As for the similarity measure, the standard Euclidean distance cannot compare two time series of different lengths. By contrast, Dynamic Time Warping (DTW) [1] can measure the similarity between two time series Q and X, even if they have different lengths.

However, DTW suffers from its huge time complexity of $O(|Q||X|)$, where $|Q|$ presents the length of Q. Thus, efficient time-series similarity search based on DTW has been an important challenge: First, Kim *et al.* [2] and Keogh *et al.* [3] proposed some branch-and-bound techniques. Then, Rakthanmanon *et al.* [4] devised a UCR suite algorithm that combines various known branch-and-bound techniques. Choi *et al.* [5] devised another pruning-based acceleration method. Although these pruning-based solutions exactly find the most similar time series

© The Author(s), under exclusive license to Springer Nature Switzerland AG 2023
O. Pedreira and V. Estivill-Castro (Eds.): SISAP 2023, LNCS 14289, pp. 19–32, 2023.
https://doi.org/10.1007/978-3-031-46994-7_2

whose DTW distance to the query becomes the smallest, they are known to work poorly for long time series, which slows down the execution speed.

Therefore, fast algorithms that approximate the DTW has been desired. Locality-Sensitive Hashing (LSH) is a known framework to realize fast approximate similarity search. However, only a few practical LSHs currently exist for time series such as [6,7]. Thus, this research theme still has much room to evolve. Specifically, Luo *et al.* [8] proposed a method called Sketch, Shingle & Hash (SSH) that approximates time series similarity search based on DTW. SSH relies on hashing and surely makes the execution time shorter than the exact UCR suite algorithm [4]. SSH extracts a set of quantized subsequences from a given time series and then searches similar time series by means of hash-based set similarity search called Min-Hash [9].

However, Min-Hash does not care about the location of set elements (i.e., quantized subsequences) in the time series, so that hash collisions have a rather weak correlation with DTW. In this paper, to strengthen the correlation between hash collisions and DTW, we propose a new method named Section Min-Hash (SMH) that also allows position shifts required by DTW. SMH selects multiple sections within the time series and generates the hash values from all the sections. In this way, SMH can recognize for which part of the time series a hash value is computed.

2   DTW

DTW is a distance between two time series. Let $X = (x_1, x_2, ..., x_m)$ and $Y = (y_1, y_2, ..., y_n)$ be two time series of length m and n. Here, x_i is an arithmetic value and corresponds to the i-th element of X. The DTW distance between X and Y is defined as the minimum cost when aligning X and Y under three constraints that are explained later. Unlike the Euclidean distance, DTW allows matching between time series with positional shifts. An alignment that satisfies the three constraints is referred to as a warping path $P(X,Y) = ((i_1, j_1), (i_2, j_2), ..., (i_K, j_K))$ which is a collection of index pairs. K is the length of $P(X,Y)$. Here, for $1 \leq k \leq K$, i_k and j_k are the indices of time series data X and Y. The warping path must satisfy the following three conditions:

Boundary Condition: The first and last points of X and Y must be aligned to each other. In other words, $P_1 = (1,1)$, and $P_K = (m,n)$.

Monotonicity Condition: The alignment between points should be monotonically non-decreasing. That is, $i_{k-1} \leq i_k$ and $j_{k-1} \leq j_k$ for $2 \leq k \leq K$.

Continuity condition: The alignment cannot skip points. That is, $i_{k-1} \leq i_k \leq i_{k-1} + 1$ and $j_{k-1} \leq j_k \leq j_{k-1} + 1$.

The cost $V(P)$ of the warping path P is defined as the sum of distances between the matched points, i.e., $V(P) = \sqrt{\sum_{k=1}^{K} d(x_{i_k}, y_{j_k})^2}$.

Formally, DTW is defined as the cost of the minimum-cost warping path. Let $A(X, Y)$ be the set of all warping paths between time series X and Y. The DTW between time series X and Y can be expressed as follows:

$$DTW(X, Y) = \min_{P \in A(X, Y)} V(P). \tag{1}$$

The minimum-cost warping path can be computed using dynamic programming. The time complexity of DTW is $O(mn)$, since the distances between x_i ($1 \leq i \leq m$) in X and y_j ($1 \leq j \leq n$) in Y need be exhaustively calculated beforehand in order to quantitate $V(P)$.

For real applications, **warping constraints** are often used to restrict the range of warping and to prevent excessive stretching or compression in the alignment. The Sakoe-Chiba band is a famous warping constraint. The Sakoe-Chiba band demands any index pair (i_k, j_k) in the warping path to satisfy $|i_k - j_k| \leq b$, where b is a positive parameter. Under the Sakoe-Chiba band, the time complexity of DTW reduces to $O(\max\{m, n\} \times b)$ that equals the number of possible index pairs, However, the longer the data, the longer the computation time.

3   SSH (Sketch, Shingle & Hash) [8]

To overcome the large time-complexity of DTW, Luo et al. [8] proposed an algorithm called SSH (Sketch, Shingle & Hash) that approximates the nearest neighbor search based on the DTW distance. Given a query time series Q, the goal is to search similar time series with small DTW distances to Q from the time series database D. SSH makes use of hashing to narrow down promising candidates in D and reduces the number of DTW calculations by computing the DTW distance only for the candidates.

For a time series $X \in D$, SSH first gathers quantized subsequences from X and represents X as a histogram S_X of quantized subsequences. Then, a hash value for X is derived by applying Min-Hash [9] to S_X. Below, we explain these two procedures in details.

3.1   Construction of a Histogram S_X of Quantized Subsequences

Let $X = (x_1, x_2, ..., x_m)$ be a time series of length m.

First, SSH segments subsequences of length w from X by sliding a window of length w along X and by setting the stride to δ. Here, the stride is a parameter that controls how far the window advances at one step. As a result, SSH acquires $\frac{(m-w+1)}{\delta}$ subsequences. To simplify the exposition, we denote $\frac{(m-w+1)}{\delta}$ by N and expresses a set of extracted subsequences as $\{X_S^{(1)}, X_S^{(2)}, \cdots, X_S^{(N)}\}$. For example, the i-th subsequences $X_S^{(i)}$ consists of w elements in X for $1 \leq i \leq N$ as follows.

$$X_S^{(i)} = (x_{(i-1)*\delta+1}, x_{(i-1)*\delta+2}, ..., x_{(i-1)*\delta+w}). \tag{2}$$

Next, SSH creates an N-bit sequence by mapping each subsequence to 0 or 1. Concretely, it computes an inner product between $X_S^{(i)}$ and a randomized filter vector r and determines the sign $B_X^{(i)}$ of the inner product. That is, $B_X^{(i)} = 1$ if $r \cdot X_S^{(i)} \geq 0$ and $B_X^{(i)} = 0$ if $r \cdot X_S^{(i)} < 0$. Here, the filter vector r is created by concatenating w random values chosen from a one-dimensional normal distribution with mean 0 and variance σ^2. Thus, r becomes a w-dimensional vector $r = (r_1, r_2, ..., r_w)$. This mapping of $X_S^{(i)}$ to a binary value is the equivalent with the LSH for the cosine similarity [10]. Therefore, for two time series X and Y, it holds that $P[B_X^{(i)} = B_Y^{(i)}] = 1 - \frac{\theta(X_S^{(i)}, Y_S^{(i)})}{\pi}$, where $\theta(X_S^{(i)}, Y_S^{(i)})$ represents the angle between $X_S^{(i)}$ and $Y_S^{(i)}$.

So far, we have obtained the N-bit sequence $B_X = (B_X^{(1)}, B_X^{(2)}, ..., B_X^{(N)})$. By extracting n consecutive bits from B_X, we can generate an n-gram. In total, the N-bit sequence B_X contains $N - n + 1$ n-grams. Obviously, an n-gram corresponds to a feature for n time-series subsequences in X. In other words, the n-gram quantizes the n subsequences in X.

Finally, SSH constructs a histogram S_X for these $N - n + 1$ n-grams. S_X holds 2^n bins that is equal to the number of possible bit patterns for n-digit binary sequences. After all, the histogram S_X represents a time series X as a multiset that contains n time series subsequences.

3.2 Hash Value for the Histogram of S_X of n-Grams

Because the histogram S_X forms a multiset of n-grams, SSH computes the hash value for S_X by applying the weighted Min-Hash. The weighted Min-Hash assures theoretically that the probability of hash collision between two multisets equals their weighted Jaccard similarity. SSH adopts the well-known Consistent Weighted Sampling [11] as a hash function for the weighted Min-Hash.

3.3 Approximate k-NN Search

This subsection describes how SSH realizes approximate k-NN search based on the DTW distance for a query time series Q. As a preprocessing, SSH prepares l hash functions $mh_1, mh_2, \cdots, mh_l$ and l hash tables $T_1, T_2, \cdots, T_l$. Then, for all the time series X_i in the database $D = \{X_i | 1 \leq i \leq |D|\}$, SSH computes l hash values $mh_1(X_i), ..., mh_l(X_i)$ and register X_i into the l hash tables according to the hash values. When the query Q arrives, SSH computes l hash values $mh_1(Q), ..., mh_l(Q)$ and examines the l hash buckets over the l tables. SSH treats the group of time series in the l buckets as promising candidates in D. SSH actually computes the DTW distance only for the promising candidates and selects the top-k time series of them that have the least DTW distance to Q. These top-k time series constitute the approximate result. To decide the top-k time series from the promising candidates, SSH relies on the pruning-based UCR suite algorithm [4].

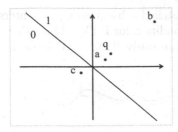

Fig. 1. Relation between distance between subsequences and their bit values

4   Drawbacks of SSH

This section states two drawbacks of SSH that motivate our research: (1) The n-grams have a weak correlation with DTW and (2) the hash values discard the temporal information of time series subsequences completely.

Let us begin with the first problem. For a pair of time series X and Y, suppose that their n-grams become the identical. Namely, $B_X^{(i)}, B_X^{(i+1)}, ..., B_X^{(i+n-1)} = B_Y^{(j)}, B_Y^{(j+1)}, ..., B_Y^{(j+n-1)}$. From this premise, we have $B_X^{(i+k)} = B_Y^{(j+k)}$ for $1 \leq k \leq n$. However, the fact $B_X^{(i+k)} = B_Y^{(j+k)}$ reveals that the subsequences $X_S^{(i+k)}$ in X and the subsequences $Y_S^{(j+k)}$ in Y have similar vector orientations, but does not assure that $X_S^{(i+k)}$ and $Y_S^{(j+k)}$ are located closely in the w-dimensional vector space. By contrast, even if $B_X^{(i+k)} \neq B_Y^{(j+k)}$, it can happen that $X_S^{(i+k)}$ and $Y_S^{(j+k)}$ are closely located. See Fig. 1 for example: q is a time series subsequences in Q. The diagonal line represents a boundary plane decided by the filter vector r. This plane has a normal vector r. In the figure, even if q and c are time subsequences closely located in the vector space, their bit values are different. On the contrary, even if q and b are very distant, their bit values grow the same. Since DTW accumulates element-wise Euclidean distances after all, the angle between two vectors used in SSH is less related to DTW than their Euclidean distance.

Furthermore, the n-grams in SSH are not adaptive to the data distribution, because the random filter vector r is chosen independently of the database D. As a result, there is a risk that the signs of $r \cdot X_S^{(i)}$ will be extremely biased toward either 0 or 1. In such cases, every time series in D is going to have a skewed histogram of n-grams, that obstructs the diversity of hash values. So that, the hash function will fail to narrow down promising candidates properly.

Let us advance to the second problem that the hash values completely discard the temporal information of time series subsequences. Due to this nature, the hash collision $mh(Q) = mh(X)$ can match two subsequences that are temporally far away and that break the specified warping constraint like the Sakoe-Chiba band. Furthermore, when multiple hash functions collide between Q and X, they might not conform to the DTW as illustrated in Fig. 2. In Fig. 2, the hash collisions between Q and X happen twice for the two hash functions mh_1 and

mh_2. However, the two matched subsequence pairs intersect with each other and break the monotonicity condition for DTW. Thus, they can never appear in the same warping path simultaneously. Thus, SSH loses fidelity to DTW.

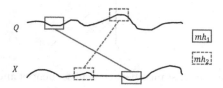

Fig. 2. Two hash collisions that break monotonicity condition

5   Proposed Solution

SSH has two drawbacks: (1) The quantization of subsequences, that is, n-grams, neither has a strong correlation with DTW nor is adaptive to the database D and (2) because hash values discard the temporal information of time series subsequences, hash collisions can match two time subsequence which are unlikely to appear in the optimal alignment corresponding to the DTW distance.

Our research aims to correct the two drawbacks in SSH. Particularly, we proposes a new method named BSecH (Bag-of-sketch Section Hash). In order to increase the correlation with DTW, BSecH quantizes time series subsequences based on the Euclidean distance rather than the vector direction. In addition, we also develop a novel Section Min-Hash (SMH). Section Min-Hash explicitly specifies multiple sections in a given time series and computes a hash value for each section, so that the hash values may link to the temporal locations in the time series. In the subsequence, Sect. 5.1 explains the quantization of time series subsequences based on the Euclidean distance. Section 5.2 discusses Section Min-Hash.

5.1   Quantization of Subsequences Based on Euclidean Distance

To quantize time series subsequences, BSecH brings the Bag-of-Visual Words (BoVW) designed for image processing into time-series analysis. The original BoVW gathers all the local features, e.g. SIFT [12] in all the images in the image database and convert them into quantized local features termed Visual Words. Thus, the BoVW represents an image as a histogram of visual words. Before us, Reference [13] has already made use of the BoVW for time series processing. This previous work tried hard to imitate the original BoVW a much as possible. In order to obtain local features, it calculates the gradient between near points in the given time series, whereas the original BoVW computes the gradient between neighbor pixels. Unlike [13], BSecH gives the highest priority to be consistent with DTW and quantizes raw time series subsequences without caring about the gradient between near points.

Given a time series $X = (x_1, x_2, ..., x_m)$ of length m, BSecH first utilizes the sliding window to segment N subsequences of length w, i.e., $\{X_S^{(1)}, X_S^{(2)}, \cdots, X_S^{(N)}\}$ completely in the same way as SSH, where $N = \frac{(m-w+1)}{\delta}$.

Next, by extracting time series subsequences from all the time series in D in this way, BSecH constructs a pool P of time series subsequences of length w. Then, BSecH treats a time series subsequence of length w as a w-dimensional vector and clusters all the subsequences in P with the k-means algorithm. Let $C_1, C_2, \cdots, C_{nc}$ be the derived clusters, where the number of clusters is denoted by nc. Through the clustering, a time series subsequence is quantized to the cluster ID that it belongs to. After the clustering, a time series X that holds N subsequences is associated with the N cluster IDs $C_X^{(1)}, C_X^{(2)}, ..., C_X^{(N)}$ where $C_X^{(i)}$ symbolizes a cluster that $X_S^{(i)}$ belongs to. Finally, BSecH represents X as a histogram S_X of cluster IDs. Note that SSH bundles n consecutive time series subsequences and quantizes them once to an n-gram, whereas BSecH quantizes a single subsequence to a cluster ID.

The quantization process in BSecH have two merits as follows. (1) Thanks to the k-means, the quantization is adaptive to the data distribution in D. Thus, the histogram S_X also becomes adaptive to the data distribution. (2) Supported by the k-means, two time-series subsequences quantized into the same cluster ID are expected to take a short Euclidean distance in the w-dimensional space. We consider that this property is crucial to approximate of DTW, because the DTW between two time series Q and X usually consists of multiple Euclidean distances between time-series subsequences(in low-dimensional subspaces) as illustrated in Fig. 3, especially if Q is similar to X under some warping constraint.

Fig. 3. DTW alignment includes Euclidean distance between subsequences

5.2   Section Min-Hash (SMH)

Section Min-Hash (SMH) aims to link the hash values of a time series X to the temporal locations in X to prevent the hash collisions from matching temporarily distant time series subsequences that are apparently not included in the alignment corresponding to the DTW distance. Roughly speaking, SMH selects

t sections of length d in the temporal space, where $t(>1)$ is a number of sections, and outputs t hash values which are derived from the t sections. d, the width of sections, and t, the number of sections are parameters that should be configured by users.

First, we explain how to select of t sections. Like the Sakoe-Chiba band, SMH assumes that every time series has roughly the equal length. Let $N_{\max}$ be the maximum number of subsequences obtained from a single time series in D.

1. We divide a finite integer interval $[1, N_{\max}]$ into t segments of equal length. The length of a segment grows $\frac{N_{\max}}{t}$.
2. For $1 \leq i \leq t$, we select the i-th start time u_i from the i-th segment randomly and decides the i-th temporal section $[u_i, u_i + d - 1]$. Though u_i is chosen randomly, u_i may not depend on a specific time series. Namely, u_i must be common for all the time-series data.

For a time series X, SMH outputs t hash values in the next way. After quantizing the subsequences, X is represented as an ordered sequence of cluster IDs $\{C_X^{(1)}, C_X^{(2)}, ..., C_X^{(N)}\}$. For $1 \leq i \leq t$, the i-th hash value of X denoted by $mh^i(X)$ is computed by applying mh, the hash function for Min-Hash, to the set of cluster IDs $\{C_X^{(u_i)}, C_X^{(u_i+1)}, ..., C_X^{(u_i+d-1)}\}$ that X holds in the temporal section $[u_i, u_i + d - 1]$.

Figure 4 illustrates the above procedure. First, the interval $[1, N_{\max}]$ is divided into t pieces separated by the vertical black lines. Next, the i-th temporal section $[u_i, u_i + d - 1]$ shown as a dotted red rectangle is determined from the i-th segment. Finally, the hash value $mh^i(X)$ is computed for the sets of cluster IDs that emerge in the dotted red rectangle.

Very significantly, SMH guarantees that, when $mh^i(Q) = mh^i(X)$, the time gap between the matched subsequences never exceeds $d\delta$. Though BSecH makes t times as many hash values as SSH, BSecH is not inferior to SSH in terms of the execution time experimentally. This is because the hash collisions are harder to occur in BSecH than in SSH: In BSecH, not only the hash value but also the section index i must coincide before a hash collision takes place.

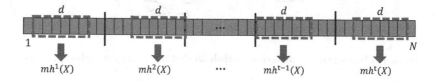

Fig. 4. Section Min-Hash

5.3   Approximate k-NN Search

BSecH makes l hash functions $mh_1, mh_2, \cdots, mh_l$ in the same way as SSH. Each hash function manages different t sections, since BSecH randomly chooses their starting times. For $1 \leq j \leq l$, a hash function mh_j outputs t hash values for a single time series X. In total, BSecH creates $t \times l = tl$ hash values for X. Let $mh_j^i(X)$ be the i-th hash value for X computed by the function mh_j.

BSecH constructs l hash tables $T_1, T_2, \cdots, T_l$. Then, for any X in D, we compute the tl hash values and register X to the l hash tables. X is to be inserted to the table T_j t times, as mh_j generates t hash values for X. To distinguish the t sections, we implement a bucket in T_j as t lists. For example, if $mh_j^i(X) = v$, X will be added to the i-th list that spans from the bucket $b(v)$ in T_j.

When a query time series Q arrives, BSecH computes the tl hash values for Q in the same way. Then, by scanning the corresponding tl lists over the l tables, BSecH identifies a group R of promising candidates in D. If a time series X belongs to R, X satisfies $mh_j^i(Q) = mh_j^i(X)$ for some pair of i and j.

Finally, by using the pruning-based UCR suite, BSecH derives the top-k times series in R that has the smallest DTW distance to Q and returns them as the approximate result. For the UCR suite, we prioritize time series that experienced more hash collisions with Q in calculating the DTW distance. Here, the frequency of hash collisions for X is defined as the number of index pairs (i, j) satisfying $mh_j^i(Q) = mh_j^i(X)$. Since such time series are expected to have small DTW distances to Q, this strategy rapidly decreases the lower bound of DTW distance for the pruning and improves the pruning efficiency. We would emphasize that SSH does not rank the time series stored in the promising candidates as BSecH. Thus, BSecH tries to bring out the potential of UCR suite more actively than SSH.

5.4 Related Works

Here, we review the attempt to utilize LSH to search similar time series. While we process uni-variate time series, Yu et al. [6] devised an LSH to approximate the DTW for multi-variate time series. PSEUDo [14] modifies [6], so that the LSH hash function may be updated based on the relevance feedback from the users. ChainLink [15] extends SSH to search subsequences similar to the query time series from the database of time series. Though ChainLink accelerates the generation of hash values by means of distributed processing, it only considers the Euclidean distance. Astefanoaei et al. [7] developed an LSH for 2D trajectories.

Other than hash-based methods, PQDTW [16] computes approximate DTW distances based on Product Quantization. PQDTW is similar to BSecH in quantizing time series subsequences with the k-means-like clustering. Like BSecH, STS3 [17] represents a time series as a set. However, by neglecting time series subsequences, STS3 loses the high frequency component in time series.

6   Experiments

We compared SSH and BSecH experimentally. The experimental platform is a PC with an Intel(R) Core(TM) i7-10700 CPU @2.9 GHz $\times$ 8 and 16 GB RAM.

All algorithms were implemented in C++. As for SSH, we used the C++ program provided by the inventor of SSH [18].

We used the real dataset ECG that was also used to evaluate SSH in the original SSH paper [8]. The ECG dataset consists of 20,140,000 ECG data points. After applying z-score normalization, we divide the ECG data into sequences of length 1024. 4000 time series were chosen from these sequences to form the time series database D. As for the query time series, we selected multiple time series of length 1024 other than the database D.

6.1   Performance Comparison with SSH

We examine the quality of 20-NN (Nearest Neighbor) search. Given a query time series Q, we evaluate an approximate result returned by either of SSH or BSecH with the recall rate. The recall rate is defined as the ratio of the exact top-20 time series with the smallest DTW distances to Q covered by the approximate result. We report the average value taken over 20 different queries. In this subsection. the Sakoe-Chiba band is set to $b = 10$.

For SSH, we used the same parameters as recommended in [8] for the ECG dataset: $w = 80$, $\delta = 3$, and $n = 15$. For BSecH, we set the parameters regarding the histogram construction as $w = 80$, $\delta = 1$, $nc = 1024$. With respect to Section Min-Hash, $t = 10$ and $d = 20$.

To examine how much correlated the hash collisions are with DTW distances, we investigated the relation between the recall and the number of DTW calculations. Recall that, both BSecH and SSH calculate the DTW distance only for the promising candidates in D which experiences at least one hash collision with Q. In Fig. 5(a), the X-axis presents the number of DTW calculations and the Y-axis shows the recall. If a higher recall is obtained without computing the DTW many times, the hash values are judged as more faithful to the DTW. In addition to BSecH and SSH, Fig. 5(a) displays also the results for the two versions (i) the one that couples the quantization based on the filter vector in SSH with Section Min-Hash and (ii) the combination of BoVW in BSecH with the conventional Min-Hash adopted in SSH. The legend for the version (i) is "SMH" and that for the version (ii) is"BoVW". To align the value range of DTW calculations, the number l of hash functions, is set to 20 for BSecH and SMH and 60 for BoVW and SSH.

BSecH achieved by far a higher recall than SSH. For instance, when DTW is calculated 500 times, the recall for BSecH is about three times as large as SSH. In addition, both BoVW and SMH consistently achieved higher recall rates than SSH. Therefore, both the quantization of subsequences based on Euclidean distance and Section Min-Hash are useful by themselves.

Next, Fig. 5(b) plots the execution time and the corresponding recall for the 20 different queries. Here, $l = 10$ for BSecH and $l = 20$ for SSH. The setting $l = 20$ is recommended by the original SSH paper. The X-axis represents the execution time, and the Y-axis shows the recall rates. Two vertical lines represent the average execution times of SSH and BSecH, while the surrounding dashed lines indicate the standard deviations. Remarkably, BSecH makes the execution

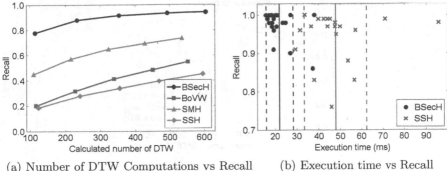

(a) Number of DTW Computations vs Recall   (b) Execution time vs Recall

Fig. 5. Comparison between BSecH and SSH

times shorter than SSH while achieving higher recall rates. The average execution time becomes 21.99 ms for BSecH and grows to 47.86 ms for SSH. Thus, BSccH halves the execution time as compared with SSH and improves the recall rates simultaneously.

In addition, the variance of execution times becomes smaller for BSecH than for SSH. SSH has a risk of increasing the variance, as it relies on a random filter vector: The random filter vector maps n time series subsequences into n-grams, while ignoring the data distribution. As a result, it can happen that most of the time series subsequences are assigned to only a few kinds of n-grams. This situation causes too many hash collisions that accompany too many DTW computations and lengthens the execution time of SSH. The rightmost red points in Fig 5(b) are instances of such cases. By contrast, the k-means method in BSecH to quantize time series subsequences tends to generate clusters of similar sizes and stabilizes the execution time.

Overall, BSecH outperforms SSH in terms of shorter execution times and higher recall rates, while reducing the variance of execution times.

6.2 Effect of Parameters in BSecH

We report how the section width d and the window size w affect the search performance in BSecH.

Effect of Section Width d in SMH: While fixing the Sakoe-Chiba band b to 10, we vary d from 10 to 40 in the interval of 10. Figure 6 shows how d affects the recall. The recall grows the highest when $d = 20$. It decreases gradually, as d moves away from 20.

Next, we increase b to 20. Figure 7(a) shows that the recall becomes the highest for $d = 30$ and decreases gradually as d moves away from 30. Then, we increases b further to 50, where d is chosen from 50, 60, 70, and 100. See Fig. 7(b). Similarly, the recall gets the highest for $d = 60$ and decreases gradually as d moves away from 60.

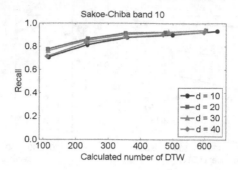

Fig. 6. Recall Rate for Various d Values when the Sakoe-Chiba band $b = 10$

These results suggest that the optimal value of d depends on the Sakoe-Chiba band b. In particular, we recommend to set d to $b + 10$.

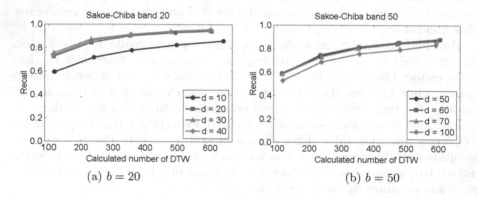

(a) $b = 20$ (b) $b = 50$

Fig. 7. Recall Rate for Various d Values

Effect of Window Size w: For two values of Sakoe-Chiba band $b = 10$ and $b = 50$, we vary the window size w from 40 to 100 in the interval of 20. Figure 8 shows how w affects the recall. For both b values, the recall betters as w increases and grows the highest for $w = 80$ and $w = 100$. These results show that the optimal window size does not depend on b.

w controls the length of time series subsequences. In BSecH, the hash collision occurs when two time series Q and X share time series subsequences that have a small Euclidean distance. BSecH considers that Q and X grow more similar if their hash values collide more often. Thus, Fig. 8 tells that, there surely exists the optimal length of time series subsequences for the strategy to approximate the DTW with the sum of Euclidean distances between matched time series subsequences.

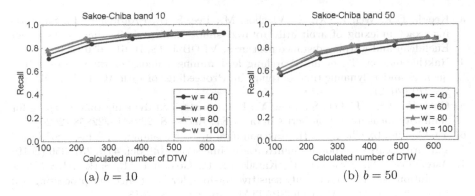

Fig. 8. Recall Rate for various w Values

7   Conclusions

We propose an algorithm BSecH for approximate time-series similarity search based on DTW. BSecH overcomes the two drawbacks of the previous hash-based method SSH: (1) The quantization of time series subsequences does not has a strong correlation with DTW, since it reflects their vector orientations that are ignored by DTW. (2) Since the hash values discard the temporal locations of subsequences, a hash collision can match two time series subsequences that are temporally far apart and unlikely to emerge in the optimal alignment for DTW. BSecH solves the first problem by clustering subsequences based on their Euclidean distances and settles down the second problem by devising a new Section Min-Hash (SMH). SMH explicitly specifies multiple temporal sections in time series and computes a hash value for each of the sections, so that the hash values may link to the temporal locations in the time series. Our experimental results demonstrate that BSecH outperforms SSH in terms of search accuracy and execution time both. We also show that the optimal section width for SMH is affected strongly by the Sakoe-Chiba band warping constraint. One future work is to evaluate BSecH on another dataset except for the ECG dataset.

Acknowledgments. This work was supported by JSPS KAKENHI Grant Number JP21K11901, 2023.

References

1. Berndt, D.J., Clifford, J.: Using dynamic time warping to find patterns in time series. In: Proceedings of 3rd International Conference on Knowledge Discovery and Data Mining, pp. 359–370 (1994)
2. Kim, S.W., Park, S., Chu, W.W.: An index-based approach for similarity search supporting time warping in large sequence databases. In: Proceedings of 17th ICDE, pp. 607–614 (2001)

3. Keogh, E.J., Wei, L., Xi, X., Vlachos, M., Lee, S.-H., Protopapas, P.: Support-ing exact indexing of arbitrarily rotated shapes and periodic time series under Euclidean and warping distance measures. VLDB J. **18**(3), 611–630 (2008)
4. Rakthanmanon, T., et al.: Searching and mining trillions of time series subse-quences under dynamic time warping. In: Proceedings of 18th ACM SIGKDD, pp. 26–270 (2012)
5. Choi, W., Cho, J., Lee, S., Jung, Y.: Fast constrained dynamic time warping for similarity measure of time series data. IEEE Access **8**, 222841–222858 (2020)
6. Yu, C., Luo, L., Chan, L.L.H., Rakthanmanon, T., Nutanong, S.: A fast LSH-based similarity search method for multivariate time series. Inf. Sci. **476**, 337–356 (2019)
7. Astefanoaei, M., Cesaretti, P., Katsikouli, P., Goswami, M., Sarkar, R.: Multi-resolution sketches and locality sensitive hashing for fast trajectory processing. In: Proceedings of 26th ACM SIGSPATIAL, pp. 279–288 (2018)
8. Luo, C., Shrivastava, A.: SSH (sketch, shingle, & hash) for indexing massive-scale time series. In: Proceedings of the NIPS 2016 Time Series Workshop, pp. 38–58 (2016)
9. Broder, A.Z., Charikar, M., Frieze, A.M., Mitzenmacher, M.: Min-wise independent permutations. In: Proceedings of 30th ACM STOC, pp. 327–336 (1998)
10. Charikar, M.S.: Similarity estimation techniques from rounding algorithms. In: Proceedings of 34th ACM STOC, pp. 380–388 (2002)
11. Ioffe, S.: Improved consistent sampling, weighted minhash and L1 sketching. In: 2010 IEEE ICDM, pp. 246–255 (2010)
12. Lowe, D.G.: Distinctive image features from scale-invariant keypoints. Int. J. Com-put. Vision **60**, 91–110 (2004)
13. Bailly, A., Malinowski, S., Tavenard, R., Chapel, L., Guyet, T.: Dense bag-of-temporal-SIFT-Words for time series classification. In: Advanced Analysis and Learning on Temporal Data, pp. 17–30 (2016)
14. Yu, Y., Kruyff, D., Jiao, J., Becker, T., Behrisch, M.: PSEUDo: interactive pat-tern search in multivariate time series with locality-sensitive hashing and relevance feedback. IEEE Trans. Visual. Comput. Graph. **29**(1), 33–42 (2023)
15. Alghamdi, N.A., Zhang, L., Zhang, H., Rundensteiner, E.A., Eltabakh, M.Y.: ChainLink: indexing big time series data For long subsequence matching. In: Pro-ceedings of 36th ICDE, pp. 529–540 (2020)
16. Zhang, H., Dong, Y., Li, J., Xu, D.: Dynamic time warping under product quantiza-tion, with applications to time-series data similarity search. IEEE Internet Things J. **9**(14), 11814–11826 (2022)
17. Peng, J., Wang, H., Li. J., Gao, H.: Set-based similarity search for time series. In: Proceedings of ACM SIGMOD 2016, pp. 2039–2052 (2016)
18. Luo, C.: ssh. www.github.com/rackingroll/ssh (2017)

Mutual k-Nearest Neighbor Graph for Data Analysis: Application to Metric Space Clustering

Edgar Chavez[1]([✉]) [iD], Stephane Marchand-Maillet[2] [iD], and Adolfo J. Quiroz[3] [iD]

[1] CICESE, Ensenada, Mexico
elchavez@cicese.mx
[2] University of Geneva, Geneva, Switzerland
stephane.marchand-maillet@unige.ch
[3] Universidad de los Andes, Bogotá, Colombia
aj.quiroz1079@uniandes.edu.co

Abstract. In this paper, we delve into the Mutual k-Nearest Neighbor Graph (mkNNG) and its significance in clustering and outlier detection. We present a rigorous mathematical framework elucidating its application and highlight its role in the success of various clustering algorithms. Building on Brito et al.'s findings, which link the connected components of the mkNNG to clusters under specific density bounds, we explore its relevance in the context of a wide range of density functions.

Keywords: Mutual k-nearest neighbor · Neighborhood graph · Cluster analysis · Connectivity

1 Introduction

Clustering demands a nuanced understanding of data and a varied toolset tailored to each unique problem. Successfully clustering hinges on formalizing intuitions into clear theorems. We aim to develop a robust mathematical framework for mkNNG clustering use and explain the success of algorithms that use this principle.

Our literature review underscores the recurring emphasis on the mkNN concept, and our study provides insights into its application in clustering.

This paper bridges intuition and mathematics, advancing clustering techniques. We offer a robust exploration of the Mutual k-Nearest Neighbor Graph, laying groundwork for improved clustering and outlier detection. The formalization is covered in Sect. 3.

2 The mkNN Graph in Clustering

Gowda and Krishna were, to our knowledge, the first to propose using the mkNN relationship for agglomerative clustering in [8]. Starting with $k = 1$, fine-grained

O. Pedreira and V. Estivill-Castro (Eds.): SISAP 2023, LNCS 14289, pp. 33–40, 2023.
https://doi.org/10.1007/978-3-031-46994-7_3

clusters are created and merged with increasing k values until only one cluster remains. This was later refined with probabilistic arguments in [5] under a density hypothesis; a detailed discussion on this is deferred to Sect. 3.

The approach in [11] offers a unique mkNN relationship, where each point connects to exactly k-neighbors, proceeding from closest to farther pairs. Once a point connects to its allotted neighbors, it's excluded from further links. Unlike other mkNN methods, distant points could form a mutual kNN relationship. Due to the necessity of computing, storing, and sorting distance pairs, this method is super-quadratic in memory and storage. The clustering approach identifies dense clusters first, claiming to detect clusters of varying densities.

In [2], the authors introduce CMUNE, emphasizing linkage based on mkNN. Dense areas are identified by measuring common points in their k-nearest neighbor neighborhoods. Seed points from these dense regions then connect unclustered points, excluding those in sparse areas. Further refinements are discussed in [1].

In [9], a regressor based on the mkNN relationship is introduced. For a given query x and value of K, it outputs the expected value of values linked to the Mutual K-nearest neighbors of x. If the mkNN result for x is empty, indicating an outlier, the regressor doesn't produce a value.

[17] details the construction of an mkNN graph, drawing parallels with earlier works [5,8]. The aim is cluster detection via the graph's connected components. Spurious connections, possibly arising from noise or outliers, are pruned by weighting the graph's edges based on an affinity measure. Edges below a set affinity threshold are discarded, and clusters are recognized from the graph's residual components.

In [14], a survey on using mkNN for cluster detection is presented. The authors' proposed algorithm hinges on two concepts: extending the mkNN relation to point groups and incorporating density considerations.

[1] unveils DenMune, a clustering algorithm leveraging mkNN to discern centrality and density, as required in the Density Peak-like algorithm [13]. Points are categorized into strong, weak, and noise points based on their reverse kNN count ($\mathcal{R}^k_x(x)$). The DenMune algorithm operates through a voting framework where points garner votes via their membership in the kNN of other points. High-vote recipients are tagged as dense or seed points, while noise points are excluded. Post noise-point removal, the points segregate into dense (seeds) and low-density (non-seeds). Clusters primarily form around seed points, with the low-density points accommodated subsequently. Remaining weak points are then aligned to the most suitable cluster. Remarkably, the authors use mkNN consistently for both density estimation and membership assessment.

A salient feature of the literature review is the dual application of the mkNN test: determining cluster membership and performing three-tier density estimation (dense, weakly-dense, sparse) as illustrated in [1]. While some algorithms appear intricate, they yield compelling experimental results. Our goal is to present a theoretically sound understanding of these heuristics, introducing a streamlined tool for clustering tasks.

3   Data Model

We consider a N-sized D-dimensional dataset $\mathcal{X} = \{x_1, \cdots, x_N\}$ from domain $\Omega \subseteq \mathbb{R}^D$. From the perspective of the continuous domain Ω, a statistical generative model defines the dataset $\mathcal{X}$ as a N-sized i.i.d sample of a probability density function (pdf) $f_{\mathcal{X}} : \Omega \rightarrow \mathbb{R}^+$ with respect to the Lebesgue measure on Ω. $\{x_i\}_{i=1}^N$ is then one realization of a set of N independent random variables $\{X_i\}_{i=1}^N$ identically distributed according to this pdf ($X_i \sim f_{\mathcal{X}}$, $i \in [\![N]\!]$)

Let $\Xi \subseteq \Omega$ be the support of $f_{\mathcal{X}}$ defined as the closure of the set where $f_{\mathcal{X}}$ is non-zero ($\Xi = \mathsf{supp}(f_{\mathcal{X}}) = \overline{\{x \in \Omega \,\|\, f_{\mathcal{X}}(x) \neq 0\}}$). The fact that Ξ is not connected indicates the presence of localized structures in the domain Ω such as (continuous) *clusters*. (Continuous) clustering is therefore defined as the labeling of connected components of the set Ξ. By (discrete) extension, given $\mathcal{X}$ as a sample of Ξ, clustering is the decision for every pair $(x_i, x_j) \in \mathcal{X} \times \mathcal{X}$ whether both data belong to the same connected component of Ξ or not (see Definition 3).

Assuming now that (Ω, d) is a metric space equipped with distance function $d : \Omega \times \Omega \rightarrow \mathbb{R}$, then neighborhood systems may be defined over $\mathcal{X}$.

Definition 1 (Mutual *k-nearest neighbor relationship*). *If $x_j \in \mathcal{V}_{\mathcal{X}}^k(x_i)$ and $x_i \in \mathcal{V}_{\mathcal{X}}^k(x_j)$ then x_i and x_j are said to be* mutual k-nearest neighbors.

Based on the mutual k-nearest neighbor relationship, one can then build the non-directed *mutual k-nearest neighbor* (mkNN) graph $G_k = (\mathcal{X}, \mathcal{E}_k)$ where data $\mathcal{X}$ serves as nodes and an edge $(x_i, x_j) \in \mathcal{E}_k$ exists if x_i and x_j are mutual k-nearest neighbors. In this paper, we wish to argue for the interest in exploiting the mkNN graph and use data clustering as a natural application domain where using the mkNN is beneficial.

Here, we are particularly interested in the connectivity of G_k given dataset $\mathcal{X}$.

Definition 2 ($k_{D,N}$). *Given a dataset $\mathcal{X} \subset \Omega$, $k_{D,N}$ is the smallest integer k such that G_k is connected.*

Authors in [5] provide us with Theorem 1 bridging the continuous and discrete domains.

Theorem 1 (rephrased from [5], Thm 2.1).
Let the i.i.d sample $\mathcal{X} = \{x_1, \cdots, x_N\} \subset \Omega \subseteq \mathbb{R}^D$ come from a distribution P with support $\Xi \subseteq \Omega$ and density $f_{\mathcal{X}}$. Assume that Ξ is connected and grid compatible, and that for constants a_1 and a_2, we have, on Ξ, $0 < a_1 \leqslant f_{\mathcal{X}}(x) \leqslant a_2$. Then, there exists a constant c such that, almost surely, $k_{D,N} \leqslant c \log N$, for large enough N.

The proof of the theorem is detailed in [5]. Constants $0 < a_1 \leqslant a_2$ make sure that values of $f_{\mathcal{X}}$ have finite non-zero extremes. Similarly, the *grid compatibility* criterion makes sure that Ξ contains a (dominated) connected cover of cubes of finite side length, each with a sufficient intersection with Ξ. Theorem 1 relates the connectivity of the mutual kNN graph (via $k_{D,N}$) and the continuity (via grid compatibility) of the local density (via constants a_1 and a_2 and "large enough N"). According to this theorem, the mutual kNN graph is connected in parts where the density of the dataset allows to estimate a support that is grid compatible and where this density is bounded away from 0 and Infinity.

That is, the size $k \geqslant k_{D,N}$ to consider for G_k to be connected (and therefore this discrete structure to represent faithfully the connected continuous domain) is bounded by a sub-linear function of N. As the local density at x_i may be estimated by $\hat{f}_X(x_i) = \frac{k}{\text{vol}(\mathcal{V}_X^k)}$, Theorem 1 states formally that as long as this local estimated density is large enough, G_k remains connected. We will exploit that interpretation to use the mkNN graph to perform data clustering. Since we rely on the formal model given by Theorem 1, we give data clustering a formal and simple definition.

Definition 3 (Data cluster). *Given dataset X, data clusters are defined as connected components of the mutual nearest neighbor graph G_k built on S, as the result of the data filtering operation*

Theorem 1 provides guarantees that data clusters defined as above asymptotically represent the connected subsets of the support of the continuous density f_X. However, since no parametric model is used for f_X, this definition accommodates arbitrary distributions, i.e arbitrary clusters shapes and arrangements.

4   Data Analysis with the Mutual k-Nearest Neighbor Graph

Theorem 1 formally relates the local estimated density of X to the mkNN graph connectivity. This creates a natural link to clustering operations. Reversing the argument, we state that parts of low estimated density represent potential *fords* thru which spurious (weak) connections in the mkNN graph may appear. Hence we propose to remove these spurious low-density parts matching the "bounded away from 0 and Infinity" continuous counterpart argument. By construction, the resulting data subset guarantees a minimal local density, which by Theorem 1 guarantees mkNN graph connectivity. Starting from dataset X, we propose a generic clustering algorithm below.

Algorithm 1. Data clustering based on mkNN graph

1: **procedure** DataAnalysis(X, k)
2:     Remove low density parts of the dataset $X \rightarrow$ subset S ▷ Data filtering
3:     Build G_k as mkNN graph over S ▷ Structure analysis
4:     Connected components of G_k represent the pieces of connected support for f_X

Data Filtering. Filtering low-density regions is a staple in the *denoizing* domain. In many contexts, sparse regions within datasets are interpreted as low-likelihood samples, which are typically smoothed over by MSE-based regression models. However, in this study, our primary concern revolves around the density $f_{\mathcal{X}}$, especially under the parameters delineated in Theorem 1 and subsequent discussions.

From this perspective, the objective becomes highlighting areas where the estimated density not only reaches a minimal threshold (a_1) but also stands as statistically reliable, backed by adequate local samples. Notably, due to the intrinsic property $\int_\Omega f_{\mathcal{X}} = 1$, identifying "low-density" zones becomes a comparative exercise, relative to the entire dataset's distribution. Such an approach closely aligns with the objective of outlier detection; an outlier's definition invariably hinges on its relation to the surrounding data [4,10].

Given this, our preliminary strategy entails employing outlier detection for data filtration, sidestepping traditional methods like KDE [15] which tends to falter in sparsely populated datasets. It's worth noting that contemporary density estimation techniques, such as generative normalizing flows [12], might find applicability in this scenario.

From the input dataset $\mathcal{X}$, the filtration process yields $\mathcal{S}$. This step intriguingly presents a dual relationship with its successor. Even though filtering frequently mandates a predefined k value, techniques spanning outlier detection to density estimation can potentially help pinpoint an optimal k.

Complexity: The construction of a compressed cover tree to solve the kNN exhibits a time complexity of $O(c_m(\mathcal{X})^8 c((\mathcal{X})^2 N \log(N)))$, where c_m and c represent expansion constants pertinent to the dataset. When tasked with deriving the k nearest neighbor table for a novel point, q, the time complexity is set at $O(c(\mathcal{X} \cup q)^2 \log(k)[c_m(\mathcal{X})^{10} \log N + c(\mathcal{X} \cup q)k])$. This infers that formulating the k nearest neighbor table remains a sub-quadratic challenge. The detailed analysis can be found in the recent PhD thesis [6]. With the $N \times k$ k-nearest neighbor table established, the determination of whether the k-nearest neighbor relationship is mutual or otherwise is achievable through a single table traversal, bearing a complexity of $O(Nk^2)$. Hence, the derivation of the mutual k-nearest neighbor graph retains a subquadratic nature.

5   Illustrative Experiments

5.1   k-Means Convex Model

To illustrate visually the ability of our toolset to naturally handle clusters of arbitrary shapes, we visualize the process with the 2D "Two Moons" dataset generated by `sklearn` ($N = 10'000$, $\sigma_{noise} = 0.1$). Due to its convex Gaussian model, k-means cannot separate this data into 2 clusters (Fig. 1[top left]) whereas our procedure (e.g with $k = 10$ and $\tau_{LOF} = 1.1$, Fig. 1 [others]) robustly splits the dataset.

One can easily picture the underlying dendrogram from the evolution of the partition. At k = 1000, connectivity overrules low density regions and the mkNN then shows a unique connected component, leading to a high FP rate.

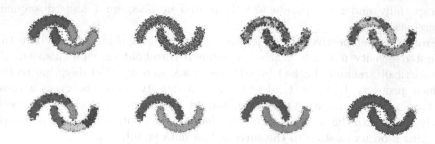

Fig. 1. Two Moons partitions. [top left] k-means partition. [others] our procedure fixing $\tau_{LOF} = 1.1$ and evolving $k \in \{1, 4, 5, 6, 7, 10, 1000\}$ in reading order

Figure 2 pinpoints the fact that this behavior is prevented by the initial filtering. If we remove this step, a large connected component appears in the mkNN graph from k = 4.

Fig. 2. Two Moons partitions without prior filtering and evolving $k \in \{1, 2, 4, 5\}$ in reading order

5.2 Density-Based Clustering Model

Now, one strong bias brought by the `sklearn` "Two Moons" dataset is that the two clusters are of approximate equal densities. When modifying the generation to obtain contrasted densities, one falls in a setup known to be adverse to the density-based clustering algorithms, of which DBSCAN [7] is a major representative.

As shown in Fig. 3, k-means fails again to identify proper clusters due to its convex model. DBSCAN appears very difficult to tune for such data since the high density pushes towards a small `Eps` radius which creates an empty ball, not passing the `MinPts` threshold in low density regions. In contrast, such a setup is accepted by our mkNN procedure with the same parameters as before.

Fig. 3. Adapted Two Moons dataset with varying densities. [left] k-means partition. [center] DBSCAN partition. [right] mkNN-based partition

Discussion: The filtering step, to be useful, should retain most data from the original dataset. Finding the proper filtering value is a problem in itself, it should be related to *fords* detection, as discussed earlier. There are many hidden parameters in the clustering procedure we outline in Algorithm 4, derived from the theorems. This dependency goes also to the bounds of density function and the α value in the grid-compatibility constant in Theorem 1. Rather than giving a clustering algorithm, we have discussed the grounding of successful clustering strategies, proposing tools to build a dedicated connectivity-based clustering algorithm for datasets adhering to precise hypothesis.

We were able to replicate clustering experiments, as reported in [1] with a simpler procedure more amenable to analysis. Also following the conclusions of the cited research, the proposed tools are useful in low-dimensional datasets. They propose to use non-linear embeddings to two dimensions, using for example t-SNE [16]. As analyzed in [3], handling high dimensional datasets is riddle with problems derived from the phenomenon of concentration of measure which generates the so-called curse of dimensionality, one of its many forms is the phenomenon of hubness and the increase of the in-degree in metric graphs, with consequences in the connectivity of the mkNNG.

Further and maybe most importantly, no vector computation is required. Hence, this toolset may operate in metric spaces that are not necessarily vector spaces and where the data is given by similarity, known to respect the metric conditions. Again, this is to be contrasted with the popular k-means algorithm operating in vector spaces, as illustrated in Fig. 1.

Acknowledgments. This work is partly funded by the Swiss National Science Foundation under grant number 207509 "Structural Intrinsic Dimensionality".

References

1. Abbas, M., El-Zoghabi, A., Shoukry, A.: DenMune: density peak based clustering using mutual nearest neighbors. Pattern Recogn. **109**, 107589 (2021)
2. Abbas, M.A., Shoukry, A.A.: CMUNE: a clustering using mutual nearest neighbors algorithm. In: 2012 11th International Conference on Information Science, Signal Processing and their Applications (ISSPA), pp. 1192–1197 (2012)
3. Angiulli, F.: On the behavior of intrinsically high-dimensional spaces: distances, direct and reverse nearest neighbors, and hubness. J. Mach. Learn. Res. **18**(170), 1–60 (2018)

4. Breunig, M.M., Kriegel, H.P., Ng, R.T., Sander, J.: LOF: identifying density-based local outliers. SIGMOD Rec. **29**(2), 93–104 (2000)
5. Brito, M., Chávez, E., Quiroz, A., Yukich, J.: Connectivity of the mutual k-nearest-neighbor graph in clustering and outlier detection. Stat. Probabil. Lett. **35**(1), 33–42 (1997)
6. Elkin, Y.: A new compressed cover tree for k-nearest neighbour search and the stable-under-noise mergegram of a point cloud. The University of Liverpool, United Kingdom (2022)
7. Ester, M., Kriegel, H.P., Sander, J., Xu, X., et al.: A density-based algorithm for discovering clusters in large spatial databases with noise. In: 2nd ACM International Conference on Knowledge Discovery and Data Mining (KDD), pp. 226–231 (1996)
8. Gowda, K.C., Krishna, G.: Agglomerative clustering using the concept of mutual nearest neighbourhood. Pattern Recogn. **10**(2), 105–112 (1978)
9. Guyader, A., Hengartner, N.: On the mutual nearest neighbors estimate in regression. J. Mach. Learn. Res. **14**(37), 2361–2376 (2013)
10. Hodge, V.J., Austin, J.: A survey of outlier detection methodologies. Artif. Intell. Rev. **22**, 2004 (2004)
11. Hu, Z., Bhatnagar, R.: Clustering algorithm based on mutual k-nearest neighbor relationships. Stat. Anal. Data Min. ASA Data Sci. J. **5**(2), 100–113 (2012)
12. Rezende, D.J., Mohamed, S.: Variational inference with normalizing flows. CoRR arXiv:1505.05770 (2016)
13. Rodriguez, A., Laio, A.: Clustering by fast search and find of density peaks. Science **344**(6191), 1492–1496 (2014)
14. Ros, F., Guillaume, S.: Munec: a mutual neighbor-based clustering algorithm. Inf. Sci. **486**, 148–170 (2019)
15. Terrell, G.R., Scott, D.W.: Variable kernel density estimation. Ann. Stat. **20**(3), 1236–1265 (1992)
16. Van Der Maaten, L.: Accelerating t-SNE using tree-based algorithms. J. Mach. Learn. Res. **15**(1), 3221–3245 (2014)
17. Zhang, H., Kiranyaz, S., Gabbouj, M.: Data clustering based on community structure in mutual k-nearest neighbor graph. In: 41st International Conference on Telecommunications and Signal Processing (TSP), pp. 1–7 (2018)

An Alternating Optimization Scheme for Binary Sketches for Cosine Similarity Search

Erik Thordsen$^{(\boxtimes)}$ (iD) and Erich Schubert (iD)

TU Dortmund University, Otto-Hahn-Straße 14, 44227 Dortmund, Germany
`erik.thordsen@tu-dortmund.de`

Abstract. Searching for similar objects in intrinsically high-dimensional data sets is a challenging task. Sketches have been proposed for faster similarity search using linear scans. Binary sketches are one such approach to find a good mapping from the original data space to bit strings of a fixed length. These bit strings can be compared efficiently using only few XOR and bit count operations, replacing costly similarity computations with an inexpensive approximation. We propose a new scheme to initialize and improve binary sketches for similarity search on the unit sphere, i.e., for cosine similarity. Our optimization iteratively improves the quality of the sketches with a form of orthogonalization. We provide empirical evidence that the quality of the sketches has a peak beyond which it is not correlated to neither bit independence nor bit balance, which contradicts a previous hypothesis in the literature. Regularization in the form of noise added to the training data can turn the peak into a plateau and applying the optimization in a stochastic fashion, i.e., training on smaller subsets of the data, allows for rapid initialization.

1 Introduction

Similarity search in large and high-dimensional data sets poses two major problems. Firstly, due to the high intrinsic dimensionality, indexing approaches are difficult and for too many dimensions degenerate to the case of a linear search in the Euclidean case as per the Nearest Neighbor Indexing Theorem [17]. Secondly, linear search is almost infeasible for many applications on very large data sets, where many such searches are necessary. In dense Euclidean space, to further complicate the issue, every distance or similarity computation is increasingly costly as the number of representation dimensions grows. Yet, when allowing for *approximate* results, i.e., not always returning correct neighbors, some of these problems can be lessened. Locality-sensitive hashing (LSH) [5] and sketching are two techniques for approximate search. LSH puts similar items into the same "hash buckets" with high probability, allowing to filter out vast parts of the data set based on comparably cheap hash functions. Alternatively, the bucket assignments can be used as features in a thus dimensionally reduced data set.

O. Pedreira and V. Estivill-Castro (Eds.): SISAP 2023, LNCS 14289, pp. 41–55, 2023.
https://doi.org/10.1007/978-3-031-46994-7_4

Sketching is a general technique for compressing data by mapping each sample onto a smaller representation like bit strings while ideally approximating certain properties by proxy, such as similarity or distance. These compressed representations can be used in a classical index or to reduce the linear search cost.

In this paper we consider similarity search using cosine similarity, although the algorithm potentially extends to inner product similarity, and we propose methods for binary sketching. Binary sketching can be seen as a special case of both sketching and LSH where each "hash function" has only two buckets. The resulting representation is a bit string for each sample. This representation is interesting because distances of bit strings can be computed efficiently using fast CPU instructions like `popcount`. To obtain such binary sketches, simple geometric expressions have been used in the literature where a 1 is assigned if a sample is inside a specific volume and a 0 otherwise (occasionally also -1 [1]). The volumes used as these "hash functions" can be hyperballs, hypercubes, half-spaces, or simple combinations (intersection or difference) thereof [11]. When these volumes are scaled or located such that approximately half of the data set is assigned a 1, the volumes and corresponding bits are called "balanced", otherwise "unbalanced". Balanced bits have been assumed to produce bit strings providing better recall values for approximate search whilst also leading to bit-strings of higher intrinsical dimension [12]. In the literature, the typical approach to balance bit assignments generated from half-spaces/hyperplanes is to add an affine bias such that exactly half the samples are assigned a 1. Yet, in the context of cosine similarity search, using half-spaces induced by non-affine hyperplanes are a natural choice and we limit this paper on non-affine hyperplanes.

Segmenting the unit sphere into cells with non-affine hyperplanes has a long history in the literature. Already in the 19th century, Schläfli derived the precise number of cells on the unit sphere $C(n, d)$ induced by n random hyperplanes in d dimensions in general position, i.e., oriented such that every intersection k hyperplanes is k-codimensional. The formula with proof and further details is given in [15, p. 299] as

$$C(n,d) = 2 \sum_{k=0}^{d-1} \binom{n-1}{k} \tag{1}$$

Whenever $d \geq n$ the number of cells equals 2^n and when $2d \geq n$ then there are at least 2^{n-1} cells. Since the number of cells only increases for larger n, we can in any other case give a lower bound of 2^{2d-1} cells. In any case, the number of cells is likely much larger than the number of samples in high dimensional data sets and with a number of hyperplanes proportional to d.

Accordingly, when intending to use non-affine hyperplane tessellation (occasionally denoted as conical tessellation [15]) for binary sketching, we can expect (almost) no identical bit vectors in practical settings. The goal, thus, is rather to find hyperplanes, such that the Hamming distance on bit strings is correlated to the distance on the hypersphere – at least for "small" distances as we are only interested in finding the nearest neighbors. Entirely random hyperplanes – originally proposed for cosine similarity search by Charikar [3] – are (almost) optimal if the data is uniformly distributed [13]. But on non-uniformly distributed, i.e., clustered or otherwise structured data, they can lead to a suboptimal distribution of cells. The resulting high pairwise Pearson correlation of bits (bit correlation) has been observed to lead to decreased filtering quality when selecting candidates for nearest-neighbor search [11], where bit correlation, bit balance, etc., are evaluated on the columns of the $|X| \times B$ matrix of bit strings of length B. Yet, only little research has been done on how to find a good set of hyperplanes, such that the bit correlation is minimal – at least for the case of cosine similarity search. Balu et al. [1] propose the use of geometrically orthogonal hyperplanes for which bits are greedily flipped to improve the correlation of samples with their "reconstruction" (sum of normal vectors where bit is 1 minus sum of normal vectors where bit is 0). This slightly accounts for the data distribution but is neither a proper optimization, nor does it work when all bit strings are approximately equal. It is also constrained to the bit string length equaling the data dimension. Mic et al. [11] introduced an algorithm that oversamples the number of required hyperplanes and discards all but a well-performing subset using a clique-based approximation algorithm.

In this paper, we intend to fill this gap by providing a fast initialization algorithm that iteratively adds the best hyperplane from a set of candidate hyperplanes and an alternating optimization algorithm that iteratively minimizes the maximum (or mean) pairwise bit similarity – a different yet similar measure to bit correlation. The algorithm does not immediately enforce bit balance yet empirically increases it. Not using an affine bias eliminates a degree of freedom for the hyperplanes, whereby it suffices to rotate the hyperplanes around the origin. The iterative process allows for budgeted training, a dynamic data set, and varying bit string lengths. The update step can be used in multiple "flavors", since either a single or multiple hyperplanes can be updated in each iteration based on the most similar or all other hyperplanes. The sole drawback is that by orthogonalizing all bits (in bit assignments, not geometrically), we increase the intrinsic dimensionality of the representation, making them more difficult to index, yet, the lower length necessary for a similar recall alleviates this.

In Sect. 2 we introduce the alternating optimization algorithm. Section 3 then focuses on our initialization, which is optional to the algorithm but allows for shorter training times. This section also includes a visual example of the initialization and the optimization algorithm. In Sect. 4 we provide empirical results based on a large real world data set. Lastly we close in Sect. 5 with a summary of the paper and an outlook on potential expansions of this approach.

2   Alternating Optimization of Binary Sketches

The alternating optimization algorithm introduced here is called "Hyperplane-based Iteratively Optimized Binarization" (HIOB). In each iteration, we update one or multiple hyperplanes – represented by their normal vectors – by rotating them such that their bit similarity with other hyperplanes decreases. For a pair of hyperplanes, we aim to equalize the number of samples assigned 00, 01, 10, and 11. By computing the number of samples having the same or opposite bits, we obtain an approximation of the "sample density per radian" and can compute an ideal angle to rotate one of the two hyperplanes. For that, we assume that the distribution in each of these four segments is approximately uniform. Instead of immediately rotating one of the normal vectors, we compute the tangent of the rotation angle, which yields an additive displacement vector in the tangent hyperplane. Using displacement vectors in the tangent hyperplane allows to aggregate multiple updates in one step, i.e., to rotate a single hyperplane towards or away from multiple other hyperplanes at once, similar to the mean gradient in a gradient-descent algorithm. The idea of this process is displayed in Fig. 1.

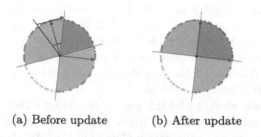

(a) Before update   (b) After update

Fig. 1. An example to motivate the update step. On uniform data, each of the differently colored areas is of approximately the same size. From the difference in area (obtained from the angle between the normal vectors), we can derive the optimal angular change. By calculating the angular change as an additive vector in the tangent space, as displayed in (a), we allow aggregating multiple changes in one step. Afterwards the bit assignments corresponding to the hyperplanes are independent, i.e., equal in size, as displayed in (b).

Let p and q be two normal vectors and let $X \subset \mathbb{R}^d$ be a set of samples with $|X| = N$. We denote the number of samples assigned 1 by both hyperplanes as

$$\mathbf{11}_{p,q} := |\{x \in X \mid \langle x, p \rangle \geq 0 \wedge \langle x, q \rangle \geq 0\}| \tag{2}$$

We similarly denote the numbers of samples for the other three possible assignments with $\mathbf{01}_{p,q}$, $\mathbf{10}_{p,q}$, and $\mathbf{00}_{p,q}$. We obtain the fractions of samples with equal and different bits for p and q with

$$f_{p=q} := \frac{\mathbf{00}_{p,q} + \mathbf{11}_{p,q}}{N} \text{ and } f_{p \neq q} := 1 - f_{p=q} \tag{3}$$

with which we can compute the angular change

$$\alpha_{p,q} := \frac{(f_{p \neq q} - f_{p=q}) \pi}{2} \tag{4}$$

The displacement vector for p using q is then defined as

$$\delta_p(q, s) := \tan(s \cdot \alpha_{p,q}) \frac{q - p \langle p, q \rangle}{\|q - p \langle p, q \rangle\|} \tag{5}$$

where $s \in (0, 1]$ is a scaling factor to control the speed of this process, resembling a learning rate. On uniformly distributed infinite data, the definition of $\delta_p(q, 1)$ is optimal, yet for non-uniformly distributed data, most of the samples may be clustered in a narrow cone. Were we to use $\delta_p(q, 1)$, we would likely skip over all samples in each update step and simply flip the bits on the assignments for p. To accommodate for that case and help convergence, the scaling factor can be reduced to a smaller value like 0.1. Choosing a smaller s monotonously improves the optimization quality, i.e., pairwise bit independence, but increases the computational cost, since the optimum is only achieved asymptotically. If the hyperplane for p should be updated using m other hyperplanes, we advise to change the scale to s/m and sum the displacement vectors, which results in the average displacement for the set of hyperplanes.

Our optimization aims at decreasing the maximum and average bit similarity which we define as

$$\mathcal{S}_{p,q} := 2 \left| f_{p=q} - \tfrac{1}{2} \right| \tag{6}$$

While $f_{p=q}$ is essentially the Simple Matching Coefficient [18], $\mathcal{S}_{p,q}$ is a sort of maximum over $f_{p=q}$ and $f_{p \neq q}$. For equal and inverted bit strings, the bit similarity is 1 and for independent balanced bit vectors it is 0. It is a fast approximation of the absolute Pearson correlation that only requires a single Hamming distance on the bit strings. The more balanced the bits are, the closer the two definitions align and the target of having 50% equal bits in each pair of bit strings is intended to push the bit strings towards balanced without forcing it.

The resulting algorithm is rather simple. In each iteration, we decide which hyperplanes to update and what hyperplanes to use for the update. We then compute all necessary displacement vectors, add them to the normal vectors and normalize to unit length. We then have to recompute the bit assignments for the updated vectors. To improve the run time, we propose to use a stochastic approach by only considering a subsample of the full data set when comparing bit strings and exchanging that subsample at fixed intervals similar to stochastic gradient descent. We observed that the overall quality does not suffer from this approach, and it makes the run time of the optimization algorithm independent of the data set size. To create the random subsets, we used permutations generated from multiple random chained modulo cycles on $\{0, \ldots, 2^n - 1\}$ with cycle-walking [2]. In that way, all samples are used for training once before reusing some. Additionally, storing the pairwise bit similarities in a matrix and only updating values affected by an update further speeds up the algorithm.

In our experiments, two "modes of operation" proved to be very useful: The first mode is to find the two most similar hyperplanes (the "worst offenders") and update one of them – chosen at random – using the other with a scale appropriate for the dataset (0.1 turned out to be very useful in all cases, but lower values should be used when the average bit similarity does not decrease). The random choice between the two hyperplanes avoids oscillating between two states if they are selected repeatedly. The second mode is to update all hyperplanes using all other hyperplanes at once. This mode requires to compute two displacement vectors for each pair of hyperplanes, which is why this method is rather costly.

Yet, even a small number of iterations (<10) produced a good starting point for further optimization even when initializing with uniformly random normal vectors. This second mode of operation, however, struggles to converge on a good final result, since at some point single bit assignments are changed. The discrete steps tend to be comparatively large and affect other bit correlations too much. We, hence, propose to use the first mode of operations until a desired result is achieved. Due to the problems with discrete steps in the pairwise bit similarity, other modes of operations investigated (e.g., always updating the "worst" hyperplane using all other hyperplanes) performed worse than the first mode. Pseudocodes for the two proposed modes are provided in Algorithm 1 and Algorithm 2.

Algorithm 1 Updates one of the two normal vectors with the "worst" bit similarity by adding the displacement vector induced by the other normal vector.

1: **function** UPDATEARGMAX($\mathcal{N} \subset \mathbb{R}^b, s \in \mathbb{R}$)
2:     $p, q \leftarrow \arg\max_{p,q \in \mathcal{N}} S_{p,q}$ ▷ Choose "worst" pair of hyperplanes
3:     $p', q' \leftarrow$ either p, q or q, p uniform at random   ▷ Shuffle p and q
4:     $p'' \leftarrow p' + \delta_{p'}(q', s)$ ▷ Add displacement by q'
5:     $\mathcal{N} \leftarrow \mathcal{N} \setminus \{p'\} \cup \{p''/\|p''\|\}$ ▷ Add new normalized vector to result
6:     **return** $\mathcal{N}$

Algorithm 2 Updates all normal vectors at once by using the displacement vectors induced by all other normal vectors.

1: **function** UPDATEBROAD($\mathcal{N} \subset \mathbb{R}^b, s \in \mathbb{R}$)
2:     $s' \leftarrow s/|\mathcal{N}|$ ▷ Compute effective scale
3:     $\mathcal{N}' \leftarrow \{\}$
4:     **for** $p \in \mathcal{N}$ **do**
5:         $p' \leftarrow p + \sum_{q \in \mathcal{N}} \delta_p(q, s')$ ▷ Add displacement by all other normals
6:         $\mathcal{N}' \leftarrow \mathcal{N}' \cup \{p'/\|p'\|\}$ ▷ Add new normalized vector to result
7:     **return** $\mathcal{N}'$

This optimization process iteratively but not necessarily monotonously reduces pairwise bit similarity, bit correlation and balances most bit assignments, i.e., results in approximately 50% samples assigned 1 for most hyperplanes. Yet, the quality in terms of indexing as, e.g., measured by the $k@n$-recall using the Hamming distance as a proxy for the cosine similarity, is only improved up to some point, after which further iterations begin to reduce the recall again. Empirical evidence for these claims are discussed in Sect. 4.

3   RANSAC-Style Initialization

As discussed in Sect. 1, hyperplanes generated from normal vectors sampled uniformly at random from the unit hypersphere do not account for the data distribution. Whilst we could run our optimization algorithm starting with entirely random normal vectors, that unnecessarily increases the number of iterations. We instead propose an initialization that is inspired by the random sample consensus, RANSAC [4]. We iteratively choose the next normal vector from a set of P random pairs sampled uniformly at random from the data. For each of these pairs we take the normalized difference as normal vector p and compute the corresponding bit assignments. Afterwards we consider the bit similarity $S_{p,q}$ to all previously chosen normal vectors q. From the normal vectors corresponding to all of these pairs, we choose the one with the smallest maximum bit similarity to any previous hyperplane. To further speed up this process, we do not compute the initialization on all samples but rather on a subsample of much smaller size. A pseudocode of this algorithm is displayed in Algorithm 3. Even for large data sets ($|X| > 1$M), parameters such as $M = 2000$ and $P = 200$ sufficed to get decent initializations in our experiments, rendering the initialization mostly invariant of data set size as well.

Algorithm 3 RANSAC-style initialization given a data set X, a number of hyperplanes b to produce, a subsample size M, and a number of pairs P to select candidates from. Returns a set of hyperplanes as normal vectors $\mathcal{N}$.

1: **function** RANSACINITIALIZATION($X \subset \mathbb{R}^d, b \in \mathbb{N}, M \in \mathbb{N}, P \in \mathbb{N}$)

2:     $X' \leftarrow M$ samples from X uniform at random

3:     $\mathcal{N} \leftarrow \left\{ \frac{x-y}{\|x-y\|} \right\}$ where $x, y \in X'$ ▷ *First normal vector*

4:     **for** $i \leftarrow 2, \ldots, b$ **do**

5:         $\mathcal{C} \leftarrow \left\{ \frac{x_i - y_i}{\|x_i - y_i\|} \mid i \in \{1, \ldots, P\}, x_i, y_i \in X' \right\}$ ▷ *Candidate normal vectors*

6:         $p^* \leftarrow \arg\min_{p \in \mathcal{C}} \max_{q \in \mathcal{N}} S_{p,q}$ ▷ *Choose best normal vector*

7:         $\mathcal{N} \leftarrow \mathcal{N} \cup \{p^*\}$ ▷ *Add to final normal vectors*

8:     **return** $\mathcal{N}$

Figure 2 shows examples of the resulting hyperplanes of an entirely random state, of the initialization proposed here, and of an optimized state. The planes are represented by the projected great circles that are obtained by intersecting the planes with the unit sphere. These great circles are the dividing line between the subspaces assigned 0 and 1. The entirely random planes as proposed by Charikar [3] do not consider the data distribution and, hence, do not focus on splitting the clusters apart and often partition empty space. The initialized planes clearly better divide each of the clusters, yet, e.g., also by chance have one of the planes mimic an "equator" which is mostly meaningless. The optimized state rotated this plane to help divide the lower left cluster and the division of the clusters appears more homogeneous. To support this intuition, Fig. 2 shows the distribution of cell sizes and also the mean cell size for each of these plane sets. The optimized state has on average the smallest number of samples per cell on the unit sphere which should lead to a better quality in terms of indexing.

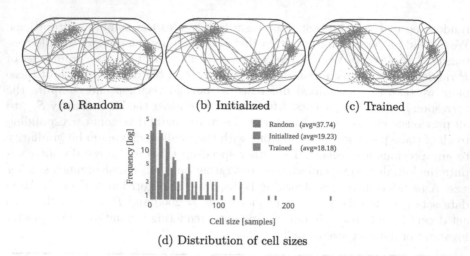

(a) Random   (b) Initialized   (c) Trained

(d) Distribution of cell sizes

Fig. 2. Cells created by 16 (a) random, (b) RANSAC-style initialized, and (c) alternating optimization trained hyperplanes for the displayed data set of 2000 samples on the three-dimensional unit sphere projected with Natural Earth projection [7]. The distribution of samples per cell is displayed in (d).

4 Evaluation

In our evaluation, we inspect how the optimization algorithm (HIOB) affects the bit correlation/similarity, the bit balance, and the quality in terms of the $k@n$-recall when using the Hamming distance on the bit strings as a stand-in for the cosine similarity. For that, we implemented the algorithm in Rust and added Python-bindings which can be accessed on GitHub.[1] The implementation uses arrays to store the pairwise bit similarity and current bit assignments, and is otherwise a straightforward implementation of Algorithms 1, 2 and 3. Aside from HIOB, we added functions to query objects from a data set using the binarization and a brute-force approach. The general idea of the query functions is to select a set of candidates using the nearest neighbors as per Hamming distance on the bit strings and refine the candidates to the number of required neighbors by evaluating the cosine similarity on all candidates. We further evaluated how filtering the candidates with a different set of longer bit strings affects the query time. We evaluated our implementation on subsets of the LAION5B data set [16] provided by the SISAP 2023 LAION2B challenge [19].[2] The data sets contain normalized vectors with 768 dimensions – embeddings from a deep neural network for images and texts – with varying sample sizes of 100K, 300K, 10M, 30M, and 100M together with a query set of 10K vectors with precomputed 100-nearest neighbors for validation.

---

[1] https://github.com/eth42/hiob.
[2] https://sisap-challenges.github.io/datasets/.

Running HIOB with RANSAC-style initialization and scale $s = 0.1$ for up to 50K iterations of Algorithm 1 (our empirical optimum in terms of $k@n$-recall) on a 32-core machine took less than 10 min even for the largest subset of 100M vectors and bit string lengths of up to 2048. We used the stochastic approach with 10K samples and 128 iterations per batch. These values were not tuned and are not sensitive based on our experience.

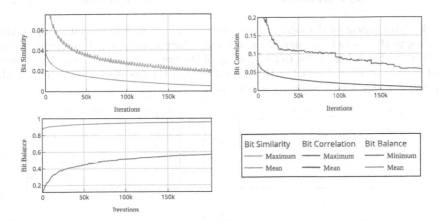

Fig. 3. Bit similarity, correlation and balance over HIOB iterations starting from a RANSAC-style initialization on the 10M subset and for 256 bit sketches.

Figure 3 displays how the bit similarity, bit correlation, and bit balance (ranging from 0 for pure to 1 for half-0-half-1) change over the number of iterations. As can be seen from the plot, HIOB decreased bit similarity and correlation and increased bit balance up to some local optimum, at which small discrete steps produce a bit of noise. The resulting increased independence of bits does not only lead to more homogeneous cells on the hypersphere, w.r.t. the data distribution, but should in theory also increase the intrinsic dimensionality of the bit strings. We evaluated the local intrinsic dimensionality (LID) using the ABID estimator [20,21] on the 200-nearest neighbor bit strings of the bit strings generated from the query set. The resulting mean LID-estimates are displayed in Fig. 4. It is unclear why the mean ABID estimates drop for the 10M subset, yet the change in absolute value is miniscule compared to the increase on the 100K subset. We interpret these results as the indexability of bit strings on the 100K subset decreasing while remaining almost unchanged for the 10M subset, perhaps due to an already high-LID initialization. Different tested indexes like those implemented in Hnswlib [9] and FAISS [8] did not execute the queries faster for comparable recall than brute-force on the bit strings, which would suggest a large intrinsic dimensionality.

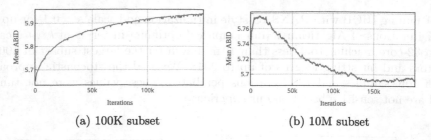

Fig. 4. Development of the local intrinsic dimensionality around the bit strings corresponding to the queries over iterations of HIOB. The low value of ≤ 6 compared to the
length of the bit strings is due to the sparsity of the space and the discrete distribution
of bit strings. It is an artefact of the estimator. The relative size of estimates should
be indicative of spatial complexity nonetheless.

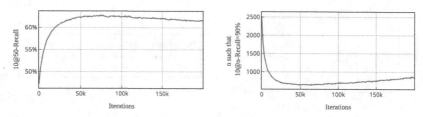

Fig. 5. The 10@50-recall and the number of candidates required for a 10@n-recall of
90% over iterations of HIOB on the 10M subset. The best results are achieved around
50K iterations (approximately 30K for the 100K subset not visualized here), which
contradicts the hypothesis that better bit balance and bit independence correlate with
filtering quality of binary sketches.

Whilst the results so far were within the expected spectrum, we observed
that the k@n-recall decreases when running HIOB too long. As shown in Fig. 5,
the k@n-recall decreased beyond roughly 50K iterations (30K iterations for
100K subset, plot omitted) even though bit balance increases and bit correlation
decreases. This result contradicts the previous hypothesis in the literature, that
the achievable recall during indexing with binary sketches is positively affected
by larger bit balance and lower bit correlation [12]. We assume that the shape of
the cells on the hypersphere is an important and so far neglected factor, although
we do not have an efficient test to verify or contradict that hypothesis. In an
attempt to regularize HIOB on the shape of cells, we added normally distributed
random noise to each subsample drawn in the stochastic approach. We hoped
it would force the hyperplanes to be less coaligned and consequentially produce
more "compact" cells. Although the von Mises-Fisher distribution would be a
better choice for spherical data, the normal distribution with subsequent normalization is isotropic on the sphere as well and much faster to compute. Adding
full-dimensional univariate normal noise with a varying standard deviation σ led
to the average angle between hyperplanes approaching $\pi/2$ for increased σ. In
principle this should provide more homogeneous cells, yet, the data distribution

is less well represented. Figure 6 displays how the added noise affects the recall on the produced bit strings. For a small amount of noise, the recall is improved and the peak in recall is "stabilized" such that it does not decrease after additional iterations. Adding too much noise rapidly decreases the recall since the data distribution is not represented by the hyperplanes well enough. The optimal choice of standard deviation in terms of recall depends on both the number of samples and the number of hyperplanes as can be seen in Fig. 7. We could, however, not find any tangible relation between the data and the optimal amount of noise, yet. The optimal amount of noise must for now be evaluated experimentally. Further insights in the topic or a better regularization are required.

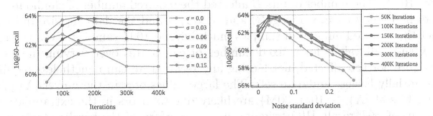

Fig. 6. Change in recall over varying iterations and standard deviation of the noise added to stochastic HIOB subsamples for 256 bits. A small amount of noise improves the recall and stabilizes the peak in recall.

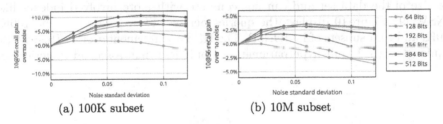

(a) 100K subset   (b) 10M subset

Fig. 7. Change in recall when using varying amounts of noise in the stochastic HIOB approach. All values were computed after 300K iterations. The ideal noise depends on both numbers of bits and samples.

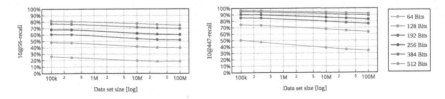

Fig. 8. Dependence of the 10@56- and 10@447-recall over varying data set size. Above $10M$ samples, the recall is almost unaffected by additional data.

We further explore the scalability of the approach. When considering different subsets of the LAION5B data set, the k@n-recall of optimized hyperplanes changed quite little over varying data set sizes as displayed in Fig. 8. This reinforces the claim that HIOB properly fits the hyperplanes to the data distribution. Aside from that, we evaluated how many candidates are necessary to achieve a certain recall for k-neighbor search and observed, that the recall-over-n curves approximately followed the model $r(n) = a/\left(1 + \left(\frac{n}{b}\right)^c\right)$ with the inverse $n(r) = b\sqrt[c]{\frac{a}{r} - 1}$. Using least-squares to fit the model to observed values from a grid search over comparably small n values, good values of n can be extrapolated. Figure 9 displays the extrapolation and the required n for a 10@n-recall of 90%. The approximation error of the recall over n curves were negligible (RMSE below 10^{-3}). The number of bits B affected the required number of candidates n somewhere between $n \propto 1/\log(B)$ and $n \propto 1/B$ and the data set size did not affect the required number of samples much beyond $10M$ samples. Even though the k@n-recall values for a fixed number of candidates over varying data set sizes were not too affected, not loosing any recall at all can nonetheless require substantially larger candidate sets. The large values of n in the n-over-$|X|$ plot for 64 bits at $|X| \in \{10M, 30M\}$ are likely due to errors in the extrapolation. In terms of k@k-recall, HIOB significantly outperformed the baseline (and only other entry) in the SISAP challenge [19] task B by Santoyo et al. [14], by achieving comparable recall while using only 192 bits where the baseline used 1024 bits.

The computational cost of the linear search on the bit strings is linear in the size of the data set and can not compete with more involved indexes like HNSW [9]. Figure 10 displays the queries per second over recall on the 100K, 300K, and 10M subsets of LAION5B. The HNSW implementation in Hnswlib [9] requires – with default hyperparameters – only less than a second (Queries per

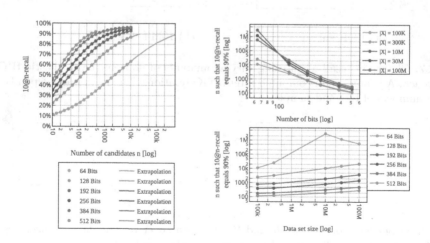

Fig. 9. Example of the extrapolated 10@n-recall over n for the $100M$ subset and plots generated from the extrapolated values for n over number of bits and data set size such that a constant 10@n-recall of 90% is achieved.

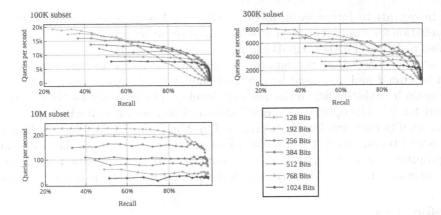

Fig. 10. Queries per second using brute-force search on the HIOB optimized bit strings over 10@n-recall where n is the varied variable for each trace. 10 000 queries were performed, so, e.g., 200 queries per second correspond to 50 s.

second >10^4) for a recall beyond 90%, which is at best comparable to the brute-force performance on up to 100K samples. Yet, the build times of HNSW can be in the hours and increase with data set size. Considering the results published by the SISAP challenge [19], stochastic HIOB sketches can be optimized and up to millions of linear search queries can be evaluated before the fastest approaches (in terms of query time only) have been built. The linear search query time on HIOB can further outperform some of the submitted indices, although that is likely in part due to a better optimized implementation. If a decently fast index during querying with a rapid build time and comparably low memory footprint is desired, stochastic HIOB can be preferred.

5   Conclusion

In this paper we introduced an optimization algorithm to improve bit indepen-dence and bit balance for binary sketches on the unit sphere. We gave empirical evidence that the algorithm improves both of these values and that the algo-rithm is able to improve the filtering quality of binary sketches in the context of similarity search. Our experiments further highlight that bit independence and balance are not entirely correlated to filtering quality, which diminishes if the optimization iterates too long. Adding small amounts of noise allows to avoid decreasing quality when training for longer, yet, the optimal parameterization of the noise remains an open problem. This observation contradicts a previous hypothesis in the literature [12]. Using a stochastic approach to the optimiza-tion and a newly introduced initialization allows for a fast optimization of the binary sketches, with running time invariant to data set size. Yet, brute-force querying with these optimized sketches is not capable of outperforming state-of-the-art approaches like HNSW [9] and the increased bit independence makes the binarized data difficult to index.

In regards to future work, further insight on how to regularize hyperplane tessellation for similarity search is required and resulting regularizations could be added to HIOB. Further, HIOB can most likely be sped-up by using a heuristic to estimate whether or not bits can be flipped. The angle to each hyperplane can be stored and decreased by the total change in angle after each update to maintain bounds, similar to the heuristic developed by Hamerly for the k-Means algorithm [6]. The optimization could automatically be stopped when a peak in some quality measure, like those proposed by Mic et al. [10], is achieved. Lastly, to extend the approach towards Euclidean distances, affine hyperplanes and an appropriately modified update routine could be introduced like rotating around the intersection point of hyperplanes in the plane spun by their normal vectors.

References

1. Balu, R., Furon, T., Jégou, H.: Beyond "project and sign" for cosine estimation with binary codes. In: IEEE International Conference Acoustics, Speech and Signal Processing, ICASSP, pp. 6884–6888 (2014). https://doi.org/10.1109/ICASSP.2014.6854934
2. Black, J., Rogaway, P.: Ciphers with arbitrary finite domains. In: Topics in Cryptology, CT-RSA, pp. 114–130 (2002). https://doi.org/10.1007/3-540-45760-7_9
3. Charikar, M.: Similarity estimation techniques from rounding algorithms. In: Symposium Theory of Computing, pp. 380–388 (2002). https://doi.org/10.1145/509907.509965
4. Fischler, M.A., Bolles, R.C.: Random sample consensus: a paradigm for model fitting with applications to image analysis and automated cartography. Commun. ACM **24**(6), 381–395 (1981). https://doi.org/10.1145/358669.358692
5. Gionis, A., Indyk, P., Motwani, R.: Similarity search in high dimensions via hashing. In: Very Large Data Bases, VLDB, pp. 518–529 (1999). https://doi.org/10.5555/645925.671516
6. Hamerly, G.: Making k-means even faster. In: Proceedings of SIAM Data Mining, SDM, pp. 130–140 (2010). https://doi.org/10.1137/1.9781611972801.12
7. Jenny, B., Patterson, T., Hurni, L.: Flex projector-interactive software for designing world map projections. Cartographic Perspect. **59**, 12–27 (2008). https://doi.org/10.14714/CP59.245
8. Johnson, J., Douze, M., Jégou, H.: Billion-scale similarity search with GPUs. IEEE Trans. Big Data **7**(3), 535–547 (2019). https://doi.org/10.1109/TBDATA.2019.2921572
9. Malkov, Y.A., Yashunin, D.A.: Efficient and robust approximate nearest neighbor search using hierarchical navigable small world graphs. IEEE Trans. Pattern Anal. Mach. Intell. **42**(4), 824–836 (2018). https://doi.org/10.1109/TPAMI.2018.2889473
10. Mic, V., Novak, D., Vadicamo, L., Zezula, P.: Selecting sketches for similarity search. In: Advance Databases and Information Systems, ADBIS, pp. 127–141 (2018). https://doi.org/10.1007/978-3-319-98398-1_9
11. Mic, V., Novak, D., Zezula, P.: Improving sketches for similarity search. In: Proceedings of MEMICS, pp. 130–140 (2015)
12. Mic, V., Novak, D., Zezula, P.: Sketches with unbalanced bits for similarity search. In: Similarity Search and Applications, SISAP, pp. 53–63 (2017). https://doi.org/10.1007/978-3-319-68474-1_4

13. Plan, Y., Vershynin, R.: Dimension reduction by random hyperplane tessellations. Discret. Comput. Geom. **51**(2), 438–461 (2014). https://doi.org/10.1007/s00454-013-9561-6
14. Santoyo, F., Chávez, E., Tellez, E.S.: A compressed index for hamming distances. In: Similarity Search and Applications, SISAP, pp. 113–126 (2014). https://doi.org/10.1007/978-3-319-11988-5_11
15. Schneider, R., Weil, W.: Stochastic and integral geometry (2008). https://doi.org/10.1007/978-3-540-78859-1
16. Schuhmann, C., et al.: LAION-5B: an open large-scale dataset for training next generation image-text models. In: NeurIPS (2022)
17. Shaft, U., Ramakrishnan, R.: Theory of nearest neighbors indexability. ACM Trans. Database Syst. **31**(3), 814–838 (2006). https://doi.org/10.1145/1166074.1166077
18. Sokal, R.R., Michener, C.D.: A statiscal method for evaluating systematic relationships. Univ. Kansas Sci. Bull. **38**(22), 1409–1438 (1958)
19. Tellez, E.S., Aumüller, M., Chavez, E.: Overview of the SISAP 2023 indexing challenges. In: Pedreira, O., Estivill-Castro, V. (eds.) SISAP 2023, LNCS, vol. 14289, pp. 255–264. Springer, Cham (2023). https://doi.org/10.1007/978-3-031-46994-7_21
20. Thordsen, E., Schubert, E.: ABID: angle based intrinsic dimensionality. In: Similarity Search and Applications, SISAP, pp. 218–232 (2020). https://doi.org/10.1007/978-3-030-60936-8_17
21. Thordsen, E., Schubert, E.: ABID: angle based intrinsic dimensionality - theory and analysis. Inf. Syst. **108**, 101989 (2022). https://doi.org/10.1016/j.is.2022.101989

Unbiased Similarity Estimators Using Samples

Conrado Martínez[1]([✉]) [iD], Alfredo Viola[2] [iD], and Jun Wang[1] [iD]

[1] Department of Computer Science, Universitat Politècnica de Catalunya, Barcelona,
Spain
conrado@cs.upc.edu, jun.wang@estudiantat.upc.edu
[2] Instituto de Computación, Universidad de la República, Montevideo, Uruguay
viola@fing.edu.uy

Abstract. Computing a similarity measure (or a distance) between two
complex objects is a fundamental building block for a huge number of
applications in a wide variety of domains. Since many tasks involve computing such similarities among many pairs of objects, many algorithmic
techniques and data structures have been devised in the past to reduce
the number of similarity computations and to reduce the complexity of
computing the similarity (e.g., dimension-reduction techniques). In this
paper, we focus on computing the similarity of two sets and show that
computing the similarity of two random samples drawn from the respective sets leads to an (asymptotically) unbiased estimator of the true
similarity, with relative standard error going to zero as the size of the
involved sets grows, and of course at a much lower computational cost
as we compute the similarity of the significantly smaller samples. While
this result has been known for a long time since Broder's seminal paper
(Broder, 1997) for the Jaccard similarity index, we show here that the
result also holds for many other similarity measures, such as the well-
known cosine similarity, Sørensen-Dice, the first and second Kulczynski
coefficients, etc.

1 Introduction

There are numerous applications where we need to evaluate the similarity (or the
distance) among many pairs of complex objects, for example to locate the best
matches to some target or to group a collection of objects into clusters of similar
objects. In order to reduce the complexity of such tasks, several approaches have
been proposed in the literature, and we can classify them, roughly speaking, in
two families: 1) one is trying to reduce the total number of similarity/distance
evaluations exploiting properties of the metric space (most notably the triangle
inequality) and organizing the information into a suitable data structure, such as

This work has been supported by funds from the MOTION Project (Project
PID2020-112581GB-C21) of the Spanish Ministry of Science & Innovation
MCIN/AEI/10.13039/501100011033.

O. Pedreira and V. Estivill-Castro (Eds.): SISAP 2023, LNCS 14289, pp. 56–63, 2023.
https://doi.org/10.1007/978-3-031-46994-7_5

vantage-point trees, Burkhard-Keller trees, GHTs or GNATs (see for instance [2, 3,9]); 2) a second approach is to use a different similarity or distance measure, one that is much simpler to evaluate while approximating the real similarity well enough, effectively reducing the dimensionality of the search space; this is, for example, the approach in the paper by Broder [1] and the approach taken here. There are many more techniques for dimensionality reduction and even those which combine both approaches, like *Locality-Sensitive Hashing* (LSH) (see for example [8,9]).

In this work, we focus on the problem of estimating the similarity $\sigma(A, B)$ of two sets A and B. It is often the case that we will need to sort the two sets and scan them once sorted to compute their intersection, their union or their symmetric difference in order to compute their similarity $\sigma(A, B)$. Moreover, this has often to be repeated many times between different pairs of sets (the sorting step can be done just once for each set). In this scenario, when the similarity of some set A with many others has to be computed, it makes sense instead to preprocess A in linear time to extract a sample S_A of significantly smaller size ($|S_A| \ll |A|$) which will be used to do the similarity evaluation. This is also the idea behind *minhashing* (see [8] and references therein), which guarantees that the probability that two sets have the same *minhash* is equal to their Jaccard similarity. In our case, we propose computing $\sigma(S_A, S_B)$[1], and we provide a formal proof that $\sigma(S_A, S_B)$ is an unbiased estimator of $\sigma(A, B)$, for several different similarity measures σ, including Jaccard and the cosine similarity, but many others as well.

Suppose we have two random samples S_A and S_B of A and B, respectively. Our goal is to prove that $\sigma(S_A, S_B)$ is an unbiased estimator or otherwise show how to correct the bias. Moreover, the accuracy (as measured by the standard relative error $\sqrt{\mathbb{V}\{X\}}/\mathbb{E}\{X\}$) of the estimation will depend on the size of the samples, and our goal is to quantify it in precise terms. A detailed knowledge of how the accuracy depends on the size of the samples is henceforth fundamental to obtain the desired compromise between accuracy and computational efficiency.

We will consider in this work many different similarity measures, presented in Table 1. For more detailed information about these measures, see for instance [4].

Table 1. Several similarity measures between two sets A and B

Jaccard	$J(A, B) = \frac{|A \cap B|}{|A \cup B|}$	Cosine	$\cos(A, B) = \frac{|A \cap B|}{\sqrt{|A| \cdot |B|}}$
Sørensen-Dice	$SD(A, B) = 2\frac{|A \cap B|}{|A| + |B|}$	Correlation	$\operatorname{corr}(A, B) = \cos^2(A, B) = \frac{|A \cap B|^2}{|A| \cdot |B|}$
Kulczynski 1	$K_1(A, B) = \frac{|A \cap B|}{|A \triangle B|}$	Kulczynski 2	$K_2(A, B) = \frac{1}{2}\left(\frac{|A \cap B|}{|A|} + \frac{|A \cap B|}{|B|}\right)$
Simpson	$\operatorname{Simpson}(A, B) = \frac{|A \cap B|}{\min(|A|, |B|)}$	Braun-Blanquet	$BB(A, B) = \frac{|A \cap B|}{\max(|A|, |B|)}$
Containment	$c(A, B) = \frac{|A \cap B|}{|A|}$		

---

[1] The random samples S_A and S_B need a final "filtering" phase before they can be used to compute the similarity $\sigma(S_A, S_B)$.

The main results of this paper can be summarized as follows: 1) For all the similarity measures in Table 1, the similarity of the random samples (after some appropriate "filtering") is an asymptotically unbiased estimator of the similarity of the corresponding sets. We have extended the results of [1] by giving general tools which help to establish the same result for many measures, eventually they could be used for some not considered here; 2) The standard relative error of the estimations goes to 0 as the size of the sets grows; something that cannot happen unless the size of the samples grows with the size of the sets. While the result is far from surprising, no previous work in the literature addressed the quantification of the standard error of the estimations, in particular, this hadn't been studied when the size of the samples is a function of the size of the sets.

Structure of the Paper. In Sect. 2 we will briefly review Affirmative Sampling, the random sampling algorithm which we assume will be used to draw random samples from the sets; it is our choice because the size of the returned samples grows with the (unknown) size of the set from which we sample, without prior knowledge of the set. This is useful in contexts in which we are presented the sets in an on-line fashion or when we have actually multi-sets (for example, text documents) and we measure their similarity in terms of the underlying sets of distinct elements (in the example, the *vocabularies* of the text documents).

In Sect. 3 we present the main results of this paper, namely, that the similarity of samples is an unbiased estimator of the similarity of the sets, for each one of the similarity measures shown in Table 1. The formal proofs of our results appear in the full version [7], not here, due to space constraints.

After that, we present in a short section (Sect. 4) the results of a small empirical study that we have conducted, showing significant accordance with the theoretical results of the previous section. We close in Sect. 5 with some final remarks and a discussion about future developments of this line of research.

2 Sampling

Let $A \subseteq \mathcal{U}$ be a finite subset of the domain $\mathcal{U}$ (also finite, but potentially extremely large). Assume that to every element $x \in \mathcal{U}$ we have assigned a random number $h(x) \in [0, 1]$, the outcome of an independent draw from a uniformly distributed random variable in $[0, 1]$. Then for any τ, $0 < \tau < 1$, the subset $A^{\geq \tau} = \{x \in A \mid h(x) \geq \tau\} \subseteq A$ is a random sample of A. Any of the n elements of A has exactly the same probability of belonging to $A^{\geq \tau}$ as any other element in A. Likewise, if we consider the subset A_k of A with the k elements of A with larger (smaller) value of h then A_k is also a random sample. In particular, if $\tau = \min\{h(x) \mid x \in A_k\}$ is the minimum h value in A_k then $A_k = A^{\geq \tau}$.

In practice, to obtain such random samples we can use a hash value $h(x)$ for each element, as presented in [1]; under pragmatic assumptions we can safely neglect the probability of collisions. Looking at the hash values as real numbers in $[0, 1]$, for any $x \in \mathcal{U}$ and any value $z \in [0, 1]$ we assume that $\Pr\{h(x) \leq z\} = z$, for a reasonably well-chosen hash function h; that is, the hash values are uniformly distributed.

If the size n of the set A from which we want to draw a sample is known beforehand, then we can take $k = k(n)$ and draw a sample of size k as described above. But even if the size of the set A were not known in advance (and we do not want to incur the costs in time and space to compute it) we can still draw random samples of variable size, growing with the (unknown) size n of the set. This can be accomplished thanks to Affirmative Sampling (AS) [6], an easy and practical alternative to just keeping the k elements with largest (smallest) hash values in A. AS also uses a fixed parameter k but produces samples of size $\geq k$ (unless $n < k$). The expected size $\mathcal{S} = |\mathsf{S}|$ of the random sample S produced by AS is $\mathbb{E}\{\mathcal{S}\} = k\ln(n/k) + \text{l.o.t.}$ with $n = |A|$. Easy variants of AS will produce random samples of size $\Theta(n^\alpha)$ for a fixed given α, $0 < \alpha < 1$ (see [6]).

Whether we use fixed-size samples or variable-size samples, as long as they are random we can make inferences about the "population" (the set A) from the sample S, for instance, about the proportion of elements in A that satisfy a certain property P. Let $n_P = |A_P|$, with A_P the subset of elements satisfying P, and n the number of elements in A. Denote $\vartheta_P := \frac{n_P}{n}$ the fraction of elements that satisfy P. Let us assume in the computations below and for the remaining of the paper that $n \geq \mathcal{S} \geq k \geq 2$, that is, that the sampling algorithm will return at least $k \geq 2$ elements. Otherwise, if the set contains less than k distinct elements, the sample contains **all** elements in A and their relevant statistics, and we can answer queries exactly.

If we have a random sample S of A of size $\mathcal{S}$ and $\mathsf{S}_P = \mathsf{S} \cap A_P$ then it is well known that the proportion $\hat{\vartheta}_P = \mathcal{S}_P/\mathcal{S}$ is an unbiased estimator of ϑ_P, even when $\mathcal{S}$ is a random variable and not a fixed value (see, for example, [6] and references therein). Quite intuitively, the accuracy of the estimator will depend on the size of the sample. The result giving its variance can be found in many places, however the size $\mathcal{S}$ of the sample is assumed fixed there. The more general statement given here, when $\mathcal{S}$ is a random variable, can be found, together with its proof, in [6].

Lemma 1. *The random variable $\hat{\vartheta}_P := \mathcal{S}_P/\mathcal{S}$, where $\mathcal{S}$ is the size of the random sample S and $\mathcal{S}_P$ is the number of elements in S that satisfy P, is an unbiased estimator of $\vartheta_P := n_P/n$, that is, $\mathbb{E}\left\{\hat{\vartheta}_P\right\} = \vartheta_P$, assuming that $n \geq \mathcal{S} > 0$. Moreover,*

$$\mathbb{V}\left\{\hat{\vartheta}_P\right\} = \frac{n_P(n - n_P)}{n(n - 1)} \cdot \left(\mathbb{E}\left\{\frac{1}{\mathcal{S}}\right\} - \frac{1}{n}\right).$$

If the behavior of the random variable $\mathcal{S} = |\mathsf{S}|$ is smooth enough[2] and $\mathbb{E}\{\mathcal{S}\} \to \infty$ when $n \to \infty$ then the accuracy of the estimator $\hat{\vartheta}_P$ will improve, as the variance will decrease and tend to 0 as $n \to \infty$.

---

[2] One can prove that in many random sampling schemes, in particular for AS, we have $\mathbb{E}\{1/\mathcal{S}\} = \mathcal{O}(1/\mathbb{E}\{\mathcal{S}\})$.

3   Estimating Similarity

Consider now two sets A and B and two random samples S_A and S_B, respectively, such that $\mathsf{S}_A = A^{\geq \tau_{\mathsf{S}_A}}$ and $\mathsf{S}_B = B^{\geq \tau_{\mathsf{S}_B}}$. Looking at Table 1 we quickly notice that many of the measures are of the form $|C_P|/|C|$, for some set C built from A and B, and some subset C_P of C of elements that satisfy a certain property P. For example, for the Jaccard similarity we have $C = A \cup B$ and the property P is "x belongs to both A and B" ($C_P = A \cap B$). To apply Lemma 1 we need to figure out how to obtain a random sample S_C of C out of the given random samples S_A and S_B, and how to find the elements in S_C which satisfy P. Our random sampling scheme guarantees that $\mathsf{S}_A = \{x \in A \mid h(x) \geq \tau_{\mathsf{S}_A}\}$ and $\mathsf{S}_B = \{x \in B \mid h(x) \geq \tau_{\mathsf{S}_B}\}$. Given two sets X and Y, let $\tau^*(X,Y) := \max\{\tau_{X \setminus Y}, \tau_{X \cap Y}, \tau_{Y \setminus X}\}$; we will take the convention that $\tau_\emptyset = 0$. Let $\tau = \tau^*(\mathsf{S}_A, \mathsf{S}_B)$. Then we can show that

$$\mathsf{S}_A^{\geq \tau} \cup \mathsf{S}_B^{\geq \tau} = (\mathsf{S}_A \cup \mathsf{S}_B)^{\geq \tau} = (A^{\geq \tau} \cup B^{\geq \tau}) = (A \cup B)^{\geq \tau},$$

that is: "filtering" S_A and S_B according to the largest threshold $\tau = \tau^*(\mathsf{S}_A, \mathsf{S}_B)$ and taking the union of these filtered samples we get a random sample of $A \cup B$. The filtering trick can also be used to produce a random sample of $A + B$ from random samples S_A and S_B, for $A \times B$, etc. This allows us to prove our first main result (its proof appears in the full version [7] of this paper).

Theorem 1. *Let σ be any of the similarity measures: Jaccard, Sørensen-Dice, containment coefficient, Kulczynski 2 (second Kulczynski coefficient) or correlation coefficient. Let S_A and S_B be random samples of A and B such that $\mathsf{S}_A = A^{\geq \tau_{\mathsf{S}_A}}$ and $\mathsf{S}_B = B^{\geq \tau_{\mathsf{S}_B}}$, and let $\tau = \tau^*(\mathsf{S}_A, \mathsf{S}_B) = \max(\tau_{\mathsf{S}_A \setminus \mathsf{S}_B}, \tau_{\mathsf{S}_B \setminus \mathsf{S}_A}, \tau_{\mathsf{S}_A \cap \mathsf{S}_B})$. Then $\hat{\sigma} = \sigma(\mathsf{S}_A^{\geq \tau}, \mathsf{S}_B^{\geq \tau})$ is an unbiased estimator of $\sigma(A,B)$, that is,*

$$\mathbb{E}\left\{\sigma(\mathsf{S}_A^{\geq \tau}, \mathsf{S}_B^{\geq \tau})\right\} = \sigma(A,B).$$

Moreover,

$$\mathbb{V}\left\{\sigma(\mathsf{S}_A^{\geq \tau}, \mathsf{S}_B^{\geq \tau})\right\} \sim \sigma(A,B) \cdot (1 - \sigma(A,B)) \cdot \mathcal{O}\left(\mathbb{E}\left\{\frac{1}{\min(|\mathsf{S}_A|, |\mathsf{S}_B|)}\right\}\right),$$

which implies that $\mathbb{V}\{\hat{\sigma}\} \to 0$ if Affirmative Sampling is used to draw the samples, since then $\mathbb{E}\{1/\min(|\mathsf{S}_A|, |\mathsf{S}_B|)\} \to 0$, if $|A|, |B| \to \infty$.

The same result holds for Simpson and Braun-Blanquet measures, provided that we know which of $|A|$ and $|B|$ is smaller (larger). If the sizes of A and B are unknown, we cannot assume that $|\mathsf{S}_A|$ and $|\mathsf{S}_B|$ have the same relation since these sizes are, in general, random variables. If we put $\min(|\mathsf{S}_A|, |\mathsf{S}_B|)$ ($\max(|\mathsf{S}_A|, |\mathsf{S}_B|)$, resp.) in the denominator of the Simpson estimator (Braun-Blanquet, resp.), we will have a bias, albeit the experiments suggest it is not very significant (see Sect. 4). The formal result, as well as a detailed discussion about the bias that we get when not knowing the sizes of A and B, can be found in the full version of the paper [7].

Last, but not least, neither the cosine similarity nor Kulczynksi 1 fit into the framework of Theorem 1. However, both can be written as functions of other similarity measures for which we have unbiased estimators. Namely, for the cosine similarity, we have $\cos(A, B) = \sqrt{\mathrm{corr}(A,B)}$. For Kulczynksi 1 we can write

$$K_1(A, B) = \frac{|A \cap B|}{|A \triangle B|} = \frac{1}{\frac{|A \cup B|}{|A \cap B|} - 1} = \frac{J(A,B)}{1 - J(A,B)}.$$

That is, these measures σ' are such that $\sigma'(A, B) = f(\sigma(A, B))$ for a similarity measure which can be estimated without bias: $\mathbb{E}\left\{\sigma(\mathsf{S}_A^{\geq \tau}, \mathsf{S}_B^{\geq \tau})\right\} = \sigma(A, B)$. In general, given a random variable X, $\mathbb{E}\left\{f(X)\right\} \neq f(\mathbb{E}\left\{X\right\})$. However, if f is a smooth function in $(0, 1)$, that is, $f \in C^{\infty}(0, 1)$, and the size of the samples grows with the size of the sets then we will have[3]

$$\mathbb{E}\left\{\sigma'(\mathsf{S}_A^{\geq \tau}, \mathsf{S}_B^{\geq \tau})\right\} = \mathbb{E}\left\{f(\sigma(\mathsf{S}_A^{\geq \tau}, \mathsf{S}_B^{\geq \tau}))\right\} \sim f\left(\mathbb{E}\left\{\sigma(\mathsf{S}_A^{\geq \tau}, \mathsf{S}_B^{\geq \tau})\right\}\right)$$

$$= f(\sigma(A, B)) = \sigma'(A, B).$$

We can therefore obtain asymptotically unbiased estimators for cosine and Kulczynski 1 (first Kulczynski coefficient) using $f(x) = \sqrt{x}$ and $\sigma = \mathrm{corr}$ for the former, and $f(x) = x/(1 - x)$ and $\sigma = \mathrm{Jaccard}$ for the latter. The proof that $\mathbb{E}\left\{f(X)\right\} \sim f(\mathbb{E}\left\{X\right\})$ under the appropriate hypotheses is given in [7].

The estimator for these similarity measures is only asymptotically unbiased, and that's because all the central moments of order $r \geq 2$ of $\sigma(A, B)$ tend to 0 as the size of the sets grows; however f or any of its derivatives might be very large, in particular when $\sigma(A, B) \to 0$ or $\sigma(A, B) \to 1$, and then the bias can be significant when the asymptotic regime hasn't been reached yet.

The variance of $\sigma'(\mathsf{S}_A^{\geq \tau}, fsB) = f(\sigma(\mathsf{S}_A^{\geq \tau}, \mathsf{S}_B^{\geq \tau}))$ can also be found using the same technique that we have used to establish that the estimator is asymptotically unbiased, by considering the Taylor series expansion of f^2. It is easy to show then that we have

$$\mathbb{V}\left\{\sigma'(\mathsf{S}_A^{\geq \tau}, \mathsf{S}_B^{\geq \tau})\right\} = \frac{1}{2} \frac{d^2}{dx^2} f^2(x)\Bigg|_{x=\sigma(A,B)} \cdot \mathbb{V}\left\{\sigma(\mathsf{S}_A^{\geq \tau}, \mathsf{S}_B^{\geq \tau})\right\} + \text{l.o.t.},$$

and since $\mathbb{V}\left\{\sigma(\mathsf{S}_A^{\geq \tau}, \mathsf{S}_B^{\geq \tau})\right\} \to 0$ when $\min(|A|, |B|) \to \infty$, we know that the variance $\mathbb{V}\left\{\sigma'(\mathsf{S}_A^{\geq \tau}, \mathsf{S}_B^{\geq \tau})\right\} \to 0$ too (and at which rate). However, $(f^2(x))''$ might be very large (actually tend to ∞) for $\sigma(A, B) \to 1$ or $\sigma(A, B) \to 0$, and then the variance of $\sigma'(\mathsf{S}_A^{\geq \tau}, \mathsf{S}_B^{\geq \tau})$ will be non-negligible in practical settings when the similarity of the two sets is very close to 0 or to 1. For example, for the cosine similarity we have $f(x) = \sqrt{x}$ and $(f^2)'' = 0$ which entails a very low variance of the estimator $\cos(\mathsf{S}_A^{\geq \tau}, \mathsf{S}_B^{\geq \tau})$ even if the similarity of the two sets is close to 0 or to 1. However, for the first Kulczynski coefficient, we have $f(x) = x/(1 - x)$ giving $(f^2)'' = (4x + 2)/(1 - x)^4$. Hence, when the

---

[3] $a_n \sim b_n$ means that $\lim_{n \to \infty} a_n/b_n = 1$.

similarity $J(A, B)$ of the two sets is very close to 1 we have $K_1(A, B) \to \infty$ and $\mathbb{V}\left\{K_1(\mathsf{S}_A^{\geq \tau}, \mathsf{S}_B^{\geq \tau})\right\} \sim \frac{3}{(1-J(A,B))^3} \cdot \mathcal{O}\left(1/\left(k \ln((|A| \cup |B|)/k)\right)\right)$, which will be quite large unless $|A| \cup |B|$ is really huge. This phenomenon shows clearly in the plot Fig. 1f given in next section.

4 Experimental Results

We have conducted a small experimental study with the aim to show representative examples of the good match between our theoretical findings and the estimates obtained in practice.

Due to the lack of space, we will report here only one experiment in which we work with two fixed sets $A = \{1, \dots, m\}$ and $B = \{r, \dots, r+n-1\}$, for some $r \leq m + 1$. Changing the value of r the intersection $|A \cap B|$ will run from 0 (if $r = m+1$) to $\min(m, n)$ (if $r = 1$). In particular, we have chosen $|A| = m = 1000$ and $|B| = n = 1500$. We apply the sampling algorithm $T = 10$ times on each set, with a different randomly chosen hash function each time, thus effectively producing T different estimates of the similarity $\sigma(A, B)$. The plots in Fig. 1 show the true similarity ($\sigma(A, B)$, red line), the estimates (blue dots) and the standard deviation in the T observations (length of the blue bars), as the size of the intersection varies from 0 to $\min(m, n) = 1000$ for some of the similarity measures studied in this paper.

The second experiment, reported in [7], studies how the quality of the estimates impact the application using them, in particular, we have studied how the clusterings produced by k-means change when we use estimates of the similarities instead of the real similarities.

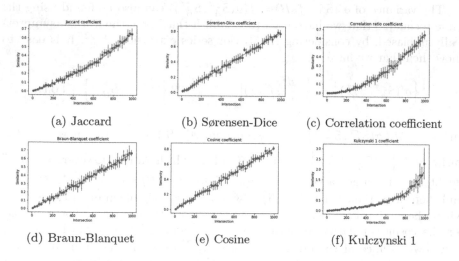

(a) Jaccard   (b) Sørensen-Dice   (c) Correlation coefficient

(d) Braun-Blanquet   (e) Cosine   (f) Kulczynski 1

Fig. 1. Empirical estimates of several similarity measures

5   Final Remarks

We show in this paper that many similarity measures between sets can be accurately estimated, going further beyond Broder's results [1] on the Jaccard and containment indices. One fundamental ingredient for (asymptotically) unbiased estimation with a high degree of accuracy is the use of variable-size sampling. Using Affirmative Sampling, the (expected) size of the samples increases with the size of the sets, and the standard relative error in the estimations goes to 0 [6].

Another line of research that we are already working on is the estimation of similarity measures between objects other than sets. For example, multisets, sequences, ... A notable example is partitions (clusterings). For instance, instead of examining all the $\binom{N}{2}$ pair of objects to compute the Rand index (a measure of similarity, see for instance, [5]) of two clusterings of a set of N objects, we can draw two random samples S and S', both containing $\ll N$ elements, form all ordered pairs combining distinct elements from S and S', and estimate the Rand index from those. It is straightforward to show that this is an unbiased estimate of the true Rand index using our techniques, and we think that they can be used to show the same for many other similarity measures between clusterings.

References

1. Broder, A.Z.: On the resemblance and containment of documents. In: Carpentieri, B., De Santis, A., Vaccaro, U., Storer, J.A. (eds.) Proceedings of the Compression and Complexity of SEQUENCES 1997, pp. 21–29. IEEE Computer Society (1997)
2. Chávez, E., Navarro, G., Baeza-Yates, R.A., Marroquín, J.L.: Searching in metric spaces. ACM Comput. Surv. **33**(3), 273–321 (2001)
3. Chen, L., Gao, Y., Song, X., Li, Z., Miao, X., Jensen, C.S.: Indexing metric spaces for exact similarity search. ACM Comput. Surv. **55**(6), 128:1–128:39 (2022). https://doi.org/10.1145/3534963
4. Deza, M.M., Deza, E.: Encyclopedia of Distances. Springer, Berlin (2009). https://doi.org/10.1007/978-3-642-00234-2
5. Hubert, L., Arabie, P.: Comparing partitions. J. Classif. **2**, 193–218 (1985)
6. Lumbroso, J., Martínez, C.: Affirmative sampling: theory and applications. In: Ward, M.D. (ed.) Proceedings of the 33$^r d$ International Meeting on Probabilistic, Combinatorial and Asymptotic Methods for the Analysis of Algorithms (AofA), vol. 225 of Leibniz International Proceedings in Informatics (LIPIcs), , Dagstuhl, Germany, pp. 12:1–12:17. Schloss Dagstuhl - Leibniz-Zentrum für Informatik (2022). https://doi.org/10.4230/LIPIcs.AofA.2022.12
7. Martínez, C., Viola, A., Wang, J.: Fast and accurate similarity estimation using samples (2023). https://doi.org/10.13140/RG.2.2.24053.14566
8. Rajaraman, A., Leskovec, J., Ullman, J.D.: Mining Massive Datasets, 3rd edn. Cambridge University Press, Cambridge (2014). https://www.mmds.org/
9. Samet, H.: Foundations of Multidimensional and Metric Data Structures. Morgan Kaufmann, Boston (2006)

Retrieve-and-Rank End-to-End Summarization of Biomedical Studies

Gianluca Moro(✉) , Luca Ragazzi , Lorenzo Valgimigli ,
and Lorenzo Molfetta

Department of Computer Science and Engineering (DISI), University of Bologna,
Via dell'Università 50, 47522 Cesena, Italy
{gianluca.moro,l.ragazzi,lorenzo.valgimigli}@unibo.it,
lorenzo.molfetta@studio.unibo.it

Abstract. An arduous biomedical task involves condensing evidence derived from multiple interrelated studies, given a context as input, to generate reviews or provide answers autonomously. We named this task context-aware multi-document summarization (CA-MDS). Existing state-of-the-art (SOTA) solutions require truncation of the input due to the high memory demands, resulting in the loss of meaningful content. To address this issue effectively, we propose a novel approach called RAM-SES, which employs a retrieve-and-rank technique for end-to-end summarization. The model acquires the ability to (i) index each document by modeling its semantic features, (ii) retrieve the most relevant ones, and (iii) generate a summary via token probability marginalization. To facilitate the evaluation, we introduce a new dataset, FAQsUMC19, which includes the synthesizing of multiple supporting papers to answer questions related to Covid-19. Our experimental findings demonstrate that RAMSES achieves notably superior ROUGE scores compared to state-of-the-art methodologies, including the establishment of a new SOTA for the generation of systematic literature reviews using Ms2. Quality observation through human evaluation indicates that our model produces more informative responses than previous leading approaches.

Keywords: Biomedical Multi-Document Summarization · Neural Semantic Representation · End-to-End Neural Retriever

1 Introduction

Given the paramount societal role of biomedicine and related natural language processing (NLP) tasks [11–15,38], aggregating information from multiple topic-related biomedical papers to help search, synthesize, and answer questions is of great interest [7]. Real-world applications require indexing, combining, and summarizing evidence from clinical trials on a research background to produce systematic literature reviews (SLRs) or answer medical inquiries. Consequently, we define such activities as *context-aware multi-document summarization* (CA-MDS) due to the presence of an input context (i.e., background or question) that

© The Author(s), under exclusive license to Springer Nature Switzerland AG 2023
O. Pedreira and V. Estivill-Castro (Eds.): SISAP 2023, LNCS 14289, pp. 64–78, 2023.
https://doi.org/10.1007/978-3-031-46994-7_6

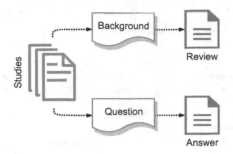

Fig. 1. The overview of biomedical CA-MDS. In our experiments, we use the input contexts (i.e., the background or question) to retrieve the salient studies and aggregate them to generate the target (Input → Output). For example, given a set of topic-related scientific papers, the goal is to select the ones more correlated to a user's input question to produce a single answer. (Color figure online)

conditions the downstream summarization task (Fig. 1). In real life, biomedical articles usually contain several thousands of words that compose lingo and complicated expressions, making understanding them a time- and labor-consuming process even for professionals. Thus, automation support for biomedical activities is practical and beneficial in facilitating knowledge acquisition.

CA-MDS solutions for biomedical applications should process all inputs without ignoring any details, reducing the risk of model hallucination, namely generating unfaithful outputs due to training on targets having facts unfounded by the source. Therefore, state-of-the-art (SOTA) models rely on sparse transformers [2], Fusion-in-Decoder strategies [18], and marginalization-based decoding [34]. However, such methods either (i) need high memory requirements that force input truncation for organizations operating in low-resource regimes [29–31,33,35], or (ii) lack end-to-end learning, reducing the potential of cooperating neural modules.

In this paper, we introduce RAMSES,[1] a retrieve-and-rank summarization approach trained via end-to-end learning to retrieve salient biomedical documents by their semantic meaning and synthesize them given an input context. RAMSES comprises a biomedical bi-encoder and a generative aggregator. The bi-encoder reads all the documents, represents their semantics via embeddings, and retrieves and scores salient documents related to an input context. Then, the aggregator is conditioned by the context along with these latent documents to decode the summary by marginalizing the token probability distribution weighted by their relevance score.

We evaluated RAMSES in two biomedical CA-MDS tasks: (i) producing SLRs on the Ms2 dataset [7] and (ii) answering frequently asked questions (FAQs) about Covid-19 in our proposed dataset FAQsUMC19. In detail, we collected 514 Covid-19 FAQs with high-quality abstractive answers written by experts. Then we augmented each instance with 30 supporting scientific papers containing the information needed to answer the question, producing 15,420 articles.

---
[1] https://disi-unibo-nlp.github.io/projects/ramses/.

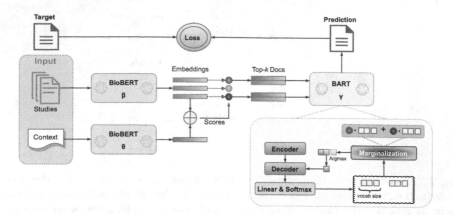

Fig. 2. The overview of RAMSES. The input is a biomedical context and multiple studies that are encoded by two different BIOBERT models. Then, we compute the relevance score of each document conditioned by the context. Finally, the top-k most salient documents are concatenated with the context and given to BART that marginalizes their token probability distribution, weighted by the relevance scores, at decoding time.

In particular, FAQSUMC19 has two essential features: (i) includes abstractive answers authored by experts, unlike other related datasets that use extractive targets [41]; (ii) is the first CA-MDS dataset for Covid-19, becoming a crucial benchmark for producing multi-document summaries to answer questions on Covid-19 with the support of updated related biomedical papers.

We perform extensive experiments, showing that RAMSES achieves new SOTA performance in the MS2 dataset and outperforms previous solutions in FAQSUMC19, whose inferred answers are also rated as of more quality by human experts.

2   Related Work

Semantic Neural Retriever Applications. The semantic representation skill exhibited by neural networks has catalyzed the emergence of groundbreaking neural methodologies in information retrieval [10] First, the algorithm BM25 has been exceeded by dense passage retrieval (DPR) [21], a remarkable neural application that has since evolved into a fundamental element within numerous neural-driven retrieval solutions [42,43]. These neural retrievers have been fused with a language model to enrich and improve input [23], generating superior models characterized by increased efficiency and improved performance [3,12]. Despite their promising results, the end-to-end application of these solutions in MDS remains unexplored.

NLP for Biomedical Documents. Much recent work in NLP has concentrated on the biomedical domain [28], including CA-MDS [7], which can decrease the burden on medical workers by highlighting and aggregating key points while reducing the amount of information to read. Previous contributions focused on the automatic generation of SLRs. In detail, cutting-edge solutions rely on three different neural architectures: (i) transformer-based models with linear complexity in the input size thanks to sparse attention [2], which concatenate the input context along with all documents in the cluster producing a single source sequence; (ii) quadratic transformers with Fusion-in-Decoder [18], which join the hidden states of documents after encoding them individually; (iii) marginalization-based decoding augmented by frozen retrievers [34], which first pinpoints salient documents w.r.t. a query and produces a single summary by summing the probability distribution of the inferred token for each document.

MDS Solutions in Other Domains. Flat approaches with MDS-specific pre-training [49] concatenate the sources in a single text, treating MDS as a single-input task. Hierarchical approaches merge document relations to obtain semantically rich representations by leveraging graph-based methods [1] and multi-head pooling and interparagraph attention [19]. Marginalization-based approaches [17] apply marginalization to the token probability distribution at the decoding time to produce a single output from many inputs. The two-stage approaches [25] adopt different strategies to rank sources before producing the summary. Unlike previous work, RAMSES is trained in end-to-end learning to retrieve relevant text from biomedical articles and marginalize the probability distribution of the latent extracted information at decoding time.

Covid-19 Datasets. With the appearance of Covid-19, thousands of articles have been published quickly. To aid experts in accessing this knowledge, large organizations collected corpora such as CORD-19 [47] and LITCOVID [6], encouraging the proposal of task-specific datasets. COVID-QA [27] study question-answering using annotated pairs extracted from 147 papers. COVID-Q [48] collects 16,690 questions about Covid-19, classifying them into 15 categories. [40] scrapped over 40 trusted websites for Covid-19 FAQs, creating a collection of 2100 questions. [45,52] proposed two datasets for the retrieval of FAQs, where user queries are semantically paired with existing FAQs. FAQSUMC19 fills this gap, introducing the first CA-MDS dataset to answer Covid-19 FAQs by summarizing multiple related studies.

Fine-grained comparisons with previous work are in Sect. 6.1.

3 Preliminary

We provide details for context-aware multi-document summarization (CA-MDS).

Definition. CA-MDS aims to compile a summary from a cluster of related articles given an input context, analogous to the query in query-focused summarization [46]. Yet, unlike answering FAQs, SLR generation does not consider questions. Thus, we define the task we face as CA-MDS. The biomedical tasks we address in this work, such as SLR generation and FAQ answering, are CA-MDS tasks because they both have an input context (i.e., the research issue in SLRs and the human question in FAQs) and many topic-related documents from which produce the output.

Problem Formulation. In the CA-MDS setting, we have $(c, \mathbf{D}, y)$, where c is the input context, $\mathbf{D}$ is the cluster of topic-related documents, and y is the target generated from $\mathbf{D}$ given c. Formally, we want to predict y from $\{c, d_1, ..., d_n | d \in \mathbf{D}\}$.

4   Method

The end-to-end learning of RAMSES allows the cooperating modules to jointly retrieve and aggregate key information from multiple sources in one output (Fig. 2).

Given the context c and the documents $\mathbf{D}$, our method first generates relevance scores on $\mathbf{D}$ with a biomedical solution based on DPR [20]:

$$p_{\beta,\theta}(d \in \mathbf{D}|c) = (Enc_\beta(d) \oplus Enc_\theta(c)) \tag{1}$$

where Enc_β and Enc_θ are two different BIOBERT-base models trained to produce a dense representation of documents and the context [39], respectively, $\oplus$ is the inner product between them, and $p(d|c)$ is the relevance score associated to the document d given c. Thus, our solution finds the most top-k relevant texts according to c. Then, given c and each $d \in$ top-k, a BART-base model [22] draws a distribution for each next output token for each d, before marginalizing:

$$p(y|c, \mathbf{D}) = \prod_z^N \sum_{d \in \text{top-}k} p_\theta(d|c) p_\gamma(y_z|d', y_{1:z-1}) \tag{2}$$

where $d' = [c, tok, d]$ is the concatenation of c and $d \in$ top-k with a special text separator token (<doc-sep>) to make the model aware of the textual boundary, N is the target length, and $p_\gamma(y_z|d', y_{1:z-1})$ is the probability of generating the target token y_z given d' and the previously generated tokens $y_{1:z-1}$.

We train our RAMSES model by minimizing the negative marginal log-likelihood of each target with the following loss function:

$$\mathcal{L} = -\sum_i \log p(y_i|c_i, \mathbf{D}_i) \tag{3}$$

End-to-End Learning. The model (Eq. 2) allows the gradient to backpropagate to all modules. For clarity, we rewrite the formula as a continuous function, as follows:

$$\text{RAMSES}(\mathbf{D}, c) = \sum_{(d_j, s_j)} B_\gamma([c, tok, d_j]) \cdot s_j \tag{4}$$

$$\text{top-}k(\mathbf{D}, c) = [(d_1, s_1), \dots, (d_k, s_k)] \tag{5}$$

$$s_j = Enc_\beta(d_j) \oplus Enc_\theta(c) \tag{6}$$

where $(d_j, s_j) \in$ top-k and B_γ is BART.

The presence of s_j in Eq. 4 allows the gradient, computed by minimizing the objective function, to reach Enc_β and Enc_θ. For this reason, the documents and context embeddings are adjusted during the training to improve the generated summary, making all modules of our solution learn jointly in an end-to-end fashion.

Table 1. The question-cluster pairs' quality. Best values are bolded.

	ROUGE			BERTScore		
	Avg	Max	Min	Avg	Max	Min
RANDOM	12.34	23.21	0.17	11.30	24.71	2.80
BM25	16.70	29.68	0.37	16.17	35.20	1.13
SUBLIMER	**20.44**	**33.31**	**2.32**	**21.11**	**36.79**	**5.75**

5 FAQsumC19 Dataset

We introduce a new dataset, FAQSUMC19, containing 514 Covid-19-related FAQs with abstractive answers written by experts, each supported by 30 abstracts of scientific articles, for a total of 15,420 documents. We obtained from the Covid-19 FAQ section on WHO[2] all available question-answer pairs. We then augmented each instance with 30 Covid-19 scientific articles strictly related to the question from the updated version of the CORD-19 dataset [47]. Specifically, we experimented with the selection of supporting articles with different information retrieval methods, such as a random baseline, BM25 [44], and SUBLIMER [38]. We used the concatenation between the question and the answer to retrieve the first 30 ranked documents regarding semantic similarity, creating a knowledge base to support the answer generation. We finally split the dataset into 464 instances for training ($\approx 90\%$) and 50 for the test ($\approx 10\%$).

To assess the quality of question-cluster pairs in our dataset, we computed the content coverage with ROUGE-1 precision [24] and BERTScore [51] of the question-answer concatenation w.r.t. each document in the cluster, and calculate the average score. We evaluate the syntactic and semantic overlap between the question and answer and the texts. Table 1 reveals that SUBLIMER achieves the best scores, as expected.

---

[2] https://www.who.int/emergencies/diseases/novel-coronavirus-2019/question-and-answers-hub.

6   Experiments

6.1   Experimental Setup

Datasets. Table 2 reports the statistics of the datasets used to test RAMSES in different biomedical tasks: **Ms2** [7] consists of 15,597 instances derived from the scientific literature. Each sample is composed of (i) the background statement, which describes the context research issue, (ii) the target statement, which is the summary to generate; and (iii) the studies, which are the abstracts of biomedical documents that contain the needed information for the research issue. **FAQsumC19** is our proposed dataset that comprises 514 Covid-19 FAQs with abstractive answers written by experts, each supported by 30 abstracts of scientific papers.

Table 2. The datasets used for evaluation (FAQsumC19 is ours). Statistics include dataset size and the average (i) number of source (S) documents per instance, (ii) number of total words in S and target (T) texts, and (iii) S-T compression ratio of words [16].

Dataset	Samples	S		T	S → T
		Docs	Words	Words	Comp
Ms2	15,597	23.30	9563.95	70.81	135.06
FAQsumC19	514	30	5635.50	139.06	40.53

Baselines. We compare RAMSES with SOTA solutions: **Bart-FiD** [7], which is BART with the Fusion-in-Decoder strategy [18], encodes all sources individually and combines their hidden states before decoding. **Led-Gaq** [7], which is LED [2] with global attention on the input query, concatenates all texts in a single input of up to 16,384 tokens. **Damen** [34], a retrieval-enhanced solution with marginalization-based decoding, discriminates important fragments of the cluster with a frozen BERT-base model and marginalizes their probability distribution during decoding. **Primera** [49], which is LED pre-trained with a multi-document summarization-specific objective, concatenates the texts with a special separator token up to 4096 tokens in size.

Evaluation Metrics. We use ROUGE-1/2/L [24] to assess fluency and informativeness. We also adopt $\mathcal{R}$ [32] as an aggregated judgment that considers the variance of the ROUGE scores. Finally, we perform qualitative analysis to bridge the superficiality of automatic evaluation measures.

Implementation. We fine-tune the models using PyTorch and the HuggingFace library, setting the seed to 42 for reproducibility. RAMSES is trained on an NVIDIA RTX 3090 GPU of 24 GB memory from an internal cluster for 1 epoch with a learning rate of 3e-5 on Ms2 and for 3 epochs with a learning rate of 1e-5 on FAQsumC19. For decoding, we use the beam search with 4 beams and the following min-max target size: 32–256 for Ms2 and 100-256 for FAQsumC19.

Table 3. Performance of models on the evaluation datasets. The best scores are in bold.

Model	Ms2				FAQsumC19			
	R-1$_{f1}$	R-2$_{f1}$	R-L$_{f1}$	$\mathcal{R}$	R-1$_{f1}$	R-2$_{f1}$	R-L$_{f1}$	$\mathcal{R}$
Baselines								
Led-Gaq	26.89	8.91	20.32	18.60	25.55	4.42	13.77	14.47
Bart-FiD	27.56	9.49	20.80	19.18	20.26	5.59	14.84	13.51
Damen	28.95	9.72	21.83	20.04	23.81	3.50	13.03	13.35
Primera	30.07	9.85	22.16	20.55	25.04	3.64	13.00	13.79
Our								
Ramses	**31.83**	**10.44**	**22.19**	**21.32**	**30.18**	**7.31**	**15.67**	**17.56**

Table 4. ROUGE F1 scores (R-1, R-2, R-L) on Ms2 on evaluating Ramses with different generator checkpoints (B and L stand for base and large, respectively) and k documents retrieved at training time. Oom means "GPU out of memory exception." The best results are bolded.

k	Ms2		
	Bart-B	Bart-L	Pegasus-L
3	31.14/9.78/21.43	31.93/10.40/21.96	27.83/7.27/18.93
6	31.12/9.88/21.68	**31.97/10.58/22.15**	28.71/8.26/19.39
9	31.81/10.30/22.13	31.00/10.17/21.76	28.61/8.32/19.35
12	31.15/10.14/21.58	31.10/10.09/6.52	27.58/7.30/18.51
15	30.94/9.82/21.39	31.72/10.53/21.90	Oom
18	31.36/10.20/21.75	31.81/10.43/21.87	Oom

6.2 Results

Table 3 reports the performance of the models in the two evaluation datasets. Ramses yields better scores, suggesting that the retrieve and rank end-to-end learning is more effective than prior SOTA approaches in both biomedical CA-MDS tasks.

The Impact of k. As our method relies on learning to select the best top-k relevant documents from the cluster, the value of k is crucial for model performance and GPU memory occupation. Therefore, we analyze the impact of k on model performance by experimenting with a different number of documents to retrieve: 3, 6, 9, 12, 15, 18. Table 4 reports a slight performance improvement as k increases until a threshold is reached (e.g., $k = 9$ for BART-base), indicating that the marginalization approach with more documents helps produce better ROUGE scores. However, a high k (i.e., $k \geq 12$) can also increase information redundancy and contradiction, lowering the final performance. Table 4 also lists the results of different models on single text summarization as the aggregator's checkpoint, such as BART and PEGASUS [50]. We notice that BART-large achieves better ROUGE scores, although PEGASUS is the largest model. However, as BART-base achieved a slightly lower result despite the noticeably fewer trainable parameters, we chose to use it for all experiments. Therefore, we tested the best checkpoint of BART-base trained with $k = 9$ with a different k at the inference time in Ms2. Table 5 reports that the best performance has been achieved with $k = 12$. Furthermore, Table 5 also shows the results on FAQSUMC19 with a different k at training time, revealing a trend similar to Ms2.

Memory Requirements. Figure 3 shows the memory complexity at the training time of RAMSES for each k. We notice that the memory occupation is linear w.r.t. k, indicating that our solution is not computationally expensive, even for large clusters.

6.3   Ablation Studies

Table 6 reports the ablation studies on Ms2 using RAMSES with BART-base and $k = 9$ with the same hyperparameter settings for all experiments.

Table 5. The results of RAMSES on Ms2 by varying k at inference time and on FAQ-SUMC19 by varying k at training time. The best scores are bolded.

k	Ms2			FAQSUMC19		
	R-1$_{f1}$	R-2$_{f1}$	R-L$_{f1}$	R-1$_{f1}$	R-2$_{f1}$	R-L$_{f1}$
3	30.98	9.75	21.48	29.32	7.05	15.71
6	31.49	10.14	21.94	28.69	6.51	15.20
9	31.81	10.30	22.13	28.74	6.83	15.44
12	**31.83**	**10.44**	**22.19**	**30.18**	**7.31**	**15.67**
15	31.73	10.36	22.10	29.14	6.65	15.26
18	31.72	10.39	22.04	29.06	6.89	15.31

Excluding the input context from the input concatenation to give to the generative aggregator (*w/o* **context**) leads to the most significant decrease in performance. Indeed, the context is the research question shared by all documents in the cluster, so it contains important information for the final summary.

Training a single model to encode both the context and documents (*w/o* **bi-encoder**), namely using a shared BIOBERT model, decreases performance. Indeed, since the context and the documents have two different purposes (i.e., we need the context to select context-related documents), two models are needed to specialize and differentiate the text representation. Despite the similarity of the two tasks, they differ for two main reasons: (i) the context is relatively shorter than the documents in the cluster, and (ii) the concept expressiveness is denser in the context than in the more verbose documents.

Removing the token separator <doc-sep> between the context and document (*w/o* **token-sep**) decreases performance. Indeed, this token is needed to make the model aware of the textual boundary between the context and the documents.

Using cosine similarity instead of the inner product (*w/o* **inner-product**) to score the documents against the input context achieves the worst results.

Freezing Enc_β (*w/o* **trained-retrieval**) decreases performance, highlighting the usefulness of end-to-end learning to allow the model to select more informative documents.

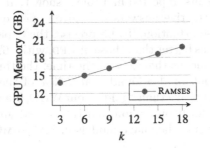

Fig. 3. The RAMSES's GPU memory requirements by varying k at training time.

Table 6. The ablation studies on MS2. We gradually remove each module of RAMSES to show the performance drop. The best scores are in bold.

	Ms2			
	R-1$_{f1}$	R-2$_{f1}$	R-L$_{f1}$	$\mathcal{R}$
RAMSES	31.83	10.44	22.19	21.32
w/o context-first	31.57	10.26	21.89	21.08
w/o trained-retrieval	31.24	10.03	21.77	20.86
w/o inner-product	31.12	10.10	21.82	20.86
w/o token-sep	31.03	10.12	21.77	20.82
w/o bi-encoder	31.47	9.88	21.52	20.79
w/o context	25.26	5.33	17.37	15.88

Switching the context with the documents in the input concatenation (*w/o* **context-first**) decreases performance, indicating that the leading context in the input helps the generative aggregator to focus better on how to join context-related information.

6.4 Human Evaluation

Considering the drawbacks of automatic metrics such as ROUGE [9], which is still the standard for evaluating text generation, we qualitatively evaluated the answers inferred from the FAQs of the entire FAQsumC19 test set with three domain experts with master's degrees in medical and biological areas.

Instructions. We gave evaluators a table, with each row containing the question and three possible answers in random order: (i) the "gold" from WHO, (ii) the prediction of RAMSES, and (iii) the prediction of LED-GAQ (the second-best model in FAQsumC19 according to $\mathcal{R}$). Each expert was asked to order the answers according to how thoroughly they answered the question, focusing primarily on the factuality. For fairness, we did not inform the evaluators about the answers' origins and the test's goal. Overall, experts completed the task in two days, reporting no difficulty ordering the 50 answers.

Results. Evaluation results, reported in Table 7, show that our method produces better informative abstractive answers to a given open question than a linear transformer with sparse attention. To be precise, the experts rated 76% of the answers of our solution as better than those of LED, with 46% agreement between the annotators (which means that 46% of the time, the three evaluators agree). Furthermore, the evaluators also found that 7.33% of the answers inferred by RAMSES are more informative than "gold" from WHO. Nevertheless, model generations are still far from being as informative as gold answers, indicating the limitations of current neural language models in FAQsumC19.

Table 7. The human evaluation on FAQsumC19 with inter-annotator agreement (IAA) using WHO ground-truth answers and the inferred ones by RAMSES and LED-GAQ. A>B means how many times the generated answer from A was scored higher than B.

	Human Evaluation		
	RAMSES>LED	RAMSES>WHO	LED>WHO
Eval 1	84%	0%	0%
Eval 2	72%	20%	14%
Eval 3	72%	2%	0%
AVG	76%	7.3%	4.7%
IAA	46%	80%	86%

7 Conclusion

In this paper, we introduced RAMSES, a retrieve-and-rank end-to-end learning solution for CA-MDS of biomedical studies. RAMSES is designed to simultaneously acquire indexing capabilities and retrieve pertinent documents to generate comprehensive summaries. Through multiple experiments on two biomedical datasets (including our proposed FAQSUMC19 to answer Covid-19 FAQs), we found that RAMSES outperforms SOTA models. This finding suggests that the integrated retrieval mechanism significantly benefits the CA-MDS task. Yet, human assessments indicate that there is still notable room for improvement, motivating further research in pursuit of novel retrieval applications within the realm of biomedical multi-document summarization.

Future works can investigate and include multimodal [36,37], cross-domain [8], and knowledge propagation [4,5,26] approaches.[3]

Acknowledgments. This research is partially supported by (i) the Complementary National Plan PNC-I.1, "Research initiatives for innovative technologies and pathways in the health and welfare sector" D.D. 931 of 06/06/2022, DARE—DigitAl lifelong pRevEntion initiative, code PNC0000002, CUP B53C22006450001, (ii) the PNRR, M4C2, FAIR—Future Artificial Intelligence Research, Spoke 8 "Pervasive AI," funded by the European Commission under the NextGeneration EU program. The authors thank the Maggioli Group for granting the Ph.D. scholarship to Luca Ragazzi and Lorenzo Valgimigli.

References

1. Amplayo, R.K., Lapata, M.: Informative and controllable opinion summarization. In: EACL, Online, April 19–23 2021, pp. 2662–2672. ACL (2021)
2. Beltagy, I., Peters, M.E., Cohan, A.: Longformer: The long-document transformer. CoRR abs/2004.05150 (2020)
3. Borgeaud, S., Mensch, A., Hoffmann, J., Cai, T., et al.: Improving language models by retrieving from trillions of tokens. In: ICML. PMLR, vol. 162, pp. 2206–2240. PMLR (2022)
4. Cerroni, W., Moro, G., Pasolini, R., Ramilli, M.: Decentralized detection of network attacks through P2P data clustering of SNMP data. Comput. Secur. **52**, 1–16 (2015). https://doi.org/10.1016/j.cose.2015.03.006
5. Cerroni, W., Moro, G., Pirini, T., Ramilli, M.: Peer-to-peer data mining classifiers for decentralized detection of network attacks. In: ADC. CRPIT, vol. 137, pp. 101–108. ACS (2013)
6. Chen, Q., Allot, A., Lu, Z.: Litcovid: an open database of COVID-19 literature. Nucleic Acids Res. **49**(Database-Issue), D1534–D1540 (2021)
7. DeYoung, J., Beltagy, I., van Zuylen, M., Kuehl, B., et al.: Ms^2: Multi-document summarization of medical studies. In: EMNLP, Punta Cana, 7–11 November, 2021, pp. 7494–7513. ACL (2021). https://doi.org/10.18653/v1/2021.emnlp-main.594

[3] https://www.maggioli.com/who-we-are/company-profile.

8. Domeniconi, G., Moro, G., Pagliarani, A., Pasolini, R.: On deep learning in cross-domain sentiment classification. In: IC3K (Volume 1), Funchal, Madeira, Portugal, November 1–3, 2017, pp. 50–60. SciTePress (2017). https://doi.org/10.5220/0006488100500060

9. Fabbri, A.R., Kryscinski, W., McCann, B., Xiong, C., et al.: Summeval: re-evaluating summarization evaluation. TACL **9**, 391–409 (2021). https://doi.org/10.1162/tacl_a_00373

10. Formal, T., Piwowarski, B., Clinchant, S.: Match your words! a study of lexical matching in neural information retrieval. In: Hagen, M., et al. (eds.) Advances in Information Retrieval: 44th European Conference on IR Research, ECIR 2022, Stavanger, Norway, April 10–14, 2022, Proceedings, Part II, pp. 120–127. Springer, Cham (2022). https://doi.org/10.1007/978-3-030-99739-7_14

11. Frisoni, G., Italiani, P., Salvatori, S., Moro, G.: Cogito Ergo *Summ*: Abstractive Summarization of Biomedical Papers via Semantic Parsing Graphs and Consistency Rewards. In: AAAI 2023, Washington, DC, USA, February 7–14, 2023. AAAI Press, Washington, DC, USA (2023)

12. Frisoni, G., Mizutani, M., Moro, G., Valgimigli, L.: Bioreader: a retrieval-enhanced text-to-text transformer for biomedical literature. In: EMNLP 2022, pp. 5770–5793. ACL, Abu Dhabi, United Arab Emirates (2022)

13. Hammoudi, Slimane, Quix, Christoph, Bernardino, Jorge (eds.): Data Management Technologies and Applications: 9th International Conference, DATA 2020, Virtual Event, July 7–9, 2020, Revised Selected Papers. Springer, Cham (2021)

14. Frisoni, G., Moro, G., Carbonaro, A.: Learning interpretable and statistically significant knowledge from unlabeled corpora of social text messages: a novel methodology of descriptive text mining. In: DATA, pp. 121–134. SciTePress (2020)

15. Frisoni, G., Moro, G., Carbonaro, A.: A survey on event extraction for natural language understanding: Riding the biomedical literature wave. IEEE Access **9**, 160721–160757 (2021). https://doi.org/10.1109/ACCESS.2021.3130956

16. Grusky, M., Naaman, M., Artzi, Y.: Newsroom: A dataset of 1.3 million summaries with diverse extractive strategies. In: NAACL (Long Papers), pp. 708–719. ACL, New Orleans, Louisiana (2018). https://doi.org/10.18653/v1/N18-1065

17. Hokamp, C., Ghalandari, D.G., Pham, N.T., Glover, J.: Dyne: Dynamic ensemble decoding for multi-document summarization. CoRR abs/2006.08748 (2020)

18. Izacard, G., Grave, E.: Leveraging passage retrieval with generative models for open domain question answering. In: EACL: Main Volume, pp. 874–880. ACL, Online (2021). https://doi.org/10.18653/v1/2021.eacl-main.74

19. Jin, H., Wang, T., Wan, X.: Multi-granularity interaction network for extractive and abstractive multi-document summarization. In: ACL, Online, July 5–10 2020, pp. 6244–6254. ACL (2020). https://doi.org/10.18653/v1/2020.acl-main.556

20. Karpukhin, V., Oğuz, B., Min, S., Lewis, P., et al.: Dense passage retrieval for open-domain question answering. arXiv preprint arXiv:2004.04906 (2020)

21. Karpukhin, V., Oguz, B., Min, S., Lewis, P.S.H., et al.: Dense passage retrieval for open-domain question answering. In: EMNLP 2020, Online, November 16–20, 2020, pp. 6769–6781. ACL (2020). https://doi.org/10.18653/v1/2020.emnlp-main.550

22. Lewis, M., Liu, Y., Goyal, N., Ghazvininejad, M., et al.: BART: denoising sequence-to-sequence pre-training for natural language generation, translation, and comprehension. In: ACL, July 5–10 2020, pp. 7871–7880 (2020). https://doi.org/10.18653/v1/2020.acl-main.703

23. Lewis, P.S.H., Perez, E., Piktus, A., Petroni, F., et al.: Retrieval-augmented generation for knowledge-intensive NLP tasks. In: NeurIPS 2020, December 6–12, 2020, virtual (2020)
24. Lin, C.Y.: ROUGE: A package for automatic evaluation of summaries. In: Text Summarization Branches Out, pp. 74–81. ACL, Barcelona, Spain (2004)
25. Liu, Y., Lapata, M.: Hierarchical transformers for multi-document summarization. In: ACL, Florence, Italy, July 28- August 2 2019, pp. 5070–5081. ACL (2019). https://doi.org/10.18653/v1/p19-1500
26. Lodi, S., Moro, G., Sartori, C.: Distributed data clustering in multi-dimensional peer-to-peer networks. In: (ADC), Brisbane, 18–22 January, 2010. CRPIT, vol. 104, pp. 171–178. ACS (2010)
27. Möller, T., Reina, A., Jayakumar, R., Pietsch, M.: Covid-qa: a question answering dataset for Covid-19 (2020)
28. Moro, G., Masseroli, M.: Gene function finding through cross-organism ensemble learning. BioData Min. **14**(1), 14 (2021)
29. Moro, G., Piscaglia, N., Ragazzi, L., Italiani, P.: Multi-language transfer learning for low-resource legal case summarization. Artif. Intell. Law **31** (2023)
30. Moro, G., Ragazzi, L.: Semantic self-segmentation for abstractive summarization of long documents in low-resource regimes. In: AAAI 2022, Virtual Event, February 22 - March 1, 2022, pp. 11085–11093. AAAI Press (2022). www.ojs.aaai.org/index.php/AAAI/article/view/21357
31. Moro, G., Ragazzi, L.: Align-then-abstract representation learning for low-resource summarization. Neurocomputing **548**, 126356 (2023). https://doi.org/10.1016/j.neucom.2023.126356
32. Moro, G., Ragazzi, L., Valgimigli, L.: Carburacy: summarization models tuning and comparison in eco-sustainable regimes with a novel carbon-aware accuracy. AAAI **37**(12), 14417–14425 (2023). https://doi.org/10.1609/aaai.v37i12.26686
33. Moro, G., Ragazzi, L., Valgimigli, L.: Graph-based abstractive summarization of extracted essential knowledge for low-resource scenario. In: ECAI 2023, Kraków, Poland, September 30 - October 4, 2023, pp. 1–9 (2023)
34. Moro, G., Ragazzi, L., Valgimigli, L., Freddi, D.: Discriminative marginalized probabilistic neural method for multi-document summarization of medical literature. In: ACL, pp. 180–189. ACL, Dublin, Ireland (May 2022). https://doi.org/10.18653/v1/2022.acl-long.15
35. Moro, G., Ragazzi, L., Valgimigli, L., Frisoni, G., Sartori, C., Marfia, G.: Efficient memory-enhanced transformer for long-document summarization in low-resource regimes. Sensors **23**(7) (2023). https://doi.org/10.3390/s23073542, www.mdpi.com/1424-8220/23/7/3542
36. Moro, G., Salvatori, S.: Deep vision-language model for efficient multi-modal similarity search in fashion retrieval, pp. 40–53 (09 2022). https://doi.org/10.1007/978-3-031-17849-8_4
37. Moro, G., Salvatori, S., Frisoni, G.: Efficient text-image semantic search: A multimodal vision-language approach for fashion retrieval. Neurocomputing **538**, 126196 (2023). https://doi.org/10.1016/j.neucom.2023.03.057
38. Moro, G., Valgimigli, L.: Efficient self-supervised metric information retrieval: A bibliography based method applied to COVID literature. Sensors **21**(19) (2021). https://doi.org/10.3390/s21196430
39. Papanikolaou, Y., Bennett, F.: Slot filling for biomedical information extraction. CoRR abs/2109.08564 (2021)
40. Poliak, A., Fleming, M., Costello, C., Murray, K.W., et al.: Collecting verified COVID-19 question answer pairs. In: NLP4COVIDEMNLP. ACL (2020)

41. Rajpurkar, P., Jia, R., Liang, P.: Know what you don't know: Unanswerable questions for squad. In: ACL 2018, Melbourne, Australia, July 15–20, 2018, pp. 784–789. ACL (2018). https://doi.org/10.18653/v1/P18-2124
42. Ren, R., Lv, S., Qu, Y., Liu, J., et al.: PAIR: leveraging passage-centric similarity relation for improving dense passage retrieval. In: ACL/IJCNLP (Findings). Findings of ACL, vol. ACL/IJCNLP 2021, pp. 2173–2183. Association for Computational Linguistics (2021)
43. Ren, R., Qu, Y., Liu, J., Zhao, W.X., et al.: Rocketqav2: A joint training method for dense passage retrieval and passage re-ranking. In: EMNLP (1), pp. 2825–2835. ACL (2021)
44. Croft, Bruce W.., van Rijsbergen, C.. J.. (eds.): SIGIR '94. Springer London, London (1994). https://doi.org/10.1007/978-1-4471-2099-5
45. Sun, S., Sedoc, J.: An analysis of bert faq retrieval models for Covid-19 infobot (2020)
46. Vig, J., Fabbri, A.R., Kryscinski, W., Wu, C., et al.: Exploring neural models for query-focused summarization. In: NAACL 2022, Seattle, WA, United States, July 10–15, 2022, pp. 1455–1468. ACL (2022). https://doi.org/10.18653/v1/2022.findings-naacl.109
47. Wang, L.L., Lo, K., Chandrasekhar, Y., Reas, R., et al.: CORD-19: the Covid-19 open research dataset. CoRR abs/2004.10706 (2020)
48. Wei, J.W., Huang, C., Vosoughi, S., Wei, J.: What are people asking about Covid-19? A question classification dataset. CoRR abs/2005.12522 (2020)
49. Xiao, W., Beltagy, I., Carenini, G., Cohan, A.: PRIMERA: Pyramid-based masked sentence pre-training for multi-document summarization. In: ACL, pp. 5245–5263. ACL, Dublin (2022). https://doi.org/10.18653/v1/2022.acl-long.360
50. Zhang, J., Zhao, Y., Saleh, M., Liu, P.J.: PEGASUS: pre-training with extracted gap-sentences for abstractive summarization. In: ICML, 13–18 July 2020. vol. 119, pp. 11328–11339. PMLR (2020)
51. Zhang, T., Kishore, V., Wu, F., Weinberger, K.Q., et al.: Bertscore: Evaluating text generation with BERT. In: ICLR, Addis Ababa, Ethiopia, April 26–30, 2020. OpenReview.net (2020)
52. Zhang, X.F., Sun, H., Yue, X., Lin, S.M., et al.: COUGH: A challenge dataset and models for COVID-19 FAQ retrieval. In: EMNLP 2021, Virtual Event, 7–11 November, 2021, pp. 3759–3769. ACL (2021)

Fine-Grained Categorization of Mobile Applications Through Semantic Similarity Techniques for Apps Classification

Elena Flondor[1](✉) ⓘ and Marc Frincu[2] ⓘ

[1] Faculty of Mathematics and Computer Science, West University of Timisoara,
bv. Vasile Parvan, Timisoara, Romania
`elena.flondor97@e-uvt.ro`
[2] School of Science and Technology, Nottingham Trent University, Cliftom Campus,
Nottingham, UK
`marc.frincu@ntu.ac.uk`

Abstract. The number of Android apps is constantly on the rise. Existing stores allow selecting apps from general named categories. To prevent miscategorization and facilitate user selection of the appropriate app, a closer examination of the categories' content is required to discover hidden subcategories of apps. Recent work focuses on exploring the granularity of the categories, but a validation of the categories' content against miscategorized apps is missing. In this research, we apply semantic similarity to apps' descriptions to uncover similarity and hierarchical clustering to search for misclassified apps. Furthermore, we apply Latent Dirichlet Allocation (LDA) algorithm to explore the existence of possible subcategories and to classify apps. Our empirical research is conducted using two data sets: 9,265 apps from Google Play Store, and 300 apps from App Store. Results confirm the existence of misclassified apps on markets and suggest the existence of multiple fine-grained categories. Our experiments outperform other LDA-based classification approaches achieving 0.61 precision. Moreover, the analysis hints the presence of misclassified apps might decrease the performance of existing classifiers.

Keywords: Semantic Similarity · Hierarchical Clustering · Latent Dirichlet Allocation · Mobile Applications

1 Introduction

The ongoing evolution of mobile devices in recent years has significantly impacted our lives. Developers of mobile apps share their products in the rapidly expanding markets of Google Play Store (GPS) [2] for Android and App Store (AS) [1] for iOS. On these platforms, developers must select a relevant category to help users find suitable apps easily. However, with thousands of apps in each category, finding apps that match consumer interests can be challenging. Moreover, some may be misclassified, making their discovery even more difficult.

O. Pedreira and V. Estivill-Castro (Eds.): SISAP 2023, LNCS 14289, pp. 79–87, 2023.
https://doi.org/10.1007/978-3-031-46994-7_7

Existing work is focused to find subcategories of apps which are hidden in the main markets' categories. Supervised and Unsupervised Machine Learning classification techniques combined with Natural Language Processing (NLP) techniques were applied to solve this challenging task. However, few of researchers mention misclassified apps on the markets [11].

We focus on validating categories' content and automating the process of identification of misclassified apps for prevention, and finding possible subcategories of apps on the markets to facilitate appropriate user app selection. We apply *semantic similarity* [10] and *hierarchical clustering* [15] to determine misclassifications, and *Latent Dirichlet Allocation* (LDA) [6] to find apps subcategories. A text-to-text *semantic similarity metric* is applied, as it outperforms traditional text similarity metrics [10]. Hierarchical clustering is applied for its capacity to represent data and to identify clusters that deviate significantly from the rest of the data. Lastly, LDA is applied to find mobile apps *subcategories* for its capacity to resume content, and to find relevant groups of words (called *topics*) which can provide insights of the main data categories. Specifically, we aim to:

- Propose an automated method for detection of miscategorized apps based on apps description using semantic similarity, hierarchical clustering, and markets classification recommendations [18,19];
- Discover hidden subcategories of apps by applying LDA on a well-defined and processed data set containing apps descriptions;
- Apply LDA to classify apps in proposed subcategories;

2   Literature Survey

This section explores existing methods of mobile apps classification based on their description corpus and reviews applied NLP techniques.

Several works are focused on exploiting the content of existing categories to extend current classification on the markets. For instance, [16,20] exploited the capacity of LDA and discovered multiple topics as an extension to the initial classification. Al-Subaihin et al. [3], extracted features from apps descriptions using the algorithm proposed in [12]. Then, a greedy hierarchical clustering algorithm was applied and the clusters, interpreted by humans, denoted that the initial categories can be extended. Ebrahimi et. al [11] encoded apps descriptions using different word embeddings and achieved the best classification results with GloVe and Support Vector Machines. Apps from Education and Health&Medical categories were manually labeled by researchers to find subcategories of apps.

The researchers applied NLP techniques to encode the descriptions: stop words removal and tokenization [11,13,20], part of speech tagging (identification of nouns, verbs, etc.) and stemming [13] or lemmatization [3,11], and adding n-grams [3]. All of these can help in boosting the performance of the classification techniques if they are combined in a proper manner.

Compared to the preceding works, our initial aim is to validate the content of the categories based on the similarity between the apps descriptions and the markets recommendations. For data processing, we combine various NLP techniques compared to previous researchers (Sect. 3.2). Then, we apply the method proposed by [10] to compute semantic similarity between apps' descriptions and fed the similarity distance matrix into hierarchical clustering algorithm [15]. Based on the distribution of the apps into clusters, we propose a new method to establish if a category presents misclassified apps and we search for subcategories of apps. Due to its promising results [20], we apply LDA to determine subcategories of apps in each category. We analyse the granularity of the categories and perform human interpretation [9] of the generated topics to validate our findings.

3 Methodology

The work discussed in Sect. 2 was adapted by combining more NLP techniques to reduce the noise of the data and to strengthen its semantic quality (Sect. 3.2). We searched for misclassified apps using *semantic similarity measures* and *hierarchical clustering* and we used LDA to find subcategories of apps included in each main category. Moreover, contrary our predecessors, we validate the content of the categories before investigating the fine-grained categories.

Data Set Gathering (Sect. 3.1) and **Data Set Processing** (Sect. 3.2) are the first two phases in our research. In the third phase we propose our method for **Categories' Content Validation** (Sect. 3.3). In the fourth phase we apply LDA in **Mining the Categories' Content** (Sect. 3.4) to discover the fine-grained categorization of the initial categories. The **Evaluation** (Sect. 3.5) phase describes the steps applied to validate our approach.

3.1 Data Set Gathering

We use a set of 9,265 Android apps to analyse fine-grained categories of GPS. It consists in English descriptions of apps from GPS between 2019-2023. They are classified in 32 categories based on the short outlines provided by the market [19]. We excluded the *Game* category as it already encompasses numerous fine-grained subcategories. The apps' descriptions were processed and LDA was applied to determine topics for subcategories (Sect. 4). To evaluate the performance of LDA in topics identification we used the data set proposed in [11]. It consists in apps crawled from AS. We selected educational apps and compared the ground-truths set by prior work (Sect. 4) with our finding.

3.2 Data Set Processing

We applied various *NLP techniques* to process the descriptions of the apps. We prepared the data to minimize the noise and to enable the algorithms to uncover latent relationships between words. We applied the following steps: uppercase characters were lowered; special characters, stop words, words with less than three characters, URLs, emojis, and numerical values were removed. Thus, we kept the most appropriate words and reduced the computational time. Then, we applied tokenization and each obtained term was labeled corresponding to its part of speech (e.g., noun, verb) to compute *semantic similarity scores* (Sect. 3.3). For subcategories discovering phase (Sect. 3.3), the remaining terms were converted to their lemma equivalent to maintain the inherent nature of words and we improved the description corpus with co-occurring word pairs of length 2, called *bi-grams* [14], as they were proven to bring improvements in topics discovering [7]. Finally, we filtered out words occurring in less than 10% and in more than 50% of the description corpus to allow LDA to build stronger relationships between the most relevant words.

3.3   Categories' Content Validation

This section proposes a method to validate the content of the categories by applying *semantic similarity measures* and *hierarchical clustering* algorithm.

Semantic Similarity Measure. To preserve the semantic similarity relations between the words, we applied the method for text similarity proposed in [10] on the apps descriptions level. The algorithm was applied to each pair of descriptions from the same category: for each part of speech from one description, we identified the one with the highest similarity (max_sim) to the other. For nouns and verbs the similarity is computed based on *Wu and Palmer similarity* [21], as it computes the degree of similarity between word senses and where the rings of synonyms occur relative to each other. As we removed cardinals (Sect. 3.2), we computed lexical similarity only for adjectives and adverbs. Therefore, a directional similarity score is given by: $sim(D_1, D_2)_{D_1} = \frac{\sum_{w_i \in D_1} max\_sim(w_i, D_2) \times idf(w_i)}{\sum_{w_i \in D_1} idf(w_i)}$, where D_1, D_2 are descriptions, w_i is the ith word in a description, and $idf(w_i)$ is the Inverse Document Frequency [17]. Both directional scores can be combined into a bidirectional similarity function: $sim(D1, D2) = \frac{sim(D1,D2)_{D_1} + sim(D1,D2)_{D_2}}{2}$. As its value ranges from 0 to 1, it allows the conversion into a normalized distance function: $dist\_sem(D_1, D_2) = 1 - sim(D_1, D_2)$ [5].

Hierarchical Clustering. We used *hierarchical clustering* [15] to investigate the granularity levels from the categories of mobile apps. Using the *similarity distance function* ($dist\_sem$), we computed the *semantic similarity distance matrix*. The shape of the matrix is given by the number of apps from a category. Let $E(D_i, D_j)$ the value of a matrix entry, where (D_i, D_j) is a pair of apps descriptions, and i, j general notations for line and column indices in the matrix. If $i = j$, we set $E(D_i, D_j) = 0$, else $E(D_i, D_j) = dist\_sem(D_i, D_j)$. We fed the similarity distance matrix obtained into the *complete linkage clustering* [15] to determine the clusters based on the maximum distance between the data points.

Possible Outliers Identification. *We define an outlier as a misclassified app* and analyse the distribution of the apps in the formed clusters to discover possible outliers. We consider the first two formed clusters to establish the existence of possible outliers. If they are highly unbalanced (e.g., a cluster contains only one or two descriptions), we compute the *semantic similarity* between the possible outlier app's description and each one of the categories recommendations proposed by the market source [18, 19]. If the maximum similarity is obtained for its main category, then it is not an outlier. The identified outliers were removed.

3.4   Mining the Categories' Content with LDA

The process continues with the discovery of different subcategories of apps within a category. We chose LDA [6], as it uses statistical models to infer topics in text data [4]. The main objective of LDA is to discover the *document topics*, within a text document. This produces a *topic-keyword* matrix. In our research, LDA is applied to the description corpus of the Android apps and the resulting topics are assumed to represent mixtures of a basic set of keywords that may describe a subcategory of apps. The number of the LDA topics, was determined based on *the highest coherence score*, as it measures the interpretability and semantic coherence of the generated topics.

3.5   Evaluation Framework

For evaluation we used *topic labeling* with *human interpretation* [9]. In the case of the second data set, we measured the performance of LDA to group mobile apps in the same category. Given the ground truth labels, we evaluated the performance of LDA by: $F2Score = \frac{(\beta^2+1) \times Precision \times Recall}{\beta^2 \times Precision + Recall}$ $(\beta = 2)$, $Precision = \frac{TP}{TP+FP}$, and $Recall = \frac{TP}{TP+FN}$. We define: TP (true positive) as the number of the apps whose label of the assigned topic matches the initial category label; TN (true negative) as the number of apps correctly classified as not belonging to inappropriate topics; FP (false positive) as the number of apps incorrectly assigned; FN (false negative) as the number of apps incorrectly excluded from appropriate topics.

4   Experiments and Results

In this section we analyse the results of outlier identification (Sect. 4.1), and the fine-grained categorization resulted using LDA (Sects. 4.2 and 4.3).

4.1   Outliers Identification Analysis

We investigated the feasibility of our approach to identify misclassified apps among the GPS data set (Sect. 3.3) and discovered two categories with outliers: *Medical* and *Weather*. Hierarchical clustering was applied for two clusters, grouping 370 samples in cluster 1 and leaving 1 sample in cluster 2 for *Medical* category. For *Weather* category, it grouped 293 in the first cluster, and left 1 in the second one. We noticed that the outlier of the *Medical* category could better fit in the *Education* category. Moreover, the maximum semantic similarity score (Sect. 3.3) between the *Medical* outlier and the categorization recommended by

GPS [19] was *0.40* for *Education*. For the *Weather* category outlier, we analysed the app and decided that it should be classified as a Game. We did not applied the same validation as in case of the other outlier, as GPS market only provides a list of games subcategories.

4.2 Google Play Store-Fine-Grained Categorization

To obtain consistent categories, we first searched for and removed possible outliers. We then applied LDA to explore the fine-grained categorization of mobile apps. For an *optimal* number of topics we investigated the behavior of *coherence scores* in relation to changes in granularity within clustering apps, which proved beneficial in practice [9]. We provided LDA with granularity levels ranging from 2 topics per category to the maximum based on apps count. We evaluated the coherence scores for each category cluster, selecting the topic count that maximized coherence scores, as topic words might define a subcategory. Given the GPS data set, the number of obtained topics sums up to 236 and hints the existence of more than 200 possible subcategories (Sect. 4.3). The subcategories coherence scores range from 0.37 (*Libraries & Demo*) to 0.58 (*Video Players & Editors*), and their average is 0.48. As the score typically ranges from 0 to 1 (Sect. 3.4), these can be promising results. The number of determined topics, might correspond to the number of existing subcategories. It ranges from 2 to 20 and does not depend on the number of the apps in each category, but on the logical relations among the keywords that exhibit higher probabilities (Sect. 3.4).

4.3 App Store Fine-Grained Categorization

To assess the ability of LDA to generate fine-grained app categories, we used the data set from [11]. We considered the *300 educational apps*, which were manually labeled with specific subcategories. We took into consideration this category as it contains misfit apps. The researchers [11] labeled them into: *Skill-based* apps, *Content-based* apps, *Function-based* apps, *Games*, and *Misfits*.

We evaluated LDA on: classification into all five types of apps (Scenario 1); classification into *Skill-based*, *Content-based*, *Function-based*, and *Games* - to analyse the effect of removing misclassified apps (Scenario 2). Furthermore, we analyse the semantic similarity between the misfits and *Education* category content recommendations and analyse their inclusion in this category. We processed the apps descriptions (Sect. 3.2) and used LDA to generate a number of topics equal to the number of subcategories. To compare our results to those obtained in [11], we labeled the topics based on contained words (Table 1) and through human interpretation: *1 - Skill-based, 2 - Function-based, 3 - Content-based, 4 - Games*. We assigned topics to each description and applied evaluation metrics (Sect. 3.5). Scenario 1 achieved the lowest performance: 0.36 *precision*, 0.38 *recall* and *f2-measure*; compared to Scenario 2: 0.614 *precision*, 0.582 *recall*, and 0.588 for *f2-measure*. However, Scenario 1 overachieved in terms of precision several classification algorithms applied in [11] on LDA vectorized descriptions (Naïve Bayes - 0.27, SVM - 0.35, Logistic Regression - 0.35). Our results show that removing misclassified apps from the categories can increase the performance of a classifier by approximately 25%. Therefore, we obtained better results compared to [11] after *Misfits* removal. We overachieved results from [11] when description was encoded using VSM (Naïve Bayes, AdaBoost, Decision Trees) and BM25 (Naïve Bayes, AdaBoost, Random Forest, KNN, SVM, etc.). For *Misfits* we applied the *bidirectional semantic similarity measure* (Sect. 3.3) between each app description and AS categories recommendations [8] to discover their most suitable main categories. E.g., we found out that: *A&A Days* is more suitable for *Productivity*, *Vuga Conf* should correspond to *Social Networking*, *Weather Saying* to *Weather*, *AR Paintings* to *Graphics & Design*, etc.

Table 1. Word topics used to establish subcategories for the markets

ID	Google Play's Education Topics	App Store's Education Topics
1	study, help, course, solution, practice, function, math, problem, data, skill	math, test, question, learn, practice, help, number, skill, exam, lesson,
2	question, test, student, class, exam, school, information, quiz, book, homework	school, feature, student, parent, information, work, access, mobile, course, service,
3	language, lesson, vocabulary, word, speak, pronunciation, course, sentence, conversation, grammar	book, study, view, available, share, version, story, record, search, experience,
4	word, feature, use, dictionary, search, time, photo, easy, share, translation, text	word, learn, game, kid, vocabulary, letter, color, picture, animal, sound,
5.	video, child, plant, world, book, play, song, educational, design, parent	–

5   Conclusion and Future Work

This paper proposed the identification of misclassified mobile apps in the markets and evaluated the performance of LDA in discovering subcategories of apps based on their descriptions. The process was done by applying NLP techniques (Sect. 3.2) and the experimental setup was based on averaged semantic similarity and hierarchical clustering to determine misclassified apps on GPS. Apps proposed in [11] were used to evaluate LDA in finding topics. We labeled the identified topics corresponding to [11] through human interpretation and evaluated LDA in classification based on the labeled descriptions. Our method for identifying misclassified apps is promising, as it found misclassified apps in the Weather and Medical categories. Moreover, our LDA based approach determined 236 possible subcategories of apps on GPS for a total of 9,265 apps. Our proposed LDA classifier out-performed LDA based classifiers applied in existing works (Sect. 4.3). Moreover, removing misclassified apps from the proposed data set, LDA can substantially improve classification process achieving a precision of 0.614.

For future work, a closer examination of the formed LDA topics is necessary to observe the fine-grained categorization of apps published on the markets. This could suggest appropriate subcategories of apps, and markets can use this to improve existing classification recommendations to avoid misclassification.

References

1. Apple app store. www.apple.com/app-store/. Accessed 18 Jun 2023
2. Google play store. www.play.google.com/store. Accessed 18 Jun 2023
3. Al-Subaihin, A.A., et al.: Clustering mobile apps based on mined textual features. In: Proceedings of the 10th ACM/IEEE International Symposium on Empirical Software Engineering and Measurement. ESEM 2016, Association for Computing Machinery, New York, NY, USA (2016). https://doi.org/10.1145/2961111.2962600
4. Al-Subaihin, A., Sarro, F., Black, S., et al.: Empirical comparison of text-based mobile apps similarity measurement techniques. Empir. Softw. Eng. **24**(6), 3290–3315 (2019). https://doi.org/10.1007/s10664-019-09726-5
5. Alcic, S., Conrad, S.: Page segmentation by web content clustering. In: Proceedings of the International Conference on Web Intelligence, Mining and Semantics. WIMS 2011, Association for Computing Machinery, New York, NY, USA (2011). https://doi.org/10.1145/1988688.1988717
6. Blei, D.M., Ng, A.Y., Jordan, M.I.: Latent dirichlet allocation. J. Mach. Learn. Res. **3**, 993–1022 (2003)
7. Bunyamin, H., Sulistiani, L.: Automatic topic clustering using latent dirichlet allocation with skip-gram model on final project abstracts. In: 2017 21st International Computer Science and Engineering Conference (ICSEC), pp. 1–5 (2017). https://doi.org/10.1109/ICSEC.2017.8443795
8. Ceci, L.: Number of available apps in the apple app store from 2008 to July 2022 (2023). www.statista.com/statistics/268251/number-of-apps-in-the-itunes-app-store-since-2008/. Accessed 18 Jun 2023
9. Chang, J., Boyd-Graber, J., Gerrish, S., Wang, C., Blei, D.: Reading tea leaves: How humans interpret topic models. In: Advances in Neural Information Processing Systems 22 (NIPS 2009), vol. 32, pp. 288–296 (2009)
10. Corley, C., Mihalcea, R.: Measuring the semantic similarity of texts. In: Proceedings of the ACL Workshop on Empirical Modeling of Semantic Equivalence and Entailment, pp. 13–18. Association for Computational Linguistics, Ann Arbor, Michigan (2005). www.aclanthology.org/W05-1203
11. Ebrahimi, F., Tushev, M., Mahmoud, A.: Classifying mobile applications using word embeddings. ACM Trans. Softw. Eng. Methodol. **31**, 1–30 (2021). https://doi.org/10.1145/3474827
12. Harman, M., Jia, Y., Zhang, Y.: App store mining and analysis: MSR for app stores. In: 2012 9th IEEE Working Conference on Mining Software Repositories (MSR), pp. 108–111 (2012). https://doi.org/10.1109/MSR.2012.6224306

13. Lavid Ben Lulu, D., Kuflik, T.: Functionality-based clustering using short textual description: helping users to find apps installed on their mobile device. In: Proceedings of the 2013 International Conference on Intelligent User Interfaces, pp. 297–306. IUI 2013, Association for Computing Machinery, New York, NY, USA (2013).https://doi.org/10.1145/2449396.2449434
14. Mikolov, T., Sutskever, I., Chen, K., Corrado, G., Dean, J.: Distributed representations of words and phrases and their compositionality (2013). https://doi.org/10.48550/ARXIV.1310.4546
15. Müllner, D.: Modern hierarchical, agglomerative clustering algorithms (2011)
16. Mokarizadeh, S., Rahman, M., Matskin, M.: Mining and analysis of apps in google play. In: Proceedings of the 9th International Conference on Web Information Systems and Technologies (BA-2013), pp. 527–535 (2013)
17. Sparck Jones, K.: A Statistical Interpretation of Term Specificity and Its Application in Retrieval, pp. 132–142. Taylor Graham Publishing, GBR (1988)
18. Store, A.A.: Choosing a category. www.developer.apple.com/app-store/categories/. Accessed 18 Jun 2023
19. Store, G.P.: Choose a category and tags for your app or game. www.support.google.com/googleplay/android-developer/answer/9859673?hl=en. Accessed 18 Jun 2023
20. Vakulenko, S., Müller, O., vom Brocke, J.: Enriching iTunes app store categories via topic modeling. In: International Conference on Information Systems (ICIS) (2014)
21. Wu, Z., Palmer, M.: Verb semantics and lexical selection. In: 32nd Annual Meeting of the Association for Computational Linguistics, pp. 133–138. Association for Computational Linguistics, Las Cruces, New Mexico, USA (1994). https://doi.org/10.3115/981732.981751www.aclanthology.org/P94-1019

Runs of Side-Sharing Tandems
in Rectangular Arrays

Shoshana Marcus[1]([✉]), Dina Sokol[2][iD], and Sarah Zelikovitz[3]

[1] Department of Mathematics and Computer Science, Kingsborough Community
College of the City University of New York, Brooklyn, NY, USA
shoshana.marcus@kbcc.cuny.edu
[2] Department of Computer and Information Science, Brooklyn College and The
Graduate Center, City University of New York, Brooklyn, NY, USA
sokol@sci.brooklyn.cuny.edu
[3] Department of Computer Science, College of Staten Island and The Graduate
Center, City University of New York, Staten Island, NY, USA
sarah.zelikovitz@csi.cuny.edu
http://www.sci.brooklyn.cuny.edu/~sokol,
http://www.cs.csi.cuny.edu/~zelikovi/

Abstract. A side-sharing tandem is a rectangular array that is composed of two adjacent non-overlapping occurrences of the same rectangular block. Furthering our understanding of side-sharing tandems should
facilitate the development of more efficient 2d pattern matching techniques and should lead to improvements in multi-dimensional compression schemes. Existing algorithms for finding side-sharing tandems are
far from optimal on 2d arrays that contain relatively few side-sharing
tandems. In this paper, we introduce the idea of a run of side-sharing
tandems, as a maximally extended chain of 2d side-sharing tandems. We
demonstrate tight upper bounds on the number of runs of side-sharing
tandems that can occur in a rectangular array. We develop an algorithm
that locates all τ runs of side-sharing tandems in an $n \times n$ input array
in $O((n^2 + \tau) \log n / \log \log n)$ time.

1 Introduction

In one dimension, a *tandem*, or *square*, is a string which consists of precisely two
consecutive occurrences of a primitive substring. For example, aa and abcabc are
squares. Squares are well-investigated objects in combinatorics of strings [1] and
form the basis of space-efficient string-matching algorithms [13] since squares are
the building blocks of larger repetitions. A *2d side-sharing tandem* is a generalization of the 1d tandem to two-dimensions. A 2d side-sharing tandem consists of
two adjacent non-overlapping occurrences of the same rectangular block, called
the *root*. This structure is depicted in Fig. 1.

Apostolico and Brimkov defined the 2d side-sharing tandem and demon-
strated that an $n \times n$ array M can contain $\Theta(n^3 \log n)$ primitively-rooted side-
sharing tandems [4]. They also presented an algorithm that locates side-sharing
tandems in time proportional to these upper bounds, i.e., $O(n^3 \log n)$ time [5].

O. Pedreira and V. Estivill-Castro (Eds.): SISAP 2023, LNCS 14289, pp. 88–102, 2023.
https://doi.org/10.1007/978-3-031-46994-7_8

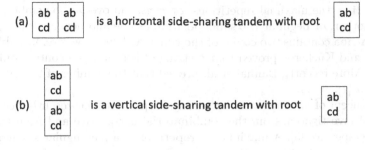

Fig. 1. Two possible configurations of a 2d side-sharing tandem

Charalampopoulos et al. [8] demonstrated $\Theta(n^3)$ bounds for the number of *distinct* primitively-rooted side-sharing tandems in an $n \times n$ array and also contributed an algorithm that runs in time proportional to these upper bounds, i.e., $\Theta(n^3)$ time. Gawrychowski et al. [15] construct an infinite family of $n \times n$ 2d arrays over the binary alphabet with $\Omega(n^3)$ distinct primitively-rooted side-sharing tandems. Both algorithms for finding side-sharing tandems [5,8] are far from optimal on a 2d array that is sparsely populated with side-sharing tandems. Thus, the goal in this work is to locate the side-sharing tandems in close to linear time, with respect to both the size of the input array M and the number of *runs* of side-sharing tandems occurring in M.

A range of applications motivate the investigation of 2d tandems. A wide variety of efficient 1d string matching techniques have not yet been extended to methods for 2d settings. As Crochemore et al. [11] noted, the essential obstacle to doing so is the unexplored different and more complex structures of 2d repetitions. Thus, it is our expectation that by furthering the understanding of 2d tandems we will precipitate the development of more efficient 2d pattern matching techniques. Just as properties of 1d tandems have enabled the speeding up of one-dimensional pattern searching algorithms and are relied on by space-efficient 1d pattern matching algorithms, discovering properties of two-dimensional tandems should open new avenues for speeding up 2d pattern matching algorithms and should enable us to design algorithms that use less working space in memory. In 1d, algorithms that compute maximal repetitions in a text have application to data compression. Likewise, the exploration of 2d runs of tandems should lead to improved compression schemes for graphics and videos.

Repetition within strings constitutes one of the most fundamental areas of string combinatorics. Repetitions are exploited in the design of efficient algorithms for string matching, data compression, and analysis of biological sequences. A maximal repetition in a string S is a substring of S that is periodic and cannot be extended at all to the right or left, e.g., ababa is a maximal repetition in the string abaababac. Maximal repetitions are important structures, as they encode all of the repetitions in a concise way. Once the set of maximal repetitions is known, repetitions of any other type (such as squares and cubes) can be extracted from it [12].

In a string, the maximal repetitions, or *runs*, can overlap, be embedded one within another, or begin at the same position as one another. For example, the run aaabaaaba contains two copies of the run aaa. Thus, it was remarkable when Kolpakov and Kucherov proved that a string of length n can contain only $O(n)$ runs [16]. More recently, Bannai et al. proved that the number of runs is strictly less than n [7].

The concept of *maximal* two-dimensional repetitions has recently been introduced and investigated, from the combinatorial perspective and from the algorithmic perspective [3]. A maximal 2d repetition is a rectangular subarray that can be decomposed into repeating non-overlapping occurrences of the same subblock horizontally and vertically that is maximally extended in all directions. Amir et el. [3] demonstrated an upper bound of $O(n^3)$ occurrences in an $n \times n$ array and introduced an algorithm that locates the maximal 2d repetitions in an $n \times n$ array M in $O(n^2 \log n + \rho)$ time, where ρ is the number of maximal 2d repetitions in M. Charalampopoulos et al. [8] tightened the bound and demonstrated that the number of maximal 2d repetitions in an $n \times n$ array is $O(n^2 \log^2 n)$. This bound proves an $O(n^2 \log^2 n)$ run-time for the algorithm of Amir et al. [3] to locate all maximal 2d repetitions in an $n \times n$ array. Gawrychowski et al. [15] construct an infinite family of $n \times n$ arrays with $\Omega(n^2 \log n)$ maximal 2d repetitions.

In 1d, once the locations of all the maximal repetitions are known, the tandems can easily be listed. In 2d, tandems contain two copies of the primitive root only horizontally *or* vertically, while repetitions have at least two copies of the primitive root horizontally *and* vertically [4]. Thus, the set of maximal 2d repetitions does not directly enable us to list the side-sharing tandems.

Amir at al. [2] demonstrate that all corner-sharing tandems in an $n \times n$ array can be located in $O(n^3 \log n)$ time. Although there can be $O(n^4)$ corner-sharing tandems in an $n \times n$ array, their algorithm represents the output with the set of maximal corner-sharing runs, which they demonstrate to be of size $O(n^3 \log n)$. They also describe an efficient algorithm for locating the approximate k-mismatch corner-sharing tandems in a 2d array.

In this paper we develop an efficient algorithm for locating side-sharing tandems in a rectangular array. For efficiency, we locate runs of tandems instead of individual tandems. In Sect. 2 we introduce the notion of a horizontal run (*h-run*), and a vertical run (*v-run*), to succinctly represent a maximally extended chain of adjacent side-sharing tandems. We prove tight upper bounds on the number of h-runs and v-runs that can possibly occur in a rectangular array. In Sect. 3 we develop an algorithm that locates all τ runs of side-sharing tandems in an $n \times n$ input array in $O((n^2 + \tau) \log n / \log \log n)$ time.

2   Runs of Side-Sharing Tandems

In this section we define runs of side-sharing tandems. Then we prove bounds on the maximum number of runs of side-sharing tandems that can occur in an $n \times n$ array.

Definition 1. *[4] A rectangular array M is* primitive *if it cannot be partitioned into non-overlapping complete occurrences of some block W.*

Furthermore, X is horizontally primitive *or h-primitive (resp., vertically primitive or v-primitive) if it cannot be represented in the form $M =$*

$$\boxed{W \cdots W}(resp., \boxed{\begin{matrix} W \\ \cdots \\ W \end{matrix}}) for\ some\ blockW \neq M. \tag{1}$$

In a 1d repetition of the form $u^n u'$ where u' is a prefix of u, u is called the period and $n + \frac{|u'|}{|u|}$ is the exponent [17]. We say that U is a horizontal prefix (resp. suffix) in rectangular array M if U is an initial (resp. ending) sequence of contiguous columns in M. A *horizontal border* of rectangular array M is a proper horizontal prefix that is also a horizontal suffix of M. We say that B is the *longest horizontal border* of M if it is the horizontal border of M that spans the largest number of columns among the horizontal borders of M.

Definition 2. *The* horizontal period, *or h-period, of an $m \times n$ array M is $n - b$ where b is the number of columns contained in the longest horizontal border of M.*

Definition 3. *[10, 19] An $m \times n$ array M with h-period p is* horizontally periodic, *or h-periodic, if $p \leq \lfloor \frac{n}{2} \rfloor$.*

The vertical period of an $m \times n$ array and vertical periodicity are defined analogously.

For example, consider the following arrays.

$$M_1 = \begin{bmatrix} a\ b\ c\ d \\ e\ f\ g\ h \end{bmatrix} \quad M_2 = \begin{bmatrix} a\ b\ c\ d\ a\ b\ c\ d\ a \\ e\ f\ g\ h\ e\ f\ g\ h\ e \\ a\ b\ c\ d\ a\ b\ c\ d\ a \end{bmatrix} \quad M_3 = \begin{bmatrix} a\ b\ c\ d\ a\ b \\ e\ f\ g\ h\ e\ f \\ a\ b\ c\ d\ a\ b \\ e\ f\ g\ h\ e\ f \end{bmatrix}$$

M_1 is primitive since it is both h-primitive and v-primitive. Although M_2 is h-periodic, it is also primitive, since it is both h-primitive and v-primitive. M_3 is h-primitive but not v-primitive.

Definition 4. *In an $m \times n$ array M, a* 2d horizontal run, *or h-run, is an h-periodic subarray M' with h-period p in which extension by one subcolumn to the right or to the left yields an array with h-period h' such that $h' > h$.*

A *2d vertical run,* or *v-run,* is defined analogously.

The h-period of an h-run is h-primitive. For 2d tandems of the configuration of Fig. 1(a), an h-run maximally extends the 2d tandem to the left and to the right. Similarly, for 2d tandems of the configuration of Fig. 1(b), a v-run maximally extends the 2d tandem upwards and downwards.

Lemma 1. *An $n \times n$ array M can contain $\Theta(n^3)$ h-runs.*

Proof. The proof is a variation of the proof in [4] for the number of side-sharing 2D tandems in a matrix. Due to space limits, the full details will be included in the journal version of this paper.

Corollary 1. *An $n \times n$ array M can contain $\Theta(n^3)$ v-runs.*

3 Algorithm

In this section we develop an efficient algorithm for locating h-runs in an $n \times n$ array M. The algorithm is easily adapted for identifying v-runs. Our algorithm achieves $O((n^2 + \tau) \log n / \log \log n)$ run-time, where τ is the size of the output set. In the extreme case, $\tau = \Theta(n^3)$, by Lemma 1.

A straightforward $O(n^3)$ time algorithm considers all $O(n^2)$ subarrays of width n, of varying heights and start rows. Within each subarray, we view the columns as metacharacters and name identical columns with the same name. Then we use a 1D linear-time algorithm (e.g., [6,16]) to find the h-runs within each subarray in $O(n)$ time.

Our more efficient algorithm iteratively identifies the h-runs of each height $1 \leq k \leq n$. First we find the h-runs of height 1 by running a 1D linear-time algorithm on each row of M. Then we iteratively identify the h-runs of each height $2 \leq k \leq n$. We link h-runs of smaller heights to obtain h-runs of larger heights, when h-runs occur in overlapping columns on adjacent rows in which the overlap is sufficiently wide to maintain h-periodicity.

We outline the steps of our algorithm for finding all h-runs in $n \times n$ array M:

1. Identify all h-runs of height 1, by locating the 1d runs on each row.
2. Find all h-runs of height 2, by linking runs on adjacent rows.
3. Go through each height $3 \leq k \leq n$ (in increasing order), and for each start row $1 \leq i \leq n - k + 1$, identify h-runs of height k by linking h-runs of smaller heights on adjacent rows.

3.1 Interval Queries

In this subsection we show how to preprocess a set of ψ intervals V so that we can efficiently answer the following kind of query. We preprocess $\forall v \in V$, $v = [a, b]$ such that $1 \leq a < n$, $1 < b \leq n$, $a < b$. This is an independent result that is itself an interesting contribution.

Definition 5. *Interval x -Intersection Query: Given an integer $x > 0$ and an interval $u = [p, q]$ with integer endpoints such that $1 \leq p < n, 1 < q \leq n, p < q$, list all intervals in V that intersect u by at least x units.*

Figure 2 demonstrates an example for the Interval x-Intersection Query.

Suppose the response to an Interval x-Intersection Query query contains ω intersecting intervals in the output list. Interval trees [20] can be used to solve this problem in $O(\omega + \log \psi)$ time. We use dynamic RMQ to solve this problem in $O(\omega \log \psi / \log \log \psi)$ time, with $O(\psi \log \psi / \log \log \psi)$ preprocessing time.

We maintain the set V in a dynamic way so that we can remove intervals from V and reinsert the intervals that were removed from V in the course of answering a query. The set V is static in the sense that only intervals that previously existed in V will be added later on.

We construct the following two arrays corresponding to V:

1. $\texttt{Endpt}[t]$, $1 \leq t \leq n$, such that $\texttt{Endpt}[t]$ contains the furthest (rightmost) endpoint of all intervals $v \in V$ that begin at location t. $\texttt{Endpt}[t] = $ largest q for all $[t, q] \in V$.
2. $\texttt{Length}[t]$, $1 \leq t \leq n$, such that $\texttt{Length}[t]$ contains the length of the longest interval $v \in V$ beginning at location t.

After we create the two arrays, we preprocess $\texttt{Endpt}$ and $\texttt{Length}$ for dynamic Range-Maximum Queries.

Now we show how to use these arrays to answer the query of interest. We begin with an empty set I. We identify an intersecting interval $\in V$ and if the intersection with $u = [p, q]$ is at least x we add it to I. We continue this process until the returned intersecting interval does not have sufficient intersection with $[p, q]$.

There are 3 ways in which an interval $[p', q'] \in V$ can intersect $[p, q]$:

 I. $p' < p$ i.e., $[p', q']$ begins before $[p, q]$.
 II. $q' > q$ i.e., $[p', q']$ ends after $[p, q]$.
III. $p < p' < q' < q$ i.e., $[p', q']$ is fully contained within $[p, q]$.

These three kinds of interval intersection are depicted in Fig. 2 and each type of intersection is shown in a different color.

To find the longest intersection of Type I., we can use range-max in the $\texttt{Endpt}$ array. For Type III., we do not know how to get the longest intersecting interval. However, if we take the range-max in the $\texttt{Length}$ array for the range $p \ldots q - x$, we will get an overlapping interval with length $\geq x$ if one exists. This suffices for answering our query, and finds intersections of both Type II. and Type III. In summary, given an interval $[p, q]$, a sufficiently long intersecting interval is obtained by taking the maximum of the following 2 results:

1. $\texttt{RangeMaximum}(\texttt{Endpt}[1 \ldots p - 1]) - p$
2. $\texttt{RangeMaximum}(\texttt{Length}[p \ldots q - x])$

Once we have identified an interval that intersects u by at least x, we add it to the set I and we remove the corresponding entry in the $\texttt{Endpt}$ or $\texttt{Length}$ arrays so that other intersecting intervals can be identified. We repeatedly take the maximum of the 2 values above until the maximum intersection is less than x. At that point we are ready to return the set I as the answer to the query and we reinsert all intervals in I to the RMQ structures for future queries on V.

Lemma 2. *We use $O(\psi \log \psi / \log \log \psi)$ time to preprocess a set of ψ intervals for Interval x-Intersection Queries.*

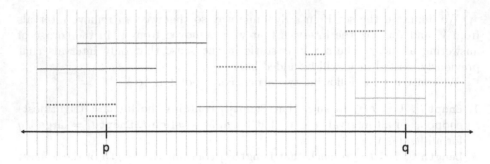

Fig. 2. Many intervals that intersect $u = [p, q]$. All intervals in the set that intersect $u = [p, q]$ by at least $x = 5$ are shown as solid lines, while the intersecting intervals that are not reported as output to the Interval x-Intersection Query are shown as dashed lines. The intersecting intervals of Type I. are red; the intersecting intervals of Type II. are yellow; the intersecting intervals of Type III. are green (Color figure online)

Proof. We preprocess the Endpt and Length arrays for dynamic RMQ. We update the Endpt and Length arrays when an interval is removed from V. We remove intervals from longest to shortest so we can store a linked list of possibilities for each array entry, sorted by decreasing interval length that they represent. These arrays remain $O(\psi)$ in size since we only insert the intervals that have been removed in the process of answering a query.

Preprocessing arrays of size ψ for dynamic RMQ requires $O(\psi \log \psi / \log \log \psi)$ time, from the lower bounds proven for dynamic RMQ in [14]. □

Lemma 3. *We can answer an Interval x-Intersection Query in $O(\omega \log \psi / \log \log \psi)$ time, where ω is the number of intersecting intervals in the output list.*

Proof. For a query with ω results, $O(\omega \log \psi / \log \log \psi)$ time suffices for $O(\omega)$ range-minimum and range-maximum queries, as well as $O(\omega)$ removals and insertions of intervals in the RMQ structures. □

3.2   Preprocessing

In addition to the preprocessing mentioned in Sect. 3.1, we precompute the least common multiple (LCM) of pairs of numbers (i, j), $1 \leq i, j \leq \frac{n}{2}$, and store them in an $\frac{n}{2} \times \frac{n}{2}$ array.

Lemma 4. *The LCM table is constructed in $O(n^2 \log n)$ time.*

Proof. Each entry in the LCM table is computed in $O(\log n)$ time [9]. Thus, the $O(n^2)$ LCM table is constructed in $O(n^2 \log n)$ time. □

3.3 Main Algorithm

Our algorithm begins by identifying the 1d runs in each row of the input array. Each run identified is output as an h-run of height 1. For each row i, $1 \leq i \leq n$, in the $n \times n$ input array M, we use a linear-time algorithm (e.g., [6,16]) to locate all runs in row i, which we call R_i. We represent each run as an interval, corresponding to the columns that it spans. We augment each interval with the period size of the run it represents.

Once we have identified the h-runs of height 1, we link together h-runs on adjacent rows that form h-runs of height 2. That is, we link runs that occur in overlapping columns in which the overlap is sufficiently wide with respect to the period sizes to maintain h-periodicity. In the following set of examples we demonstrate the different possibilities of how runs in adjacent rows may link to form a run of height 2.

Example 1. In some instances, entire runs on adjacent rows link to form a single h-run of height 2. Consider the strings cabcabcabcab and babababababa. These strings are both runs of length 12; the first has period size 3 and the second has period size 2. If these two runs occur directly above / below one another in a 2d array, they link in their entirety to form an h-run of height 2 with length 12 and h-period of size 6:

```
cabcabcabcab
babababababa
```

Example 2. In other settings, part of one run can align with part of another run or with another entire run. Consider the string cabcabcabca aligned directly above bababaaaaaa, in the same columns. cabcabcabca consists of a single run with period cab and length 11 while bababaaaaaa contains a run of length 6 with period ba as well as an overlapping run of length 6 with period a. The longer run above links with the second run below to form a single h-run of height 2 with length 6 and h-period of size 3:

```
cabcabcabca
bababaaaaaa
```

Example 3. It is possible for runs in adjacent rows to align in the same columns, yet they will not form any h-runs of height 2 since the overlap is not sufficiently long to be h-periodic. Consider the run abaababaab with period abaab and the run abababab with period ab. Even if these runs occur directly above below one another in a 2d array, they do not form an h-run of height 2.

```
abaababaab
abababab
```

Example 4. Suppose one row contains the substring abaabaaba (a run with period aba, which contains two runs with period a within it) and directly below it we find the substring bbabbbabb (a run with period bbab, which contains three runs with period b within it). Although each of these strings contains more than one run, their alignment in the array does not admit even one h-run of height 2.

```
abaabaaba
bbabbbabb
```

Example 5. One run α of width ℓ can link to several runs above (or below) it, yielding several h-runs of height 2 and widths $\leq \ell$, each with possibly different h-periods. Consider the string ababababab, which is a single run with period 2, and the string aaaaacbbbb, which contains two smaller runs (aaaaa and bbbb). When these two strings align on adjacent rows, their runs link to form 2 h-runs of height 2.

That is,
ababababab
aaaaacbbbb
contains
ababa babab
aaaaa cbbbb
and
ababababab
aaaaacbbbb

Example 6. Consider the strings aaaaaaaaaaaa and ababaaababaa. The string aaaaaaaaaaaa is a single run with period a. However, ababaaababaa contains five runs, two with period ab, two with period a, and another with period ababaa. When these two strings align on adjacent rows and their runs are linked together,

aaaaaaaaaaaa
ababaaababaa

contains the following five h-runs of height 2, shown in red:

aaaa aaaaaaaa	aaaaaaaaaaaa	aaaaaaaaaa aa	aaaaaaaaaaaa
abab aaababaa	ababaaababaa	ababaaababaa	ababaaababaa

Now we explain how to readily link two specific runs on adjacent rows if they connect to form an h-run of height 2 (in the Run Linking Problem). Then we explain how to efficiently link all the runs that can be linked on two adjacent rows to obtain all the h-runs of height 2 that span those two rows (in the Row Run Linking Problem).

Definition 6. Run Linking Problem: *Let $\alpha \in R_i$ have period p_α and exponent e_α and let $\beta \in R_{i+1}$ have period p_β and exponent e_β.[1] Identify the h-run of height 2 that is formed by linking α and β together, if they can link to form an h-run of height 2.*

In Example 1, α is the run cabcabcabcab with period $p_\alpha =$ cab and exponent $e_\alpha = 4$, while β is the run bababababababa with period $p_\beta =$ ba and exponent $e_\beta = 6$.

Observation 1 specifies a necessary (but not sufficient) condition for runs in adjacent rows to link to form an h-run of height 2; Observation 2 specifies a *necessary and sufficient* condition.

Observation 1. Suppose α and β overlap by o columns. Let $\rho = max(p_\alpha, p_\beta)$. It is only possible for α and β to link together to form an h-run of height 2 when $o \geq 2\rho$.

Observation 2. Suppose α and β overlap by o columns. Then, α and β link to form an h-run of height 2 iff o is at least twice $LCM(p_\alpha, p_\beta)$.

---

[1] Throughout the remainder of this section, α represents a run on row i with period p_α and exponent e_α and β represents a run on row $i+1$ with period p_β and exponent e_β.

To determine if α and β link to form an h-run of height 2, we first compute o, the number of columns in which they overlap. Then we see if $o \geq 2 \text{ LCM}(p_\alpha, p_\beta)$, the necessary and sufficient condition for the overlap to be h-periodic (stated in Observation 2).

Lemma 5. *The* Run Linking Problem *is solved in $O(1)$ time.*

Proof. $O(1)$ time suffices to compute o, the number of overlapping columns in α and β, by comparing their start and end columns. Then we determine if the condition specified in Observation 2 occurs, i.e., that $o \geq 2 \cdot LCM(p_\alpha, p_\beta)$. This is done in constant time by referring to the precomputed LCM table. $\square$

Now we generalize to the problem of linking *all* the runs on two adjacent rows to obtain *all* the h-runs of height 2 that span those rows.

Definition 7. *Row Run Linking Problem*
Input: *Runs on adjacent rows, R_i and R_{i+1}, $1 \leq i < n$.*
Output: *Set of $\tau_{i,2}$ h-runs of height 2 that span rows i and $i+1$.*

We determine in $O(1)$ time if a pair of runs on adjacent rows link to form an h-run. Yet we do not want to try all $O(n^2)$ possible pairs. Thus, we devise a more efficient solution to the Row Run Linking Problem. The sizes of R_i and R_{i+1} are both smaller than n since there are fewer than n runs in a string of length n [7]. In each subarray of M that has width n and height 2, we can view the subcolumns as metacharacters. From this perspective, the size of the output set $\tau_{i,2}$ is also smaller than n.

We split the set of $\tau_{i,2}$ h-runs of height 2 spanning rows i and $i+1$ into two disjoint subsets, S_1 and S_2, and search separately for h-runs in each subset. S_1 consists of h-runs in which $p_\alpha \geq p_\beta$, i.e., either the run on row i either has a larger period than the run on row $i+1$ or the runs on adjacent rows have periods that are equal in size. S_2 consists of h-runs in which $p_\alpha < p_\beta$, i.e., the run on row i has a smaller period than the run on row $i+1$. Example 1 $\in S_1$. The first three h-runs in Example 6 belong to S_2 since the smaller period is above the larger period, while the last two h-runs shown in Example 6 belong to S_1 since the periods on both rows are the same size of 1.

We use a bidirectional search to solve the Row Run Linking Problem and search separately for h-runs in each subset. We identify the h-runs $\in S_1$ by *downward extensions* of runs in row i and we identify the h-runs $\in S_2$ by *upward extensions* of runs in row $i+1$. In the remainder of this paper, we focus on the downward extension phase since the upward extension phase of the algorithm is analogous. To extend a run α in row i downward to the h-runs of height 2 that it forms, we want to find all runs β in row $i+1$ that overlap α by $\geq 2p_\alpha$ columns, which is the necessary condition of Observation 1. In Lemma 6 we consider a substring of a run α on row i which contains exactly two adjacent copies of p_α, and show that there can be at most one run β below it with a period that is shorter than p_α or of the same size as p_α. By Lemma 6 and Corollary 3, from the perspective of the run with larger period in an h-run of height 2, there are a limited number of runs on the adjacent row to possibly connect to. Thus, we always begin with the run with larger period and search either downwards or upwards for the limited number of runs on the adjacent row to consider connecting to. This is depicted in Fig. 3.

Lemma 6. *Consider a substring s of α of length $2p_\alpha$ that begins in column c_s and ends in column c_e. There exists at most one run $\beta \in R_{i+1}$ with period size $p_\beta \leq p_\alpha$ and length $\geq 2p_\alpha$ beginning in a column $\leq c_s$ and ending in a column $\geq c_e$.*

Proof. It is important to keep in mind that there is one string below α on row $i+1$. A known fact of periodicity is that two overlapping runs within one string can overlap by at most one character less than the larger period [18]. The run β on row $i + 1$ has period size $p_\beta \leq p_\alpha$ and length $\geq 2p_\alpha$. Thus, there are at least two copies of p_β directly below s and this substring is periodic in p_β and cannot be periodic in anything else.   □

Corollary 2. *Suppose $\beta \in R_{i+1}$ with period $p_\beta \leq p_\alpha$ and overlaps α by $\geq 2p_\alpha$ columns. Another run $\beta' \in R_{i+1}$ with period $p'_\beta \leq p_\alpha$ and length $\geq 2p_\alpha$ can overlap β by at most $p_\alpha - 1$ columns.*

Corollary 3. *Fewer than e_α runs β in row $i+1$ can overlap α by $\geq 2p_\alpha$ columns, with period size $p_\beta \leq p_\alpha$.*

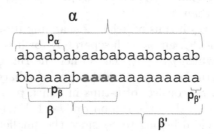

Fig. 3. A run α on row i with two overlapping runs below it on row $i+1$, β and β', that overlap by the maximum $p_\alpha - 1$ characters. The overlap between β and β' is shown in bold blue font. $p_\alpha = 5$, $p_\beta = 5$, $p_{\beta'}(Color figure online) = 1$

We sort the runs for each row, R_i, $1 \leq i \leq n$, by their period sizes, in descending order. We follow this sorted order to extend one run at a time, in order of decreasing period size. This facilitates the search for runs on the next row that have periods no longer than the periods in the runs above, as we are only looking for h-runs $\in S_1$. For each R_i, we construct and maintain the Endpt and Length arrays so that Interval x-Intersection Queries can be asked. We preprocess the arrays for dynamic RMQ, as described in Sect. 3.1. We maintain two sets of double linked lists of possible entries for each element of the Endpt and Length arrays. In one set of linked lists, possibilities are sorted by decreasing interval length (as explained in Sect. 3.1, to enable Interval x-Intersection Queries to return multiple results) and in the other we sort by decreasing period size. This way, we can remove a run from the set R_i and update the corresponding Endpt and Length arrays in $O(\log n / \log \log n)$ time. Once we have completed the downward extension of runs in a row, we reinsert the runs that have been removed in time proportional to the number of runs that are reinserted, with an $O(\log n / \log \log n)$ slowdown to update the dynamic RMQ data structures.

Definition 8. *Two-Period Overlap Query*: *Given a run $\alpha \in R_i$, $1 \leq i < n$, with period p_α and a set of runs R_{i+1} that occur on the row below α, identify the runs in the set R_{i+1} that each overlap the columns of α by at least $2p_\alpha$ columns.*

We solve the Two-Period Overlap Query as an Interval x-Intersection Query over the intervals of R_{i+1}, with parameters $x = 2p_\alpha$ and $u =$ the interval of columns that α spans.

Lemma 7. *We can answer a Two-Period Overlap Query in $O(e_\alpha \log n / \log \log n)$ time.*

Proof. From Corollary 3 we know there are fewer than e_α runs returned by the query. The time complexity is derived from this fact combined with Lemma 3, the time complexity of answering an Interval x-Intersection Query that returns up to e_α results, over a set of up to n intervals. □

In the downward extension phase, we iterate through each of the runs $\alpha \in R_i$ in decreasing order of period size. As we extend downwards each run $\alpha \in R_i$ with period p_α, we first remove any runs in R_{i+1} with period larger than p_α, which is straightforward as it follows the sorted order. Whenever we remove a run from R_{i+1}, we update the Endpt and Length arrays. (We note that in the downward extensions of all the runs in R_i, we can remove each of the runs in R_{i+1} only once.) Then we ask a Two-Period Overlap Query on α and R_{i+1} to identify the set of runs $I \subseteq R_{i+1}$ such that each run $\beta \in I$ overlaps α by at least $2p_\alpha$ columns and has period $p_\beta \leq p_\alpha$. The set I consists of the runs on row $i + 1$ that can possibly link with α to yield an h-run in S_1 (by Observation 1). For each run $\beta \in I$ returned by the Two-Period Overlap Query, we solve the Run Linking Problem and determine if α and β overlap sufficiently to form an h-run, and if so, we announce an h-run of height 2 as output.

Lemma 8. *In $O(e_\alpha \log n / \log \log n)$ time we can extend $\alpha \in R_i$ to all h-runs of height $2 \in S_1$ that α extends to.*

Proof. This follows from Corollary 3 and Lemmas 5 and 7 since the downward extensions of α are obtained by a Two-Period Overlap Query (answered in $O(e_\alpha \log n / \log \log n)$ time), and then up to e_α instances of the Run Linking Problem (each of which is answered in $O(1)$ time) on the results returned by the Two-Period Overlap Query. □

Lemma 9. *We can solve the Row Run Linking Problem in $O(n \log n / \log \log n)$ time.*

Proof. We sort R_i and R_{i+1} in linear $O(n)$ time. We construct the Endpt and Length arrays and preprocess the arrays for dynamic RMQ so that Interval x-Intersection Queries can be asked in $O(n \log n / \log \log n)$ time (as described in Sect. 3.1).

The sum of the exponents of all runs in a string of length n is $O(n)$ [16]. Thus, there are a total of $O(n)$ extensions of runs in row i to row $i + 1$. With this fact in mind, the time complexity follows naturally from Lemma 8. $\qquad\square$

Corollary 4. *All h-runs of height 2 are identified in $O(n^2 \log n / \log \log n)$ time.*

Observation 3. For integer $3 \leq k \leq n$ and $1 \leq i \leq n - k$, an h-run of height k begins on row i spanning columns c_s through c_e, $1 \leq s < n, 1 < e \leq n, s < e$, iff an h-run u of height $k - 1$ begins on row i and a run ℓ begins on row $i + k - 1$, both u and ℓ span columns c_s through c_e, and $c_e - c_s + 1 >=$ twice the LCM of the periods of u and ℓ.

Now that we have shown how to efficiently solve the Row Run Linking Problem to identify h-runs of height 2, we show how the approach generalizes for identifying all h-runs in the input array. Once we have identified the h-runs of height 2, we iteratively locate the h-runs of each height k, $3 \leq k \leq n$, beginning in each row i of the array. We link h-runs of smaller heights beginning on row i with h-runs occurring directly below them, to obtain h-runs of larger heights. For instance, to identify the h-runs of height k beginning on row i, we can link the h-runs of height $k - 1$ that begin on row i with the runs on row $i + k - 1$. By Observation 3, this is sufficient. We store each set of h-runs of different heights beginning on different rows separately from one another. We sort the h-runs in each set by their period sizes, in descending order. Then, for each set, we construct the Endpt and Length arrays and preprocess the arrays for dynamic RMQ.

Considering Corollary 4 and the $O(n)$ different heights of h-runs indicates that our algorithm identifies all h-runs in the input array in $O(n^3 \log n / \log \log n)$ time. However, a more careful analysis reduces the time complexity of our algorithm.

Let σ_α denote the sum of the exponents of h-runs of height 2 that are extensions of α and are $\in S_1$. We note that h-runs can only get narrower and not wider as we extend them downwards.

Lemma 10. *Suppose $\alpha \in R_i$ has exponent e_α, $\beta \in R_{i+1}$ has exponent e_β, and $p_\alpha \geq p_\beta$. If α and β link to form an h-run of height 2 with period $p_{\alpha\beta}$ and exponent $e_{\alpha\beta}$, then $e_{\alpha\beta} \leq e_\alpha$ and $e_{\alpha\beta} \leq e_\beta$.*

Proof. From Observation 2 we see that $p_{\alpha\beta}$ is a multiple of p_α. Thus, $p_{\alpha\beta} \geq p_\alpha$. This implies that $e_{\alpha\beta} \leq e_\alpha$. Similarly, $e_{\alpha\beta} \leq e_\beta$. $\qquad\square$

Lemma 11. *Suppose $\tau_{\alpha,2}$ h-runs of height 2 are extensions of α and are $\in S_1$. Let σ_α denote the sum of the exponents of these $\tau_{\alpha,2}$ h-runs of height 2. Then $\sigma_\alpha \leq e_\alpha + \tau_{\alpha,2}$.*

Proof. From Corollary 2, we see that the intervals returned by a Two-Period Overlap Query cannot overlap one another by more than $p_\alpha - 1$ columns. Consider the extreme case in which each result overlaps another by p_α characters at each end, and we have the maximum number of results returned, which is strictly less than e_α (by Corollary 3). Even in this case, $\sigma_\alpha \le e_\alpha + \tau_{\alpha,2}$.   □

Lemma 12. *Let τ_α denote the set of h-runs of any height that are in S_1 and are extensions of α. In $O((e_\alpha + \tau_\alpha) \log n / \log \log n)$ time, we extend run α with exponent e_α down to all h-runs that it extends to.*

Proof. By Lemma 6 and Corollary 3, a Two-Period Overlap Query on α returns fewer than e_α results. For each result, we either have new output or the exponent shrinks. When we obtain new output, we charge the $O(1)$ work to the output generated. In the case that the exponent shrinks, there are at least p_α columns of α that do not extend to an h-run of height 2. This is derived from Corollary 2, which indicates that the runs $\in R_{i+1}$ returned by a Two-Period Overlap Query on α are at a distance of at least p_α columns from one another.

It is possible to simultaneously have the effect of both obtaining output and the exponent shrinking when certain regions of α split and extend to two h-runs and other regions do not extend at all. We can charge the splitting of regions into several h-runs to the new output generated and we can charge the rejections to the overall shrinking exponent.   □

Theorem 1. *All τ h-runs in the $n \times n$ input array are identified in $O((n^2 + \tau) \log n / \log \log n)$ time.*

Proof. For any row i, $1 \le i \le n$, the sum of the exponents of all runs on the row is $O(n)$ [16]. Thus, over all extensions downwards from R_i, we can charge the rejection of results from Two-Period Overlap Queries to the combination of the sum of exponents of R_i and the number of h-runs (of all heights) that extend from R_i. Thus, the theorem generalizes Lemma 12 from the perspective of extending a single run to the perspective of extending all runs that occur on rows of the input array.   □

Funding. Shoshana Marcus was partially supported by the Professional Staff Congress of the City University of New York (PSC-CUNY) grant number 66557-00 54.

References

1. Allouche, J.P.: Algebraic combinatorics on words. Semigroup Forum **70**, 154–155 (2005). https://doi.org/10.1007/s00233-004-0146-9
2. Amir, A., Butman, A., Landau, G.M., Marcus, S., Sokol, D.: Double string tandem repeats. Algorithmica **85**(1), 170–187 (2023). https://doi.org/10.1007/s00453-022-01016-9
3. Amir, A., Landau, G.M., Marcus, S., Sokol, D.: Two-dimensional maximal repetitions. Theoret. Comput. Sci. **812**, 49–61 (2020). https://doi.org/10.1016/j.tcs.2019.07.006

4. Apostolico, A., Brimkov, V.E.: Fibonacci arrays and their two-dimensional repetitions. Theoret. Comput. Sci. **237**(1–2), 263–273 (2000). https://doi.org/10.1016/S0304-3975(98)00182--0
5. Apostolico, A., Brimkov, V.E.: Optimal discovery of repetitions in 2D. Discrete Appl. Math. **151**(1–3), 5–20 (2005). https://doi.org/10.1016/j.dam.2005.02.019
6. Bannai, H., Tomohiro, I., Inenaga, S., Nakashima, Y., Takeda, M., Tsuruta, K.: A new characterization of maximal repetitions by Lyndon trees. In: Proceedings of the Twenty-Sixth Annual ACM-SIAM Symposium on Discrete Algorithms, SODA 2015, San Diego, CA, USA, 4–6 January 2015, pp. 562–571 (2015). https://doi.org/10.1137/1.9781611973730.38
7. Bannai, H., Tomohiro, I., Inenaga, S., Nakashima, Y., Takeda, M., Tsuruta, K.: The "runs" theorem. SIAM J. Comput. **46**(5), 1501–1514 (2017). https://doi.org/10.1137/15M1011032
8. Charalampopoulos, P., Radoszewski, J., Rytter, W., Walen, T., Zuba, W.: The number of repetitions in 2D-strings. In: Grandoni, F., Herman, G., Sanders, P. (eds.) 28th Annual European Symposium on Algorithms, ESA 2020, 7–9 September 2020, Pisa, Italy (Virtual Conference). LIPIcs, vol. 173, pp. 1–18. Schloss Dagstuhl - Leibniz-Zentrum für Informatik (2020). https://doi.org/10.4230/LIPIcs.ESA.2020.32
9. Cormen, T.H., Leiserson, C.E., Rivest, R.L., Stein, C.: Introduction to algorithms. MIT press (2022)
10. Crochemore, M., Gasieniec, L., Hariharan, R., Muthukrishnan, S., Rytter, W.: A constant time optimal parallel algorithm for two-dimensional pattern matching. SIAM J. Comput. **27**(3), 668–681 (1998). https://doi.org/10.1137/S0097539795280068
11. Crochemore, M., Ilie, L., Rytter, W.: Repetitions in strings: algorithms and combinatorics. Theoret. Comput. Sci. **410**(50), 5227–5235 (2009). https://doi.org/10.1016/j.tcs.2009.08.024
12. Crochemore, M., Iliopoulos, C.S., Kubica, M., Radoszewski, J., Rytter, W., Walen, T.: Extracting powers and periods in a word from its runs structure. Theoret. Comput. Sci. **521**, 29–41 (2014). https://doi.org/10.1016/j.tcs.2013.11.018
13. Crochemore, M., Rytter, W.: Squares, cubes, and time-space efficient string searching. Algorithmica **13**(5), 405–425 (1995). https://doi.org/10.1007/BF01190846
14. Davoodi, P.: Data structures: range queries and space efficiency, Ph. D. thesis, Aarhus University (2011)
15. Gawrychowski, P., Ghazawi, S., Landau, G.M.: Lower bounds for the number of repetitions in 2D strings. In: Lecroq, T., Touzet, H. (eds.) SPIRE 2021. LNCS, vol. 12944, pp. 179–192. Springer, Cham (2021). https://doi.org/10.1007/978-3-030-86692-1_15
16. Kolpakov, R.M., Kucherov, G.: Finding maximal repetitions in a word in linear time. In: 40th Annual Symposium on Foundations of Computer Science, FOCS 1999, 17–18 October 1999, pp. 596–604. New York, NY, USA. IEEE Computer Society (1999). https://doi.org/10.1109/SFFCS.1999.814634
17. Kolpakov, R., Kucherov, G.: On maximal repetitions in words. In: Ciobanu, G., Păun, G. (eds.) FCT 1999. LNCS, vol. 1684, pp. 374–385. Springer, Heidelberg (1999). https://doi.org/10.1007/3-540-48321-7_31
18. Lothaire, M.: Applied combinatorics on words, vol. 105. Cambridge University Press (2005)
19. Marcus, S., Sokol, D.: 2D Lyndon words and applications. Algorithmica **77**(1), 116–133 (2017). https://doi.org/10.1007/s00453-015-0065-z
20. McCreight, E.M.: Priority search trees. SIAM J. Comput. **14**(2), 257–276 (1985). https://doi.org/10.1137/0214021

Turbo Scan: Fast Sequential Nearest Neighbor Search in High Dimensions

Edgar Chavez[1]([✉])(iD) and Eric S. Tellez[1,2,3](iD)

[1] CICESE, Ensenada, Baja California, Mexico
{elchavez,stellez}@cicese.mx
[2] INFOTEC, Mexico City, Mexico
[3] IxM CONACyT, Mexico City, Mexico

Abstract. This paper introduces Turbo Scan (TS), a novel k-nearest neighbor search solution tailored for high-dimensional data and specific workloads where indexing can't be efficiently amortized over time. There exist situations where the overhead of index construction isn't warranted given the few queries executed on the dataset.

Rooted in the Johnson-Lindenstrauss (JL) lemma, our approach sidesteps the need for random rotations. To validate TS's superiority, we offer in-depth algorithmic and experimental evaluations. Our findings highlight TS's unique attributes and confirm its performance, surpassing sequential scans by 1.7x at perfect recall and a remarkable 2.5x at 97% recall.

Keywords: Approximate nearest neighbor search · Online knn search · Johnson-Lindenstrauss lemma

1 Introduction

Indexing is the standard approach for addressing the challenges of high-dimensional vector search. An index is a data structure that organizes vectors to enable efficient retrieval based on their similarity or distance metric. The index typically employs techniques such as space partitioning, hashing, or tree structures to group vectors with similar properties together, reducing the search space and speeding up the search process. The use case for indexing is when the database is static, and many queries will be performed over time, amortizing the index construction cost. In the above scenario, it is also worthwhile to have an overhead in memory usage; the most efficient indexes need to store additional information about distances between objects, quantized vectors, or links between objects.

Typical indexing methods assume that the cost of building an index to search a database can be ignored as the heavy workload of queries will eventually make up for it. However, we acknowledge situations where the number of queries is relatively small, and the cost of creating an index cannot be recovered. For example, searching for data in archives can be useful, but there may be few queries. On the other hand, the nearest-neighbor graph is a critical tool for visualizing

O. Pedreira and V. Estivill-Castro (Eds.): SISAP 2023, LNCS 14289, pp. 103–110, 2023.
https://doi.org/10.1007/978-3-031-46994-7_9

high-dimensional data. The datasets that need to be visualized are usually small, high-dimensional, and tend to change quickly over time. As a result, the costs of index construction and maintenance may outweigh the benefits of improved search efficiency. In such cases, the only available option is to use brute-force or linear search.

We are unaware of other approaches in the literature for online nearest neighbor searches. In [2], the authors propose a technique that utilizes the Johnson-Lindenstrauss (JL) theorem [3] for mapping in each distance computation. This mapping has a complexity logarithmic to the dimension. The proposed method is intended for refining a small dataset of candidates at the output of a standard index.

Another interesting approach, discussed in [8], involves using a lower bound of the distance instead of the full distance. The authors survey various methods in this context and propose a novel method to compute a hierarchy of approximations of increasing complexity. It's worth noting that these lower-bound methods require a preprocessing step to filter out queries, which typically incurs a linear cost based on the database size.

In time series analysis, Dynamic Time Warping (DTW) can facilitate online searches through specific heuristics as detailed in [6]. This research employs defined bounds and early termination to efficiently identify similar sequences under DTW. However, it is vital to distinguish that time series differ from vector databases, as the former represents a singular sequence while the latter encompasses complete vectors without overlap. Additionally, determining the lower bound of distance to the nearest neighbor in similarity searches is as challenging as identifying the nearest neighbors themselves.

2   Our Proposal

The Johnson-Lindenstrauss (JL) theorem is a mathematical result related to dimensionality reduction in high-dimensional data. It was introduced by William B. Johnson and Joram Lindenstrauss in 1984. The theorem states that given a set of high-dimensional points in Euclidean space, there exists a linear transformation that can approximately embed these points into a lower-dimensional space while preserving their pairwise distances up to a certain factor. In other words, the theorem guarantees that reducing the dimensionality of a set of points is possible without losing too much information about their pairwise distances.

More formally, let's say you have a set of n points in a high-dimensional space with dimensionality d. The Johnson-Lindenstrauss theorem states that for any $\epsilon > 0$ and any positive integer k, there exists a linear transformation (or projection) that maps the n points from d-dimensional space to k-dimensional space, where k is much smaller than d, such that the pairwise distances between the points are preserved up to a factor of $(1 \pm \epsilon)$. This means that the distances between points in the lower-dimensional space are approximately the same as those between the original points in the higher-dimensional space, up to a small error factor ϵ. The JL theorem provides a theoretical basis for dimensionality reduction techniques, such as random projections.

The Johnson-Lindenstrauss lemma states that a set of n points in a high-dimensional space can be projected onto a lower-dimensional space while preserving their pairwise distances up to a factor of $(1 + \epsilon)$ with high probability, where ϵ is a small constant and the dimension of the projected space is $O(\log n/\epsilon^2)$. However, there are recall issues when using random projections for indexing. The recall is the fraction of relevant items retrieved by a search. Because random projections can introduce noise and distortion, the recall of the search results may be reduced.

2.1 Turbo Scan: A Fast Filtering-Based Approximate Similarity Search Algorithm

Our first observation is about the squared Euclidean distance[1]. Please note that we can compute this sum in any order and partially. While we advance, the computation will be approximately better in precise order. The same is true for the inner product. Then, we can compute the partial distances from the query to all the vectors in the database using only a few coordinates, i.e., following the JL lemma, $h = O(\log n)$ at each iteration, and take decisions with the partial information that yields to reduce the final computing cost.

Therefore, instead of using the JL lemma to compute a projection, we use a linear memory approach and evaluate partial distances that solve an approximate nearest neighbor query in $O(n \log_\alpha n)$ average arithmetic operations where n is the number of vectors in the dataset and α related to the required quality. We call this procedure Turbo Scan (TS).

The Algorithm in Brief. Let us explore an initial idea for enhancing the efficiency of our nearest neighbor search. Our approach involves computing partial distances between the query and the entire database but with a limited number of coordinates. We then divide the database into two halves and retain the closest half to the query based on these partial distances. We repeat this procedure, gradually increasing the number of coordinates in our partial list and selecting the top half of the database at each step. Finally, in the last stage of this process, we calculate full vector distances for only a tiny fraction of the remaining database. By employing this method, we aim to improve our search algorithm's overall speed and efficiency, thereby enhancing our system's performance.

When implementing the initial idea, it is important to consider two parameters carefully. Firstly, determining the number of coordinates to be considered in partial distance computation is crucial for computational speed and recall rate. We'll explore alternatives using a fixed database to establish a more concrete approach. Below is a detailed framework to formalize the initial idea.

---

[1] Instead of computing $\sqrt{\sum_i |u_i - v_i|^2}$ we calculate $\sum_i |u_i - v_i|^2$, which produces the same ordering of the results.

A More Detailed Description. Let $X_{d \times n}$ and $Q_{d \times m}$ be our d-dimension data and query set, respectively, in a matrix representation of n and m vectors respectively. We need some notation to access vectors and slices. The ith column vector is accessed as $X_{1..d,i}$ and an slice of length h as $X_{j..j+h-1,i}$. Our method computes $\sum_i |u_i - v_i|^2$ (or $\sum_i u_i \cdot v_i$) incrementally, for X and Q, discarding distant candidates at each iteration.

Algorithm 1 shows the Turbo Scan pseudo-code. It receives the dataset X and query q, along with hyperparameters $1 \leq h \leq d$ and $1 < \alpha \leq n$. Our algorithm takes a small slice of vectors (of length h) and computes the distance partially, filtering out many objects in the result set (P and D at lines 10–11). The procedure is repeated iteratively using different slices (lines 3–12). The candidate set is reduced by a factor or $1/\alpha$ (lines 4–8).

Algorithm 1. Turbo Scan algorithm with fixed h-sized slices and α candidate reduction parameter.

Inputs:
- $X_{d \times n}$ the input dataset.
- q a d-dimensional vector query.
- k the number of neighbors to retrieve.
- $1 \leq h \leq d$ the vector slice size, typical values are factors of $\log n$.
- $1 < \alpha \leq n$ the candidate reduction value, typical values are small integers.

Pseudo-code:
1: Let $P_{1 \times n}$ and $D_{1 \times n}$ be two coupled arrays containing identifiers and partial distances. Initialize P as an identity permutation and D as an array of zeros.
2: $j \leftarrow 1$
3: **while** $j + h - 1 \leq d$ **do**
4:     **while** $i \in P$ **do**
5:         Let $X_{j..j+h-1,i}$ be named as u.
6:         Let $q_{j..j+h-1}$ be named as q'.
7:         $D_{P_i^{-1}} \leftarrow D_{P_i^{-1}} + \sum_\ell |u_\ell - q'_\ell|^2$
8:     **end while**
9:     Partially sort D and P to keep top $\max\{k, \lceil |D|/\alpha \rceil\}$ candidates
10:     Truncate both arrays to hold only these top $\max\{k, \lceil |D|/\alpha \rceil\}$ candidates
11:     $j \leftarrow j + h$
12: **end while**

Note 1: P and D arrays can be cached by thread and reused to avoid memory allocation between queries.
Note 2: P_i^{-1} retrieves the rank of i in P.
Note 3: Partial inner product can be used modifying line 7.

Figure 1 presents an illustration of the algorithm. The dataset matrix is accessed in slices. In the first iteration, we calculate all the partial distances for the first slice. In the second iteration, there are fewer elements as many are discarded based on the α parameter; the number of iterations required can be determined by $\log_\alpha n$. The optimal value of α depends on the dataset. The number of arithmetic operations per query reduces to $O(nd/\log_\alpha n)$, which is smaller than the standard $O(dn)$ sequential scan.

Similarly, the number of iterations can also be determined by h, specifically $\lceil d/h \rceil$. Still, determining the factor by which elements are discarded is more challenging.

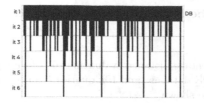

Fig. 1. Turbo Scan in brief. In each iteration we compute a more precise distance over a smaller fraction of the database

Both parameters are interrelated, but we choose to retain both since Alg. 1 and Fig. 1 demonstrate that adjusting the slices or α values can be done easily. It is worth investigating since dynamic parameters can offer certain advantages. We will experimentally examine both fixed and varying hyperparameters in the next section.

3   Experimental Results

This section presents the experimental results that characterize our approach and demonstrate its effectiveness. To conduct these experiments, we utilized the LAION-300K dataset, a subset of the 768-dimensional CLIP embeddings from the LAION2B database [7]. The LAION-300K dataset was obtained from the SISAP'2023 challenge site.[2]

We conducted two sets of experiments. In the first set, we utilized the official ten thousand queries provided by the SISAP challenge. The second set of experiments involved computing the $k$nn graph for the LAION-300K dataset, which means computing the $k$nn sets for the entire 300K vectors in the dataset. For both sets of experiments, we set the value of $k = 16$, which is a typical magnitude for multimedia information retrieval and non-linear dimensional reduction techniques.

Our benchmarks were performed on a workstation with two Intel(R) Xeon(R) Silver 4216 CPUs running at 2.10 GHz (32 cores) and equipped with 256 GB of RAM. The workstation operated on CentOS 8.5. We have provided an MIT-licensed package that includes an experimental implementation of our approach using the Julia programming language.[3] This implementation incorporates multithreading search functionality as well as various slicing and splitting strategies. The results we obtained represent the real-time duration in seconds required to solve queries in parallel.

---

[2] Available at: https://sisap-challenges.github.io/datasets/.
[3] Available at: https://github.com/sadit/SlicedSearch.jl.

3.1 Determining the Discrimination Ratio $1/\alpha$

Table 1. Required percentage distance evaluation to obtain a given recall for different slices for the LAION-300K database and 10k queries. The accumulated computing cost for the 10k queries is also shown

slice width	searchtime (seconds)	dataset review (%) for recall r						
		$r=0.25$	$r=0.5$	$r=0.75$	$r=0.9$	$r=0.95$	$r=0.99$	$r=1.0$
4	23.48	6.690	17.974	37.276	58.433	70.462	87.723	99.895
8	25.96	3.302	10.767	26.521	47.233	60.636	81.814	99.993
16	28.81	1.054	4.546	14.593	32.217	45.849	71.203	99.973
32	26.42	0.154	0.977	4.783	14.808	25.189	50.920	99.067
64	28.06	0.024	0.157	1.053	4.872	10.411	29.969	98.400
128	31.23	0.006	0.026	0.141	0.739	1.995	10.064	81.557

Table 1 shows the percentage required to be reviewed to achieve a given recall using a single iteration of our algorithm. We can estimate the expected quality using this table for a given α, i.e., each value is $100/\alpha$ since we present percentages for readability. For instance, supposing that we can expect the same properties from each iteration, i.e., the filtering power can be seen as an independently randomly distributed variable, in the first iteration, for $h = 64$, we can obtain a recall of 0.99 using $\frac{1}{\alpha} = 0.29969$ (e.g., $\alpha \approx 3.33$). Note that we can solve in $\log_{3.33} n \approx 10$ iterations with an expected recall of 0.9, i.e., 0.99^{10}. For an expected recall of 0.96, fixing $h = 128$, we obtain $\alpha = 50$ using 3 iterations.

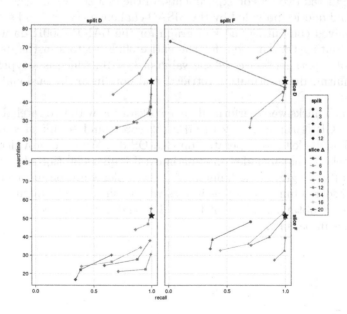

Fig. 2. Search time and recall of the 10k knn queries on the LAION-300K dataset (k = 16). Brute force performance is marked with the **black** star

3.2   Performance of the Full Turbo Scan Algorithm

Figure 2 shows the performance of solving knn queries of our Turbo Scan on the LAION-300K dataset, finishing all iterations and using $h = 10\Delta$. Each row corresponds to fixed and dynamic slicing policies (e.g., iterating with fixed-size or dynamically-sized sub-vectors). The first column shows the performance of the dynamic splitting policy. The second one fixes the split's α parameter.

Curves correspond to a single slicing Δ value, varying on split α value. The star marker corresponds to the brute-force solution of the queries. Top-left figure (DD quadrant) shows how small values of Δ give a poor performance; hence larger Δ values are better. This can be counter-intuitive but can be explained by the fast SIMD instructions available in modern hardware. This favors the splitting capabilities instead of the splicing ones. Note how the bottom-left figure (FD) decreases the search time on fixed slicing while treads quality as compared with DD; note that large fixed slicing ($\Delta = 10$) and dynamic split ($\alpha = 4$) (DF quadrant) performs better.

Regarding the fixed splitting policy, we can see higher search times due to the lack of adaptability of the policy. The top-right figure shows several high recall setups, except for the extreme hyper-parameters $\alpha = 2$ and $\Delta = 4$. The rest of the parameters yield acceptable-quality performances. The bottom-right figure (FF) describes the relatively good performance, especially for $\Delta = 10$ and 8, that achieve high recalls and are still faster than the brute force solution. Nonetheless, the DD and FD perform better in our benchmark.

Table 2. Performance of computing the all knn graph on the LAION-300K dataset ($k = 16$)

name	slice	Δ	split	α	recall	searchtime
BF					1.0	1431.0 s
TS	F	8	F	3	0.8692	1333.1 s
TS	D	16	F	3	0.9992	1699.5 s
TS	F	8	D	12	0.6182	749.4 s
TS	D	16	D	12	0.9728	1074.4 s

3.3   Computing the k Nearest Neighbor Graph

Our next experiment computes the dataset's k nearest neighbor graph. This is the input of many statistic and data science procedures, like non-linear dimensional reduction techniques like t-SNE, ISOMAP [1], or UMAP [4]; also, the spectral clustering algorithm, as described by Ng et al. [5]. In these application domains, high-quality results are required, and most users are not willing to tune and store a metric index.

Table 2 shows the performance comparison of our TS with the brute force solution. We can observe how split-F has higher search times and recall values.

The best configuration is the DD which shows high-recall and faster solution times. We avoid a large ablation study in this experiment due to the extensive computing times involved and limit ourselves to a pair of good hyper-parameters, as found in our previous experiment, see Fig. 2, showing the practical capabilities of our TS.

4   Conclusions

This paper presents Turbo Scan (TS), a JL-filtering based algorithm, tailored for approximate nearest neighbor searches in high-dimensional spaces. TS is optimized for memory, demanding a linear memory footprint per thread and executes queries with $O(n \log_\alpha n)$ operations, where n denotes dataset vector count and α indicates result quality.

TS calculates distances in partial slices, discarding elements iteratively according to partial findings. We detail the method, focusing on slice width and the α parameter, which defines the algorithm's discarding intensity (represented by the $1/\alpha$ discarding factor).

Using the LAION-300K dataset, we tested TS with both static and dynamic Δ and α values over 10k queries to construct the k nearest neighbor graph. Our findings reveal a 2–3x speedup and recall scores exceeding 0.9 compared to conventional brute force methods, signifying superior result quality and efficiency.

References

1. Anowar, F., Sadaoui, S., Selim, B.: Conceptual and empirical comparison of dimensionality reduction algorithms (PCA, KPCA, LDA, MDS, SVD, LLE, ISOMAP, LE, ICA, t-SNE). Comput. Sci. Rev. **40**, 100378 (2021)
2. Gao, J., Long, C.: High-dimensional approximate nearest neighbor search: with reliable and efficient distance comparison operations. arXiv preprint arXiv:2303.09855 (2023)
3. Johnson, W.B., Lindenstrauss, J.: Extensions of Lipschitz mappings into a Hilbert space. Contemp. Math. **26**, 189–206 (1984)
4. McInnes, L., Healy, J., Melville, J.: UMAP: uniform manifold approximation and projection for dimension reduction. arXiv preprint arXiv:1802.03426 (2020)
5. Ng, A., Jordan, M., Weiss, Y.: On spectral clustering: analysis and an algorithm. In: Advances in Neural Information Processing Systems 14 (2001)
6. Rakthanmanon, T., et al.: Addressing big data time series: mining trillions of time series subsequences under dynamic time warping. ACM Trans. Knowl. Discov. Data (TKDD) **7**(3), 1–31 (2013)
7. Schuhmann, C., et al.: LAION-5B: an open large-scale dataset for training next generation image-text models (2022)
8. Zhang, H., Dong, Y., Xu, D.: Accelerating exact nearest neighbor search in high dimensional Euclidean space via block vectors. Int. J. Intell. Syst. **37**(2), 1697–1722 (2022)

Class Representatives Selection
in Non-metric Spaces for Nearest
Prototype Classification

Jaroslav Hlaváč[1,2]($\boxtimes$) (iD), Martin Kopp[1] (iD), Jan Kohout[1,3], and Tomá Skopal[2] (iD)

[1] TD&R Data Science, Cisco Systems, Prague, Czech Republic
hlavac.jaroslav@gmail.com
[2] Faculty of Mathematics and Physics, Charles University, Prague, Czech Republic
[3] TruU, Prague, Czech Republic

Abstract. The nearest prototype classification is a less computationally intensive replacement for the k-NN method, especially when large datasets are considered. Centroids are often used as prototypes to represent whole classes in metric spaces. Selection of class prototypes in non-metric spaces is more challenging as the idea of computing centroids is not directly applicable. Instead, a set of representative objects can be used as the class prototype.

In this paper, we present CRS, a novel memory and computationally efficient method that finds a small yet representative set of objects from each class to be used as prototype. CRS leverages the similarity graph representation of each class created by the NN-Descent algorithm to pick a low number of representatives that ensure sufficient class coverage. Thanks to the graph-based approach, CRS can be applied to any space where at least a pairwise similarity can be defined. In the experimental evaluation, we demonstrate that our method outperforms the state-of-the-art techniques on multiple datasets from different domains.

Keywords: Class Representation · Nearest Prototype Classification · Prototype Selection

1 Introduction

The k-NN classifiers are often used in many application domains due to their simplicity and ability to trace the classification decision to a specific set of samples. However, their adoption is limited by high computational complexity and memory requirements. Because contemporary datasets are often huge, containing hundreds of thousands or even millions of samples, computing similarity between the classified sample and the entire dataset may be computationally intractable.

In order to decrease computational and memory requirements, the nearest prototype classification (NPC) method is commonly used, c.f. [1–3]. In NPC, each class is represented by a *prototype*, that represents typical characteristics of

© The Author(s), under exclusive license to Springer Nature Switzerland AG 2023
O. Pedreira and V. Estivill-Castro (Eds.): SISAP 2023, LNCS 14289, pp. 111–124, 2023.
https://doi.org/10.1007/978-3-031-46994-7_10

the class. The classified sample is then compared just to the prototypes instead of calculating similarity to the entire dataset. Therefore, the goal of prototype selection is to find a memory-efficient representation of classes such that classification accuracy is preserved while the number of comparisons is significantly reduced.

An intuitive prototype in a metric space can be a centriod. But even in metric spaces a centroid is often not an optimal solution because a single point does not represent the whole class well. Sometimes the centroid does not make sense and in non-metric spaces (also called distance spaces [4]) it is not defined. Such is the case in many application domains, where objects exist in space where only a pairwise (dis)similarity is defined, e.g., bioinformatics [5], biometric identification [6], or pattern recognition [7].

Our focus on non-metric spaces comes from the problem of behavioural clustering of network hosts [8], where we need to quickly assign a network host to a group of other hosts. A newly appearing network host in a computer network needs to be quickly assigned to a correct host group (or a new group must be created). The space we operate in is defined by the domains and IP addresses that the whole network has communicated with in the previous sliding time window (e.g. day). The similarity we use is expensive to compute (see [8] for details) as the dimension of the space is high and changes quickly.

Nevertheless, the problem of selecting a minimal number of representative samples is of more general interest. Only a few methods have been developed for non-metric scenarios, and to the best of our knowledge the only general (not domain-specific) approach is selection of small subset of objects to represent the whole class. The method is referred to as representative selection and the *representatives* (selected objects), are used as a prototype. Several recent methods capable of solving representatives selection on non-metric spaces exist (i.e. DS3 [9], δ-medoids [10]).

In this paper, we present a novel method to solve the problem of representative selection – Class Representatives Selection (CRS). CRS is a general method capable of selecting small yet representative subset of objects from a class to serve as its prototype. Its core idea is fast construction of an approximate reverse k-NN graph and then solving minimal vertex cover problem on that graph. Only a pairwise similarity is required to build the reverse k-NN graph, therefore application of CRS is not limited to metric spaces.

To show that CRS is general and domain-independent, we present an experimental evaluation on datasets from image recognition, document classification and network host classification, with appealing results when compared to the current state of the art. The code for CRS can be found at https://github.com/jaroslavh/ceres.

The paper is organized as follows. The related work is briefly reviewed in the next section. Section 3 formalises the representative selection as an optimization problem. The proposed method is described in detail in Sect. 4. The experimental evaluation is summarized in Sect. 5 followed by the conclusion.

2   Related Work

During the past years, significant effort has been made to represent classes in the most condensed way. The approaches could be categorized into two main groups.

The first group gathers all prototype generation methods [11], which create artificial samples to represent original classes, e.g. [12,13]. The second group contains the prototype selection methods. As the name suggests, a subset of samples from the given class is selected to represent it. Prototype selection is a well-explored field with many approaches, see, e.g. [14].

However, most of the current algorithms exploit the properties of the metric space, e.g., structured sparsity [15], l_1-norm induced selection [16] or identification of borderline objects [17].

When we leave the luxury of the metric space and focus on situations where only a pairwise similarity exists or where averaging of existing samples may create an object without meaning, there is not much previous work.

The δ-medoids [10] algorithm uses the idea of k-medoids to semi-greedily cover the space with δ-neighbourhoods, in which it then looks for an optimal medoid to represent a given neighbourhood. The main issue of this method is the selection of δ: this hyperparameter has to be fine-tuned based on the domain.

The DS3 [9] algorithm calculates the full similarity matrix and then selects representatives by a row-sparsity regularized trace minimization program which tries to minimize the rows needed to encode the whole matrix. The overall computational complexity is the most significant disadvantage of this algorithm, despite some proposed approximate estimation of the similarity matrix using only a subset of the data.

The proposed method for Class Representatives Selection (CRS) approximates the topological structure of the data by creating a reverse k-NN graph. CRS then iteratively selects nodes with the biggest reverse neighbourhoods as representatives of the data. This approach systematically minimizes the number of pairwise comparisons to reduce computational complexity while accurately representing the data.

3   Problem Formulation

In this section, we define the problem of prototype-based representation of classes and the nearest prototype classification (NPC). As we already stated in Introduction, we study the prototypes selection in general cases, including non-metric spaces. Therefore, we further assume that a class prototype is always specified as (possibly small) subset of its members.

Class Prototypes. Let T be an arbitrary space of objects for which a pairwise similarity function $s : T \times T \to \mathbb{R}$ is defined and let $X \subseteq T$ be a set of (training) samples. Let $\mathcal{C} = \{C_1, ..., C_m\}$ be a set of classes of X such that $C_i \cap C_j = \emptyset, \forall i \neq j$ and $\bigcup C_i = X$. Let $C_i = \{x_1, x_2, ..., x_n\}$ be a class of size n. For $x \in C_i$, let us denote U_x^k the k closest samples to x, i.e., the set of k samples that have the highest similarity to x in the rest of the class $C_i \setminus \{x\}$. Then the goal of the prototype selection is to find a prototype of class C_i, $R_i \subseteq C_i$ for each class such that:

$$\forall x \in C_i \; \exists \, r \in R_i \; : \; x \in U_r^k \tag{1}$$

In order to minimize computational requirements of NPC, we search for a minimal set of class representatives R_i^* for each class, which satisfies the coverage requirement (1):

$$R_i^* = \underset{|R_i|}{\arg\min} \left\{ r : \bigcup_{r \in R_i} U_r^k = C_i \right\} \tag{2}$$

Note that several sets might satisfy this coverage requirement.

Relaxed Prototypes. Finding class prototypes that fully meet the coverage requirement (1) might pose a computational burden and produce unnecessarily big prototypes. In most cases, covering the majority of the class objects while leaving out a few outliers leads to a smaller prototype that still captures the essential characteristics of a class. Motivated by this observation, we introduce a relaxed requirement for class prototypes. We say that a set $R_i \subseteq C_i$ is a representative prototype of class C_i if the following condition is met:

$$\left| \bigcup_{r \in R_i} U_r^k \cap C_i \right| \geq \epsilon \, |C_i|, \tag{3}$$

for a preset parameter $\epsilon \in (0, 1]$.

In further work, we replace the requirement (1) with its relaxed version (3) with $\epsilon = 0.95$. In case of need, the full coverage can be enforced by simply setting $\epsilon = 1$. Even in the relaxed version, we seek a prototype with minimal cardinality which satisfies (3).

Nearest Prototype Classification. Having the prototypes for all classes $\mathcal{R} = \{R_1, ..., R_m\}$, an unseen sample x is classified to the class with the most similar prototype $R^* \in \mathcal{R}$. R^* is the prototype containing representative r with the highest similarity to x.

$$r^* = \underset{r \in \bigcup R_i}{\arg\max} \, s(x, r).$$

Note that in our research we take into account only the closest representative $r*$. This choice comes from previous research [8] where 1-NN was yielded the best results.

4   Class Representatives Selection

In this section, we describe our method CRS for building the class prototypes. The entire method is composed of two steps:

1. Given a class C and a similarity measure s, a reverse k-NN graph G is constructed from objects C using the pairwise similarity s.
2. Graph G is used to select the representatives that satisfy the coverage requirement while minimizing the size of the class prototype.

The simplified scheme of the whole process is depicted in Fig. 1.

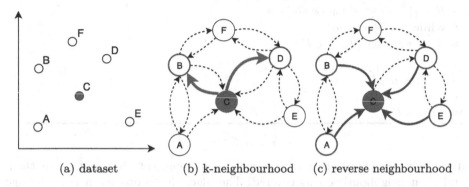

      (a) dataset   (b) k-neighbourhood   (c) reverse neighbourhood

Fig. 1. Illustration of the steps of CRS algorithm. (a) Visualization of a toy 2D class. (b) 2-NN graph created from the class. (c) Reverse graph created from the graph depicted in (b). Node C's reverse neighbourhood covers A, B, D, E and thus would be a good first choice for a representative. Depending on the coverage parameter ϵ, the node F could be considered an outlier or also added to the representation

4.1   Building the Prototype

For the purpose of building the prototype for a class C a weighted reverse k-NN graph G_C^{-1} is used. It is defined as $G_C^{-1} = (V, E, w)$, where V is the set of all objects in the class C, E is a set of edges and w is a weight vector. An edge between two nodes $v_i, v_j \in V_{i \neq j}$ exists if $v_i \in U_{v_j}^k$, while the edge weight w_{ij} is given by the similarity s between the connected nodes, $w_{ij} = s(v_i, v_j)$.

The effective construction of such graph is enabled by employing the NN-Descent [18] algorithm, a fast converging approximate method for the k-NN graph construction. It exploits the idea that "a neighbour of a neighbour is also likely to be a neighbour" to locally explore neighbouring nodes for better solutions. NN-Descent produces a k-NN graph G_C. The reverse k-NN graph G_C^{-1} is then obtained from G_C by simply reversing directions of the edges in G_C.

Omitting all edges with weight lower than τ from the reverse k-NN graph G_C^{-1} ensures that very dissimilar objects do not appear in the reverse neighborhoods:

$$(\forall y \in U_x : s(x, y) \geq \tau)$$

The selection of representatives is treated as a minimum vertex cover problem on G_C^{-1} with omitted low similarity edges. We use a greedy algorithm which

---

Algorithm 1: Pseudocode for Class Representatives Selection

Data: class $C = \{c_1, ..., c_n\}$, similarity s, coverage threshold ϵ, size of
          neighbourhood k, weight threshold τ
Result: set of selected representatives $R \subseteq C$

1 G_C = NN-Descent(C, s, k)
2 G_C^{-1} = ReverseGraph(G_C, τ)
3 $Z = C$ //set of uncovered objects
4 $R = \{\}$ //set of representatives
5 **while** $\frac{|C|-|Z|}{|C|} < \epsilon$ **do**
6      $r = \arg\max_c(|U_c|, c \in U_c)$
7      $Z = Z \setminus U_r$
8      $R = R \cup \{r\}$
9 **end**
10 **return** R

---

iteratively selects objects with maximal $|U|$ as representatives and marks them and their neighbourhood as covered. The algorithm stops when the coverage requirement (3) is met.

The whole algorithm is summarized in Algorithm 1.

4.2 Parameter Analysis

This subsection summarizes the parameters of the CRS method.

- k: number of neighbours for the k-NN graph creation. When k is high, each object covers more neighbours, but on the other hand it also increases the number of pairwise similarity calculations. This trade-off is illustrated for different values of k in Fig. 2. Due to the large impact of this parameter on properties of the produced representations and computational requirements, we further study its behaviour in more detail in a dedicated experiment in Sect. 5.
- ϵ: coverage parameter for the relaxed coverage requirement as introduced in Sect. 3. In this work, we set it to 0.95 which is a common threshold in outlier detection. It ensures that the vast majority of each class is still covered but outliers do not influence the prototypes.
- τ: threshold on weights, edges with lower weights (similarities) are pruned from the reverse neighbourhood graph G_C^{-1} (see Sect. 4.1). By default it is automatically set to approximate *homogeneity* $h(C)$ of the class C defined as:

$$h(C) = \frac{1}{|C|} \sum_{x_i, x_j \in C, i \neq j} s(x_i, x_j) \tag{4}$$

Additionally, the NN-Descent algorithm, used within the CRS method, has two more parameters that specify its behaviour during the k-NN graph creation. First, the δ_{nn} parameter which is used for early termination of the NN-Descent

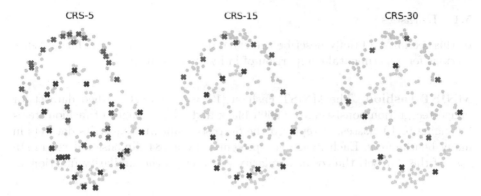

Fig. 2. In CRS the number of selected representatives and the quality of representation are both determined by k. For low ks the NN-Descent subsamples dense areas of the class too much and the information about neighbours is not propagated (CRS-5). As each object explores a bigger neighbourhood for higher k, the number of other objects it represents grows, therefore the number of representatives decreases. On the other hand, with less representatives, some information about the structure is lost, as in the case of $k = 30$

algorithm when the number of changes in the constructed graph is minimal. We set it to 0.001, as suggested by the authors of the original work [18]. Second, the sample rate ρ controls the number of reverse neighbours to be explored in each iteration of NN-Descent. Again, we set it to 0.5 to speed up the k-NN creation while not diverging too far from the optimal solution.

5   Experiments

This section presents experimental evaluation of the CRS algorithm on multiple datasets from very different domains that cover computer networks, text documents processing and image classification. First, we compare the CRS method to the state of the art techniques DS3 [9] and δ-medoids [10] on the nearest prototype classification task on different datasets. Then, we study the influence of the parameter k (which determines the number of nearest neighbours used for building the underlying k-NN graph).

We set δ in the δ-Medoids algorithm as approximate homogeneity h (see Eq. 4) calculated from random 5% of the class. Setting δ is a difficult problem not explained well in the original paper. From our experiments, homogeneity is a good estimate. The best results for DS3 we obtained with $p = \inf$ and $\alpha = 3$, while creating the full similarity matrix for the entire class. We tried $\alpha = 0.5$ which was suggested by the authors, but the algorithm always selected only one representative with much worse results. Finally, for CRS we set $\epsilon = 0.95$, $\tau = h$ (to be fair in comparison with δ-medoids). By far the most impactful parameter is k. Section 5.4 looks at the selection in depth. A good initial choice for classes with 1000 or more samples is $k = 20$ and $k = 10$ works well for smaller classes.

5.1 Datasets

In this section we briefly describe the three datasets used in the following subsections for experimental comparison of individual methods.

MNIST Fashion. The MNIST Fashion [19] is a well established dataset for image recognition consisting of 60000 black and white images of fashion items belonging to 10 classes. It replaced the overused handwritten digits datasets in many benchmarks. Each image is represented by a 784 dimensional vector. In case of this dataset, the cosine similarity was used as the similarity function s.

20Newsgroup. 20Newsgroup dataset is a benchmark dataset for text documents processing. It is composed of nearly 20 thousand newspaper documents from 20 different classes (topics). The dataset was preprocessed such that each document is represented by a TF-IDF frequency vector of dimension 130,107. We used the cosine similarity which is a common choice in the domain of text documents processing as a similarity function s.

Private Network Dataset. Network dataset is the main motivation for our research. It was collected on a corporate computer network, originally for the purpose of network host clustering based on their behaviour [8]. The work defines a specific pair-wise similarity measure for network devices based on visited network hosts which we adopt for this paper. The dataset consists of all network communication collected on more than 5000 network hosts for one day (288 5-minute windows). This dataset resides in the space of all possible hostname and port number combinations. The dimension of this space is theoretically infinite, hence we work with a similarity that treats this space as non-metric.

For the purposes of the evaluation, classes smaller than 10 members were not considered, since such small classes can be easily represented by any method. The sizes and homogeneities of the classes can be found in Table 2. In contrast to the previous datasets, the sizes and values of homogeneity of classes in the Network dataset differ significantly, as can be seen in Table 2.

5.2 Evaluation of Results

In this section we present the results for each dataset in detail. The main results are summarized in Table 1. For a more complete picture we also included results for selecting a random 5% and all 100% of the class as a prototype. When evaluating the experiments, we take into account both precision/recall of nearest prototype classification and the percentage of samples selected as prototypes. Each method was run 10 times over a 80%/20% train/test split of each dataset. The results were averaged and the standard deviations of precisions and recalls were smaller than 0.005 for all methods, which shows stability of all algorithms. The only exception was δ-medoids on Network dataset where the precisions fluctuated up to 0.015.

Table 1. Average precision/recall values for each method used on each dataset. The table also shows the percentage of the class that was selected as a prototype. CRS outperforms both DS3 and δ-medoids on all datasets. In the network dataset CRS-k10 outperforms event the full-100% baseline as CRS does not try to cover outliers (in this case network hosts being very different from the rest of the class)

Method	MNIST Fashion	20Newsgroup	Network
δ-medoids	0.765/0.737 (10.45%)	0.362/0.3 (7.57%)	0.992/0.952 (7.01%)
DS3	0.727/0.715 (0.19%)	0.291/0.291 (0.57%)	0.853/0.956 (3.04%)
random-5%	0.780/0.767 (5.0%)	0.33/0.435 (5.0%)	0.94/0.959 (5.0%)
full-100%	0.849/0.846 (100.0%)	0.56/0.548 (100.0%)	0.987/0.963 (100.0%)
CRS-k10	0.823/0.817 (11.94%)	0.45/0.391 (11.28%)	0.992/0.973 (6.53%)
CRS-k20	0.813/0.806 (8.56%)	0.377/0.329 (6.58%)	0.972/0.976 (3.21%)
CRS-k30	0.806/0.798 (6.1%)	0.344/0.295 (4.83%)	0.983/0.968 (2.38%)

Table 2. Sizes and homogeneity for each class from network dataset. Classes with size lower than 10 were removed from the dataset

Class	A	B	C	D	E	F	G	H	I	J	K	L	M	N
Size	1079	2407	75	2219	346	59	248	49	52	108	218	44	42	32
Homogeneity	0.58	0.14	0.84	0.64	0.60	0.92	0.34	0.84	0.69	0.35	0.78	0.35	0.79	1.0

As we have shown in the experiment in Sect. 5.4, CRS can be tuned by the parameter k to significantly reduce the number of representatives and maintain a high precision/recall values. The DS3 method selects a significantly lower number of representatives than any other method. However, it is at the cost of lower precision and recall values.

MNIST Fashion. The average homogeneity of a class in the MNIST Fashion dataset is 0.76. This corresponds with a slower decline of the precision and recall values as the number of representatives decreases. In Fig. 3 are the confusion matrices for the methods.

20Newsgroup. The 20Newsgroup dataset has the lowest average homogeneity $h = 0.0858$ from all the datasets. The samples are less similar on average, therefore the lower precision and recall values. Still CRS-k10 with only 11% of representatives performs quite well, compared with the other methods. Confusion matrices for one class form each subgroup are in Fig. 4.

Network Dataset. The results for data collected in real network further prove that lowering K does not lead a great decrease in performance. Again Fig. 5 shows confusion matrices for main 3 algorithms. Particularly interesting are the biggest classes A, B and D which were most difficult to cover for all algorithms.

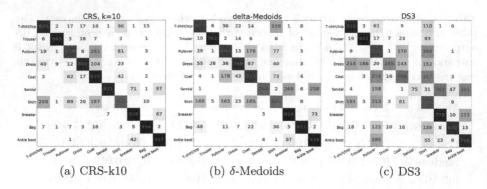

Fig. 3. Confusion matrices for each class in the MNIST Fashion dataset show the performance all 3 methods compared. The Sandal class was the hardest to represent for all methods

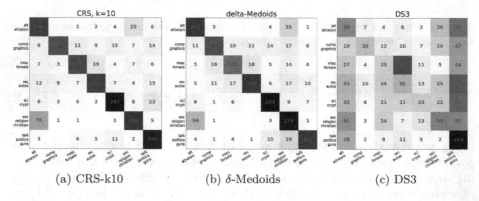

Fig. 4. Confusion matrices for each class in the 20Newsgroup dataset show the performance all 3 methods compared

For sizes of all classes see Table 2. Moreover, lower homogeneity for B is also clearly seen in the confusion matrix.

5.3 Time Efficiency

When considering the speed of the algorithms, we particularly focus on cases where the slow and expensive computation of the pairwise similarity overshadows the rest, e.g. in the case of the Network dataset. Therefore, we compare the algorithms by the relative number of similarity computations S defined as:

$$S = \frac{S_{actual}}{S_{full}}, \tag{5}$$

where S_{actual} stands for the number of comparisons made and S_{full} is the number of comparisons needed for computing the full similarity matrix.

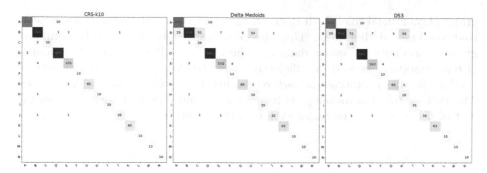

Fig. 5. Confusion matrices for each class in the Network dataset. Particularly interesting are the biggest classes A, B and D which were most difficult to cover for all algorithms. Moreover, lower homogeneity for B is also clearly seen in the confusion matrix

Table 3. The average number of similarity calculations relative to computing full similarity matrix in classes that have more than 1000 samples. For the DS3 algorithm, we always calculate the full similarity matrix; therefore, it is not included in the table

dataset	δ-Medoids		CRS-k10		CRS-k20	
	mean	std	mean	std	mean	std
MNIST Fashion	0.132	0.031	0.074	0.008	0.218	0.013
Network Dataset	0.467	0.266	0.178	0.058	0.507	0.16

We measured S for classes bigger than 1000 samples to see how the algorithms perform on big classes. In smaller classes the total differences in comparisons are not great as the full similarity matrices are smaller. Also the smaller the class, the closer are all algorithms to $S = 1$ (for CRS it can be seen in Fig. 6h). The results for big classes are in Table 3. We use DS3 with the full similarity matrix to get most accurate results, therefore $S_{DS3} = 1$.

For CRS the number of comparisons is influenced by k, sample rate ρ, and homogeneity of each class and its size. However, we use very high ρ in the NN-Descent part of CRS, which significantly increases the number of comparisons. The impact of k is discussed in detail in Sect. 5.4 and experimenting with ρ is up for further research. For δ-Medoids the number of similarity computations performed is determined by the difficulty of the class and the δ parameter. In CRS, the parameters can be set according to the similarity computations we have available to achieve the best prototypes given the time. This does not hold neither for δ-medoids nor for $DS3$.

5.4   Impact of k

When building class prototypes by the CRS method, the number of nearest neighbours (specified by the parameter k) considered for building the k-NN graph plays crucial role. With small ks, each object neighbours only few objects that are most similar to it. This also propagates into reverse neighbourhood graphs, especially for sparse datasets. Therefore, small ks increase the number of representatives needed to sufficiently cover the class. Using higher values of k produce smaller prototypes as each representative is able to cover more objects. The cost of this improvement is increased computational burden because the cost of k-NN creation increases rapidly with higher ks.

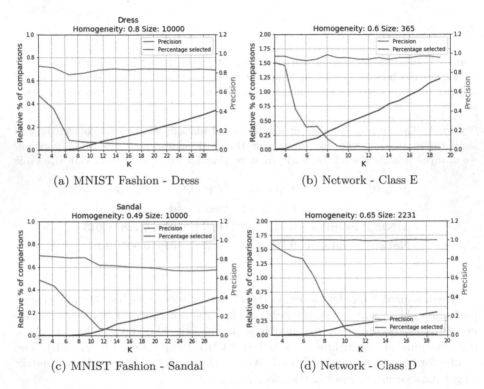

(a) MNIST Fashion - Dress   (b) Network - Class E

(c) MNIST Fashion - Sandal   (d) Network - Class D

Fig. 6. Illustration of how the selection of k influences the number of representatives and number of similarity computations. The number of representatives is in relative numbers to the size of the class. For different classes as k increases the relative number of comparisons also increases. However, the size of prototype selected decreases steeply while the precision decreases slowly (Color figure online)

Figure 6 shows trends of precision, sizes of created prototypes and numbers of similarity function evaluations depending on k for several classes that differ in their homogeneity and sizes. We can see the trade-off between computational requirements (blue line) and memory requirements (red line) as the k increases. From some point (e.g. where red line crosses the blue line), the classification precision decreases slowly. The cost limitations of building the prototype or the classification can be used to set the parameter k. If k is low, CRS selects prototypes faster, but the number of selected representatives is higher and therefore the classification cost is also higher. If the classification cost (number of similarity computations to classify an object) is more important than prototype selection, parameter k can be higher.

6 Conclusion

This paper proposes CRS, a novel method for building representations of classes, class prototypes, which are small subsets of the original classes. CRS leverages nearest neighbour graphs to map each structure of each class and identify representatives that will form the class prototype. This approach allows CRS to be applied in any space where at least pairwise similarity is defined.

The proposed method was compared to the prior art in a nearest prototype classification setup on multiple datasets from different domains. The experimental results show that the CRS method achieves superior classification quality while producing comparably compact representations of classes.

References

1. Seo, S., Bode, M., Obermayer, K.: Soft nearest prototype classification. IEEE Trans. Neural Netw. **14**(2), 390–398 (2003)
2. Schleif, F.-M., Villmann, T., Hammer, B.: Local metric adaptation for soft nearest prototype classification to classify proteomic data. In: Bloch, I., Petrosino, A., Tettamanzi, A.G.B. (eds.) WILF 2005. LNCS (LNAI), vol. 3849, pp. 290–296. Springer, Heidelberg (2006). https://doi.org/10.1007/11676935_36
3. Cervantes, A., Galván, I., Isasi, P.: An adaptive michigan approach PSO for nearest prototype classification. In: Mira, J., Álvarez, J.R. (eds.) IWINAC 2007. LNCS, vol. 4528, pp. 287–296. Springer, Heidelberg (2007). https://doi.org/10.1007/978-3-540-73055-2_31
4. Anthony, M., Ratsaby, J.: Large width nearest prototype classification on general distance spaces. Theoret. Comput. Sci. **738**, 65–79 (2018)
5. Martino, A., Giuliani, A., Rizzi, A.: Granular computing techniques for bioinformatics pattern recognition problems in non-metric spaces. In: Pedrycz, W., Chen, S.-M. (eds.) Computational Intelligence for Pattern Recognition. SCI, vol. 777, pp. 53–81. Springer, Cham (2018). https://doi.org/10.1007/978-3-319-89629-8_3
6. Becker, G.C.: Methods and apparatus for clustering templates in non-metric similarity spaces, October 12, US Patent 7,813,531 (2010)
7. Scheirer, W.J., Wilber, M.J., Eckmann, M., Boult, T.E.: Good recognition is nonmetric. Pattern Recogn. **47**(8), 2721–2731 (2014)
8. Kopp, M., Grill, M., Kohout, J.: Community-based anomaly detection. In: 2018 IEEE International Workshop on Information Forensics and Security (WIFS), pp. 1–6. IEEE (20180
9. Elhamifar, E., Sapiro, G., Sastry. S.S.: Dissimilarity-based sparse subset selection. IEEE Trans. Pattern Anal. Mach. Intell. **38**(11), 2182–2197 (2015)
10. Liebman, E., Chor, B., Stone, P.: Representative selection in nonmetric datasets. Appl. Artif. Intell. **29**(8), 807–838 (2015)
11. Triguero, I., Derrac, J., Garcia, S., Herrera, F.: A taxonomy and experimental study on prototype generation for nearest neighbor classification. IEEE Trans. Syst. Man Cybern. Part C (Appli. Rev.) **42**(1), 86–100 (2011)

12. Geva, S., Sitte, J.: Adaptive nearest neighbor pattern classification. IEEE Trans. on Neural Networks **2**, 2 (1991)
13. Xie, Q., Laszlo, C.A., Ward, R.K.: Vector quantization technique for nonparametric classifier design. IEEE Trans. Pattern Anal. Mach. Intell. **15**(12), 1326–1330 (1993)
14. Garcia, S., Derrac, J., Cano, J., Herrera, F.: Prototype selection for nearest neighbor classification: taxonomy and empirical study. IEEE Trans. Pattern Anal. Mach. Intell. **34**(3), 417–435 (2012)
15. Wang, H., Kawahara, Y., Weng, C., Yuan, J.: Representative selection with structured sparsity. Pattern Recogn. **63**, 268–278 (2017)
16. Zhang, X., Zhu, Z., Zhao, Y., Chang, D., Liu, J.: Seeing all from a few: l1-norm-induced discriminative prototype selection. IEEE Trans. Neural Netw. Learn. Syst. **30**(7), 1954–1966 (2018)
17. Olvera-López, J.A., Carrasco-Ochoa, J.A., Martínez-Trinidad, J.: Accurate and fast prototype selection based on the notion of relevant and border prototypes. J. Intell. Fuzzy Syst. **34**(5), 2923–2934 (2018)
18. Dong, W., Moses, C., Li, K.: Efficient k-nearest neighbor graph construction for generic similarity measures. In: Proceedings of the 20th International Conference on World Wide Web, pp. 577–586 (2011)
19. Xiao, H., Rasul, K., Vollgraf, R.: Fashion-mnist: a novel image dataset for benchmarking machine learning algorithms (2017)

The Dataset-Similarity-Based Approach to Select Datasets for Evaluation in Similarity Retrieval

Matheus A. L. Matiazzo[1], Vitor de Castro-Silva[1], Rafael S. Oyamada[2], and Daniel S. Kaster[1(✉)]

[1] University of Londrina, Londrina, Brazil
{vitor.castro.silva,dskaster}@uel.br
[2] University of Milan, Milan, Italy
rafael.oyamada@unimi.it

Abstract. Most papers on similarity retrieval present experiments executed on an assortion of complex datasets. However, no work focuses on analyzing the selection of datasets to evaluate the techniques proposed in the related literature. Ideally, the datasets chosen for experimental analysis should cover a variety of properties to ensure a proper evaluation; however, this is not always the case. This paper introduces the dataset-similarity-based approach, a new conceptual view of datasets that explores how they vary according to their characteristics. The approach is based on extracting a set of features from the datasets to represent them in a similarity space and analyze their distribution in this space. We present an instantiation of our approach using datasets gathered by surveying the dataset usage in papers published in relevant conferences on similarity retrieval and sample analyses. Our analyses show that datasets often used together in experiments are more similar than they seem to be at first glance, reducing the variability. The proposed representation of datasets in a similarity space allows future works to improve the choice of datasets for running experiments in similarity retrieval.

Keywords: Similarity Retrieval · Datasets · Experimental Analysis · Similarity Space of Datasets

1 Introduction

Similarity search enables retrieving similar elements from a database given one or more reference elements according to their features. An important aspect to consider when developing new methods for similarity searching is the choice of datasets to evaluate the proposed methods. There are several datasets in the field of similarity search for the most varied applications [3,9,12]. However, in

This work has been financially supported by the Brazilian funding agencies CNPq and Araucaria Foundation, and by the University of Milan, Italy.

most works in similarity searching, the selection of datasets for evaluating the proposals is solely based on experience, without a deeper variability analysis.

In the literature, there are several papers describing the underlying characteristics of datasets that make them more or less difficult for similarity search purposes. Particularly, dataset properties that might affect the performance of similarity search methods include intrinsic dimensionality [2], relative contrast [7], fractal correlation [4], and the concentration of distances [6]. However, works in the field rely on properties like these to select the datasets for experimentation without analyzing the interconnection between properties that affect the dataset variability. To the best of our knowledge, no previous research covers a wide range of these datasets, focusing on jointly analyzing their underlying characteristics for similarity search purposes.

In this paper, our goal is to survey and analyze the properties of datasets employed by researchers in the field of similarity search in order to provide a better understanding of the diversity of datasets studied. We accomplish that by gathering several datasets used in the literature and extracting an assortion of their features. The relevance of this contribution relies on considering joint properties to represent the datasets using a new conceptual view, called the *dataset-similarity-based* approach. Employing meaningful features allows us to create a similarity space of datasets, which can be used to analyze the diversity of datasets for many purposes, including selecting datasets to conduct experiments on similarity retrieval methods. We present an instantiation of proposed conceptual view based on a survey on dataset usage in works in the field over the last decade and analyses illustrating the potential of the approach. The methods used to extract such features and the surveyed datasets are publicly available.

This paper is organized as follows. Section 2 presents the background to understand the paper and related work. Section 3 describes the proposed approach, including how we define the dataset similarity space and the methodology we employed to survey the dataset usage in papers on similarity retrieval in the last decade to instantiate our approach. Section 4 presents sample analyses using our proposed dataset-similarity-space view, and in Sect. 5 we conclude.

2   Background and Related Work

The most intuitive strategy to evaluate searching algorithms is to compare their performances across datasets with different characteristics, such as cardinality and dimensionality. However, the choice of the datasets employed in the evaluation is fundamentally based on experience, which can create a bias in the choice, such as selecting a dataset only based on popularity.

Several metrics can be extracted from datasets to measure the complexity of a similarity search problem [2,10]. Existing metrics include the Relative Contrast [7], intrinsic dimensionality [2], fractal correlation [4], and the concentration of distances [6]. These metrics can characterize datasets providing valuable insights for choosing similarity search algorithms and their parameters.

For instance, Aumller and Ceccarello [2] describe an application of the local intrinsic dimensionality to measure the complexity of a dataset. The authors show that the so-called *curse of dimensionality* is not the leading cause to degrade the performance of similarity queries and that the intrinsic dimensionality of a dataset can be used to obtain better insights. On the other hand, in [14], we proposed a learned approach to map the combination of dataset metrics to the performance of proximity graph algorithms. This way, it becomes possible to predict the performance of different algorithms for a given dataset and to choose the best algorithm for a given dataset without having to run a greedy search to find a suitable algorithm configuration [16].

Understanding the complexity of datasets is also a relevant subject in other fields. In [15], Šikonja presents a study comparing complex datasets to determine if datasets and subsets are similar enough to be used together in a data mining task. The proposal consists of a methodology to compare the statistical properties of the attributes and the similarity between clusters of elements. In the machine learning field, [11] presents a study on the properties of datasets aiming at understanding the impact of these properties on the performance of classification algorithms. These properties include generic properties (e.g., dimensionality and cardinality), classification properties (e.g., class imbalance and number of classes), and neighborhood properties (e.g., number of clusters and hub score). In a similar direction, a new research field called dataset discovery has recently emerged in the literature [5]. The problem consists of finding datasets that are relevant to a given query. However, most works in this field are keyword-based [8], metadata-based [13], or context-based [1].

Therefore, to the best of our knowledge, there is no survey in the literature with an analysis of the datasets used by researchers in similarity search and no approach that considers the similarity between datasets as we propose in this work. Defining a dataset similarity space is challenging as it requires gathering a comprehensive amount of datasets and identifying a set of features to extract from them concerning a variety of aspects. This is a laborious task as datasets commonly used come from different sources scattered over several repositories, most poorly structured and often storing inconsistent versions of the same dataset. Moreover, defining a suitable set of features to extract from the datasets still demands investigation.

3   A Dataset-similarity-Based Proposal to Support Experimental Analysis

In this work, we introduce a novel view of datasets employed for experimental analysis in similarity retrieval. The choice of a set of datasets to evaluate a new method is challenging, both to cover a variety of properties of the underlying problem and to discuss the results.

Our proposal is to define a similarity space of datasets and observe how they are spread across this space to gain an understanding of their variability. We call the *dataset-similarity-based* approach such a conceptual view, which has the potential to provide a broader knowledge of datasets based on their intrinsic properties. Our proposal is based on the fact that datasets vary according to a multitude of properties that may contribute to the selection criterion for experimentation and to interpret the results. Using an assortion of features to represent datasets in a similarity space provides a joint view of how the datasets are distributed according to their properties.

3.1 Definition of the Dataset Similarity Space

This section details how we define the datasets' similarity space. We employed features comprising three categories.

1. Statistical and information-theoretical measures on the dataset attributes.
2. Measures commonly used in similarity retrieval analysis, such as the dataset cardinality and embedding dimensionality.
3. Measures of the hardness of similarity searching, including local intrinsic dimensionality, relative variance, and features derived from these complexity metrics (e.g., histograms of local intrinsic dimensionality).

For the complete list of features, refer to the implementation we developed to extract dataset features, available online[1]. We highlight that the most important point is the dataset-similarity-based conceptual view we propose. The instantiation we present herein considers a wide number of features to describe the datasets, but it is not intended to be exhaustive. Certainly, there is room for improvement following our conceptual view.

3.2 Survey of Dataset Usage in Similarity Retrieval Evaluation

To instantiate our proposal, we surveyed the datasets used by papers related to similarity search and gathered those publicly available in a centralized repository. We limited the survey scope to papers published in the International Conference on Similarity Search and Applications (SISAP), which is the principal forum on similarity search, and in the two most prestigious conferences on databases, namely the International Conference on Management of Data (SIGMOD) and the International Conference on Very Large Data Bases (VLDB). We justify such a limitation due to the need to survey the datasets used in the papers, which multiplies the effort.

The analysis covered more than one decade, comprising the years between 2008, which was the first edition of SISAP, and 2020. Every article published in any of the three conferences in the period was manually checked to verify if the main contribution is on similarity searching (e.g., an index method, a search algorithm, etc.), adding up to 146 articles. Other topics related to similarity were kept out of our scope.

Then, we searched for the dataset information presented in the papers, collecting features like the dataset's common name, the raw data type (e.g., image, text, etc.), the feature vector extracted (e.g., for images, color histogram, texture features, SIFT, etc.), the cardinality, the dimensionality, and others. We

---

[1] https://github.com/raseidi/annmf

identified 461 dataset mentions in the papers, comprising 237 distinct dataset instances. We performed a manual data integration of the instances, grouping datasets we understand as the same dataset by the context provided in the paper. Finally, we imputed missing dataset features with information about the datasets from external sources, being the original source whenever available.

From the datasets surveyed, we found 49 available for download. We generated several subsets of many of these datasets through random sampling to explore the impact of cardinality in the dataset properties, as this is a key performance factor for searching. In total, the repository built in this work is composed of 198 datasets. Then, we extracted features for all datasets in the repository.

4 Analyses in the Dataset Similarity Space

This section presents sample analyses to illustrate what can be done using our dataset-similarity-based conceptual view. We performed a Principal Component Analysis (PCA) on the features of all datasets in the built repository and generated visualizations using the two first principal components to show how they relate to each other in the similarity space.

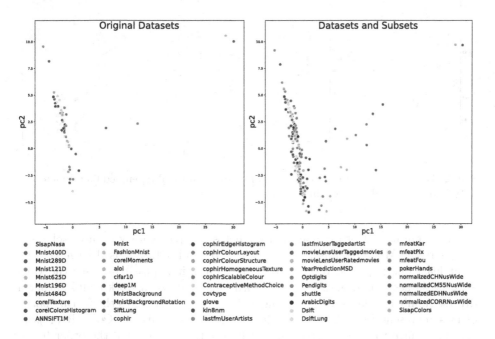

Fig. 1. Datasets and sub-datasets spreaded out in the similarity space

Figure 1 shows on the left the distribution of the 46 original (complete) datasets, and on the right, the distribution of all datasets (including the subsets). In this paper's figures, subsets of the same dataset have the same color.

It is clear that subsets promoted a significant variability in most datasets, highlighting the significance of cardinality for dataset characterization. The fact that most datasets got clustered parallel to the PC2 axis in the visualization does not indicate that they are redundant but that the PCA is highly impacted by the "outliers" *movieLensUser* and *lastfmUser* datasets.

Figure 2 shows the dataset distribution classified by ranges of cardinality, embedding dimensionality, and intrinsic dimensionality. As shown, datasets with small cardinality tend to cluster on the lower area, datasets with average and high cardinality in the middle, and datasets with very high cardinality are mainly on the upper part of the graph. Regarding dimensionality, it is clear that intrinsic dimensionality is more relevant for dataset distribution than embedding dimensionality, since intrinsic dimensionality more clearly spreads datasets. The datasets out of the main cluster are challenging datasets, such glove (Global Vectors for Word Representation) and deep1m (features derived from convolution neural networks), with high to very high cardinality and intrinsic dimensionality.

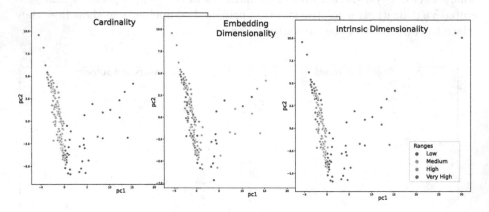

Fig. 2. Dataset distribution according to ranges of cardinality, embedding dimensionality and intrinsic dimensionality

An example of feature analysis that helps complementary dataset selection considers the MNIST dataset. Figure 3(a) shows all datasets used with MNIST in at least one paper surveyed. From this figure, it is possible to see the variance of datasets used with MNIST. With this analysis, a work could gather a diverse number of datasets by making a choice that maximizes the feature space of this representation. A viable choice would be the datasets CoPhIR or pokerHands, since the feature space would be relatively large. On the other hand, FashionMNIST would result in a less diverse dataset selection.

Another use case for the tools made available by this work is the ability to create experiments with any other datasets that other studies may use. To illustrate this usage, a comparison was made between datasets and sub-datasets created by the SIFT and Dense SIFT extractors when applied to an image dataset. While the SIFT extractor picks central points in an image and generates feature vectors based on these points, Dense SIFT leverages points over the whole image. Both extractors were applied to two datasets of fairly distinct image types: images from lung medical exams and random images from Flickr. Figure 3(b) shows the distribution of the datasets and sub-datasets created. It is clear that the most significant difference is between extractors and not between the image types, clustering the datasets by the feature regardless of the image type.

We showed sample analyses that could be done using our similarity-based conceptual view of datasets. It allows for gaining a broader knowledge of the distribution and variability of datasets employed for evaluating similarity search methods according to the datasets' intrinsic features. Potential applications of this approach include the selection of datasets for experimentation based on a varied number of properties, enhancing the variability among the chosen ones, and taking advantage of dataset similarities to guide decisions, such as parameter recommendation for indexing structures, as we did in a previous work [14].

5   Conclusion

In this work, we introduced the new dataset-similarity-based approach to handle datasets for different purposes. Our approach considers joint properties to represent the datasets as a similarity space composed of extracted features. We presented an instantiation of the approach based on a survey on how similarity search researchers have used datasets and showed analyses highlighting that it is possible to enhance data set choices when feature variance is a relevant selection criterion.

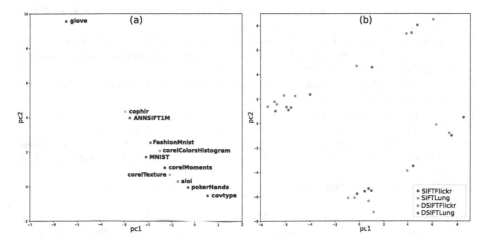

Fig. 3. MNIST and datasets used at least once with MNIST (no sub-datasets) and features SIFT and Dense SIFT extracted from different image types

Future work includes advancing the approach, for instance, by adding more features to describe the datasets and expanding the repository since the more datasets are included in the feature space, the richer the data set choice for other studies becomes. We also plan to employ the dataset-similarity-based conceptual view to support decisions or applications demanding dataset analysis.

References

1. Altaf, B., Akujuobi, U., Yu, L., Zhang, X.: Dataset recommendation via variational graph autoencoder. In: ICDM, pp. 11–20. IEEE (2019)
2. Aumüller, M., Ceccarello, M.: Benchmarking nearest neighbor search: influence of local intrinsic dimensionality and result diversity in real-world datasets. In: EDML SDM. CEUR Workshop Proceedings, vol. 2436, pp. 14–23. CEUR-WS.org (2019)
3. Bolettieri, P., et al.: CoPhIR: a test collection for content-based image retrieval. CoRR abs/0905.4627 (2009)
4. Camastra, F., Vinciarelli, A.: Estimating the intrinsic dimension of data with a fractal-based method. IEEE Trans. Pattern Anal. Mach. Intell. **24**(10), 1404–1407 (2002)
5. Chapman, A., et al.: Dataset search: a survey. VLDB J. **29**(1), 251–272 (2020)
6. François, D., Wertz, V., Verleysen, M.: The concentration of fractional distances. IEEE Trans. Knowl. Data Eng. **19**(7), 873–886 (2007)
7. He, J., Kumar, S., Chang, S.: On the difficulty of nearest neighbor search. In: ICML. icml.cc/Omnipress (2012)
8. Hendler, J.A., Holm, J., Musialek, C., Thomas, G.: US government linked open data: semantic.data.gov. IEEE Intell. Syst. **27**(3), 25–31 (2012)
9. Lecun, Y., Bottou, L., Bengio, Y., Haffner, P.: Gradient-based learning applied to document recognition. Proc. IEEE **86**(11), 2278–2324 (1998). https://doi.org/10.1109/5.726791
10. Li, W., et al.: Approximate nearest neighbor search on high dimensional data - experiments, analyses, and improvement. IEEE Trans. Knowl. Data Eng. **32**(8), 1475–1488 (2020)
11. Lorena, A.C., Garcia, L.P.F., Lehmann, J., de Souto, M.C.P., Ho, T.K.: How complex is your classification problem?: A survey on measuring classification complexity. ACM Comput. Surv. **52**(5), 1–34 (2019)
12. Lowe, D.G.: Object recognition from local scale-invariant features. In: ICCV, pp. 1150–1157. IEEE Computer Society (1999)
13. Neumaier, S., Umbrich, J., Polleres, A.: Automated quality assessment of metadata across open data portals. ACM J. Data Inf. Qual. **8**(1), 1–29 (2016)
14. Oyamada, R.S., Shimomura, L.C., Barbon, S., Jr., Kaster, D.S.: A meta-learning configuration framework for graph-based similarity search indexes. Inf. Syst. **112**, 102123 (2023)
15. Robnik-Šikonja, M.: Dataset comparison workflows. Int. J. Data Sci. **3**(2), 126–145 (2018)
16. Shimomura, L.C., Oyamada, R.S., Vieira, M.R., Kaster, D.S.: A survey on graph-based methods for similarity searches in metric spaces. Inf. Syst. **95**, 101507 (2021)

Suitability of Nearest Neighbour Indexes for Multimedia Relevance Feedback

Omar Shahbaz Khan[1]([⊠])(ID), Martin Aumüller[2](ID), and Björn Þór Jónsson[1](ID)

[1] Reykjavik University, Reykjavik, Iceland
{omark,bjorn}@ru.is
[2] IT University of Copenhagen, Copenhagen, Denmark
maau@itu.dk

Abstract. User relevance feedback (URF) is emerging as an important component of the multimedia analytics toolbox. State-of-the-art URF systems employ high-dimensional vectors of semantic features and train linear-SVM classifiers in each round of interaction. In a round, they present the user with the most confident media items, which lie furthest from the SVM plane. Due to the scale of current media collections, URF systems must be supported by a high-dimensional index. Usually, these indexes are designed for nearest-neighbour point queries, and it is not known how well they support the URF process. In this paper, we study the performance of four state-of-the-art high-dimensional indexes in the URF context. We analyse the quality of query results, compared to a sequential analysis of the collection, over a range of classifiers, showing that result quality depends (i) heavily on the quality of the SVM classifier and (ii) the index structure itself. We also consider a search-oriented workload, where the goal is to find the first relevant item for a task. The results show that the indexes perform similarly overall, despite differences in their paths to the solution. Interestingly, worse recall can lead to better application-specific performance.

Keywords: High-Dimensional Indexing · Interactive Learning · User Relevance Feedback · Multimedia Retrieval

1 Introduction

In user relevance feedback (URF), the goal is to train interactive classifiers to satisfy specific information needs based on direct feedback from the user. When interacting with a multimedia collection, the user is presented in each interaction round with a set of items from the collection and asked to judge some items as relevant and some items as non-relevant for a specific task. At the start of the URF process, the items are typically randomly sampled from the collection or retrieved using a query, but once the initial classifier has been trained the items are usually selected from the top items returned by the current version of the classifier. This interactive process continues until the user's information need is satisfied or they determine that the collection holds no items of interest. As URF

O. Pedreira and V. Estivill-Castro (Eds.): SISAP 2023, LNCS 14289, pp. 133–147, 2023.
https://doi.org/10.1007/978-3-031-46994-7_12

allows users to express and refine fuzzy information needs, it is an important component of the multimedia analytics toolbox.

A state-of-the-art multimedia URF system, such as [12], is implemented as follows. First, the multimedia items are described by semantic labels, which are produced by advanced deep-learning models and compressed using an index-based compression technique. Second, such a system builds linear-SVM classifiers, which are known to work well with few examples, and presents users in each interaction round with the multimedia items that (a) are furthest away from the resulting hyperplane, and (b) have not been seen before in the process. And third, a high-dimensional index is used to speed up the retrieval and ensure a stable response time. While the state-of-the-art has shown URF to work at scale using these elements, such furthest neighbour queries from an SVM-based hyperplane have not been studied much in the literature. In particular, to the best of our knowledge these evaluations have not been carried out in the context of URF. This paper therefore opens an investigation into the suitable choice of such a high-dimensional index for URF over multimedia collections.

Indexing high-dimensional data to support similarity queries such as finding (approximate) nearest or furthest neighbors suffer the *curse of dimensionality*. In general, this means that there are no *sublinear* time algorithms that solve these tasks *exactly* on *arbitrary data*, and a linear scan through the dataset is the best one can hope for. However, if a small loss in accuracy can be accommodated or if the data is "favorable", a large collection of scalable solutions for finding nearest neighbors is available. A solution is either provided with strong theoretical guarantees, such as *hashing-based* approaches [9] with their theoretical time guarantees, or with strong empirical evidence on the quality of the query result, such as *cluster-based* [11], *graph-based* [10,14], or *tree-based* [5] approaches. The ANN benchmarking effort [1] summarizes these approaches and demonstrates that, in practice, nearest neighbor search tasks on million-scale datasets can be solved several orders of magnitude faster than a linear scan with little loss in accuracy.

1.1 Problem Definition

We formulate the SVM-based hyperplane setting used in URF as follows. First, the distance between a point $p \in \mathbb{R}^d$ and a hyperplane $q \in \mathbb{R}^{d+1}$ is

$$d_{P2H}(p, q) = \frac{q_{d+1} + \sum_{i=1}^{d} p_i q_i}{\sqrt{\sum_{i=1}^{d} q_i^2}}$$

Second, the problem of finding furthest points from a given hyperplane in the positive direction is equivalent to finding the points in the dataset with largest (positive) distance to the hyperplane. To avoid an exhaustive scan through the dataset, the task is to build an *index* data structure over the point set $S \subseteq R^d$ that supports these furthest point queries.

Point-to-Hyperplane queries are challenging because d_{P2H} is not a distance measure in the strict sense. However, we can reduce the problem to an inner product space as follows: Append a 1 to each point in the dataset, so that both points and hyperplanes have $d + 1$ dimensions, and notice that—since the hyperplane is fixed at query time—finding the furthest neighbors according to d_{P2H} is equivalent to finding vector p that *maximizes* the *inner product* (MIPS) $d'(p, q) = \sum_{i=1}^{d+1} p_i q_i$. Solving maximum inner product search has seen significant progress in the research community and arises in particular in recommender systems [2]. The standard approach involves asymmetric transformations of data and query points [2,20]. After such a transformation, finding points that maximize the inner product becomes equivalent to finding nearest neighbors in the transformed space, which is usually Euclidean distance. However, these transformations usually lower the contrast between points. For example, Huang et al. [8] consider transformations for hashing-based *closest point to hyperplane* queries and experimentally show that finding *furthest neighbors* instead of *nearest neighbors* (under slightly different transformations) provides empirical speed-ups. In the context of graphs, Morozov and Babenko [16] show that the transformed vectors produce *worse graph indexes* than using the inner product directly.

1.2 Contributions

The current state-of-the-art large-scale URF approach, Exquisitor [12], uses the high-dimensional ANN index eCP (extended Cluster Pruning) [15]. The stated reasons for this choice are its comprehensibility, time guarantees, and ability to work with hyperplane queries using MIPS. Based on practical advances in nearest neighbor search, we evaluate the suitability of state-of-the-art high-dimensional indexing approaches for URF over multimedia collections. In particular, we inspect three diverse approaches—Annoy (Approximate Nearest Neighbor Oh Yeah) [5], IVF (Inverted File Index on k-means clustering) [11], and HNSW (Hierarchical Navigable Small World) [14]—that perform well on million-scale nearest neighbor search with regards to supporting maximum inner product queries. We evaluate these indexes along with the eCP index using an automated evaluation protocol, based on the Lifelog Search Challenge 2019 dataset [7], to simulate URF sessions with the goal of finding one relevant item. All source code for the URF evaluation is made available on GitHub[1] to provide the research community with an experimental pipeline to compare different high-dimensional indexes for URF. In the process, we make the following contributions:

- All evaluated indexes show adequate quality towards hyperplane queries in terms of recall, with HNSW achieving the best overall performance.
- Indexes that introduce variety, due to build quality or search approach, are better at solving URF tasks (eCP and Annoy). Thus, high recall does not directly translate to being the best at solving the actual URF tasks. This relates to hyperplane queries being refined throughout a URF session, so the quality of initial queries may not be well defined for the task.

---

[1] https://github.com/Ok2610/urf-indexing-eval.

- Finally, we analyze the effects of the approximation of each index. We find that eCP and IVF have a more comprehensible parameter for this than HNSW and Annoy, and eCP's setting has the least variability leading to better time estimates.

2   User Relevance Feedback

Analytical tasks for multimedia focus on discovering knowledge from the media items that reside within the ever growing multimedia collections of today. To truly uncover this knowledge it is essential to explore and search through the contents of such collections in real time. While long-running machine tasks may be capable of categorising and summarising parts of the collection for a task, typically as new knowledge is discovered the goal of the task can shift. In such situations, user relevance feedback is preferred as it allows the user to shift the classifier based on the new knowledge [22].

The most common approach to URF is to present a suggestion set S to the user, using the current classifier C. From S the user labels p items as positives and n items as negatives, which are then used to update C. There are multiple ways to determine which items to include in S. In URF the most confident items of C are presented. This is beneficial when the intention is not solely on creating a strong classifier, as the information need may not be entirely clear and may be susceptible to change throughout the session. With URF, the user may explore the collection or be more search-oriented, depending on how much refinement is put towards the classifier.

URF for content-based retrieval has been around for several decades [4, 17, 18, 23], but as multimedia collections started to rapidly expand, it became cumbersome at such scale due to the response time. Even prior to the scale issues, it was difficult to have explainable classifiers due to the feature representations of the multimedia items. Comprehension is important for the user to better understand the effects of their actions on the classifier. With improvements in deep learning, machines have become much better at discovering semantic features in multimedia contents [3, 6], making it a preferred choice of feature for URF applications. Semantic features extracted through deep nets result in sparse high-dimensional vectors. The current state of the art large-scale URF approach, Exquisitor, uses the high-dimensional feature vectors in a compressed representation together with the clustering-based ANN index eCP. The compressed representation selects the top f features and stores them in a space-efficient representation. Note that the compressed representation does not transform the feature space to ensure it remains comprehensible. The eCP index has been modified such that it works with this compressed representation. Furthermore, the index handles hyperplane queries from Exquisitor's Linear-SVM classifier [12]. In addition to the basic URF scheme, Exquisitor also employs incremental retrieval, which continues a search within the index, in case not enough items are found to be returned. This is linked to the search expansion parameter b in the eCP index, which is the number of clusters it needs to retrieve. Exquisitor has been shown capable of working with YFCC100M, a multimedia collection with nearly 100 million items, achieving subsecond response time with modest computing resources [12].

3   High-Dimensional Indexing

There exists a plethora of different approaches for solving nearest neighbor search queries. The most successful approaches can be categorized into *clustering-based*, *graph-based*, *hashing-based*, and *tree-based* approaches. For our practical evaluation, we pick approaches from each category which have not been considered for URF. We exclude locality-sensitive hashing based approaches because they were part of an earlier evaluation [12] and were shown to be inferior to using the eCP index. We review the considered approaches next.

Cluster-based approaches (IVF [11], eCP [15]). Given a dataset $S \subseteq \mathbb{R}^d$ and two parameters k and ℓ, run a clustering algorithm such as k-means to find k centroids. By associating each point with its closest centroid, the space is partitioned into k parts. The data structure that stores the centroids and the associated lists is referred to as an inverted file index (IVF). To find nearest neighbors to a query $q \in \mathbb{R}^d$, inspect the points associated to the ℓ closest centroids to q, possibly indexing the centroids for large k. The eCP index uses the same approach but uses the k initial random points as centroids to find a balanced space partition. Furthermore, eCP builds a hierarchy using the centroids.

Graph-based approaches (HNSW [14]). Given a dataset $S \subseteq \mathbb{R}^d$ and parameters k, ℓ, the goal is to build a graph $G = (V, E)$, where each point is represented by a vertex and edges exist between a point and a "diverse" set of at most k close points. Let us assume that such a graph G is given. To find the nearest neighbors of a query point q, HNSW uses a hierarchy of graphs to find a good entry point into the bottom-layer graph that indexes all points. Given such a start point, carry out a greedy hill climbing. In each round, consider the currently closest point to the query not considered before. Inspect the neighborhood and compute the distances to the query point. After each round, trim the list of current closest points (inspected and non-inspected) to ℓ, which is usually called the beam width. Terminate if all ℓ points have been considered. (Note that this is not a bound on the number of distance computations, since considered points might be trimmed.) To build the graph, order all the points and insert them one-by-one using the search algorithm, often with a smaller ℓ' than used for the queries. From the points inspected in this search, a pruned set of k points is chosen as neighbors of the inserted point (pruning might be necessary for its neighbors if the degree bound k is not met). There exist many other graph-based indexes that change details of this construction [10,21].

Tree-based approaches (Annoy [5]). Annoy builds a collection of trees based on random projections. Given a set of points S and two parameters k, ℓ, the data structure works as follows. First, a node in a tree is described by a hyperplane a that splits up a point set $S' \subseteq S$. For example, one can find the median inner product of the data points with a and split S' into two balanced subsets based on that. At the root, the whole dataset is taken into consideration, and a leaf is created as soon as the number of points at a node is below a certain threshold. Instead of a single tree, k trees are created to boost the quality of the results. Given a query point and a collection of k trees, carry out root-to-leaf-traversals in each tree for the query. Traverse the trees upwards until ℓ (unique) points are found and return the closest among these points as the result of the search.

4   Evaluation

Evaluating user relevance feedback approaches is best done through real user tests. As we are attempting to gauge the performance of hyperplane queries on different indexes in a preliminary URF setup, however, we do not need real users at this stage. Instead we employ an automated evaluation protocol, based on real-life applications [13]. The evaluation protocol consists of tasks where the objective is to find the first relevant item within the tasks relevant item set, R_t. Typically, in an automated URF session, a set number of positives and negatives are considered from the suggestion set S in each round. Based on the labelling policy used in the protocol, positives and negatives are added or replaced from their respective global sets to train the linear-SVM classifier. Positives and negatives are labelled by comparing the distances between the combined maximum feature vector of R_t and the feature vector of items in S.

Based on [13], we design the evaluation protocol with the following parameter choices; rd number of rounds in each URF session, k number of items to retrieve, s the number of suggestions to consider, p the number of positives to label and n the number of negatives to label, P and N the positive and negative sets used to train the linear SVM. rd is set to 50 to simulate a long-running session. s is set to 25, which typically would also be the value for k. Items seen in previous rounds of a URF session should not be presented to the user again, however, and in situations where all k items are seen the session ends. When the underlying URF approach is able to remove previously seen items internally to avoid returning them again, then we would set $k = s$. Since this is not the case for the tested indexes, however, the indexes are asked to return a higher number $k = 1000$ of items, from which the previously seen items are then pruned. The labelling policy of the evaluation protocol is *AccRep*, where in each round p positives and n negatives are added to P and N. If better items exists in the remaining suggestions for either P or N, they replace the weakest items in the sets [13]. In our evaluation $p = 5$ and $n = 15$ leading to 5–250 positives and 15–750 negatives for each session, with each session presenting a total of 1250 ($s \cdot rd$) items.

Dataset. The dataset used in the evaluation is from the Lifelog Search Challenge 2019. Lifelogging is the principal of recording your daily life with as much data as possible, i.e. logging biometric data, taking images throughout the day, food logs, and more. Pure lifeloggers often walk around with a miniature camera on their person that takes an image at a set interval, and thus they accumulate a large

Table 1. Build and search parameters for each index

Index	Build Parameters	Search Parameter
eCP	L: 3, $cSize$: 100	b: 64
Annoy	n_{trees}: 100	$search_k$: 10000
HNSW	M: 48, ef_C: 500	ef_S: max$(10, k)$
IVF	n_{list}: 417	n_{probe}: 64

multimedia collection. The Lifelog Search Challenge (LSC) is a live interactive retrieval challenge, where tasks are defined over a snippet of the extremely large collection. The LSC 2019 dataset represents one lifelogger's daily life across 1 month, consisting of 41,666 images. There are 24 retrieval tasks defined over this collection [7]. Semantic feature labels have been extracted using a deep net, with the top 7 features being used and the rest set to zero. At LSC multiple multimedia retrieval systems attempt to solve the tasks one by one within a time limit. A task consists of a text query, but unlike regular search challenges where the entire text is given at once, in LSC the text query is presented in parts. Every 30 s new information is added to the presented text of the current task. In the evaluation protocol this aspect is reflected when filters are used, but as our focus here is on hyperplane queries, this aspect is ignored.

Index Parameters. We use Annoy v1.17.2, HNSW from hnswlib v0.7.0 and IVF from Faiss v1.7.4. The build parameters for each index used in the experiments can be seen in Table 1. The choice of the build parameters for Annoy and HNSW are based on their settings for datasets of relevant sizes from ANN-benchmarks [1]. The eCP index it has two build parameters; L the level of its hierarchy and $cSize$ the number of items in each cluster. Note that the latter is a soft enforcement, so there is still a chance of clusters being larger or smaller. We aim to have clusters with 100 items in a 3-level hierarchy where L_l has $\sqrt{L_{l+1}}$ clusters, with $L = 3$ having all clusters. The IVF index has a single build parameter n_{list} which specifies the number of clusters to divide the items into. We select this to be similar to eCP's clusters at $L = 3$. Each index has a run-time approximation parameter that is set according to their recommendations. The effects of this parameter is crucial to understand in terms of distance computations performed, as it indicates whether an index is able to provide stable response time guarantees. All indexes were built using Euclidean distance.

4.1 Experiment 1: Hyperplane Queries

In the first experiment we investigate the recall of each index when encountering hyperplane queries. The hyperplane queries for this experiment are generated by running the evaluation protocol on the LSC dataset, leading to 1200 hyperplane queries in total (24 tasks with URF sessions of 50 rounds). The groundtruth for this experiment are the top 1000 items for each hyperplane under maximum

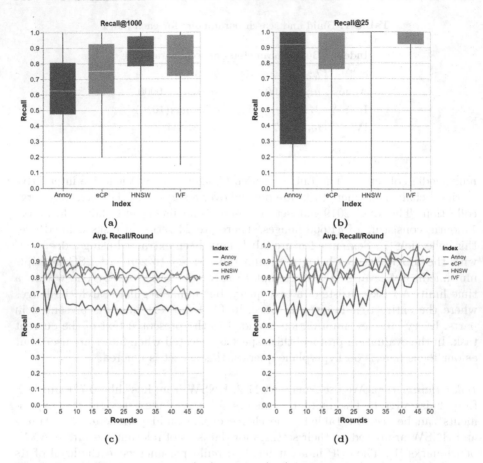

Fig. 1. Recall distribution @1000(a,c) and @25(b,d) for hyperplane queries and average recall per round. Annoy (blue), eCP (orange), HNSW (red), IVF (teal) (Color figure online)

inner product similarity obtained by an exact linear scan. Here, recall is the fraction of the k items returned by the implementation that belong to the true top k items (groundtruth) with largest inner product for a given hyperplane. The results from this experiment are depicted in Fig. 1. Figure 1a shows the recall distribution for the top 1000 items. The best-performing index is HNSW with a consistent distribution above 75% recall. The IVF index is close but has a slightly lower recall overall. eCP fares worse than IVF, which shows the extra effort in constructing the clusters is beneficial for recall, but not by much. Annoy has a similar distribution to eCP, but generally 10% lower recall on average. Figure 1b shows the recall distribution for the top 25 items, or the items that would actually be presented to the user. Here we see an increase in recall overall for all the indexes. HNSW remains at top, achieving seemingly 100% apart from some outliers. IVF is slightly worse with eCP following. Annoy has a much wider distribution, but a high median above 90% recall. Thus, the items the user sees are high-recall items for the hyperplane queries regardless of the index.[2] Figures 1c and 1d depict the average recall across all tasks per round in the URF session. As more rounds pass in a session, the hyperplane queries should become more descriptive. For the top 1000 items all indexes follow a similar pattern of starting with a high recall that falls during the initial rounds, and then settles at a lower average recall. For the top 25 items the pattern starts similarly with a drop off, but as the hyperplane queries become more refined the recall for all indexes increase. Given that the top 25 items are the ones the user sees, this behavior is desired.[3]

4.2   Experiment 2: User Relevance Feedback

From the previous experiment, it is clear that HNSW performs best in the case of recall for hyperplane queries while Annoy performs worst. When it comes to URF tasks, however, the theoretical top items for a query may not necessarily be items of interest, especially in early rounds when the hyperplane still needs to be refined. We consider two scenarios. In Scenario 1, the indexes are provided with the hyperplanes from using an exact linear scan in the evaluation protocol. In Scenario 2, the linear-SVM is trained from the individual results of an index.

Table 2 shows the round where each index managed to find the first relevant item in the top 25 in Scenario 1, where Scan represents the results from the linear scan. Scan solves the most tasks (14 out of 24), while the indexes solve 10 tasks each except for Annoy which solves 11. With Annoy solving 1 more task it shows that high recall for majority of hyperplane queries in a URF session is not always necessary. Looking at the solved tasks, none of the indexes complete a task that is not also solved by Scan, and they either solve it in the same round or 1 round after it. Notable exceptions, are tasks 11, 20 and 23 where Scan solves them 5–11 rounds earlier, and task 14 where eCP and HNSW solve the task 7 rounds prior to Scan. From these results we see how the restriction on the indexes require more rounds to solve a task, while task 14 shows that having the entire collection available can also introduce noise.

We now turn to Scenario 2. As the indexes employ a structure and approximations on the collection, a suggestion set from the same hyperplane can differ, and from that point they have different hyperplane queries throughout the session. This is depicted in Table 3 where each index has run the evaluation protocol using hyperplane queries generated from their own suggestions. By using queries defined through their own sessions they solve more and different tasks, and for the tasks they solve in common there are larger gaps between the rounds. Here, Annoy solves 15 tasks and the others solve 14. It is also worth noting that Annoy,

---

[2] This experiment was also conducted using Annoy, HNSW, and IVF built using inner product instead of Euclidean distance. In all cases, average recall @1000 was lower, while for HNSW recall @25 was improved.

[3] Similar results are observed when (roughly) targeting a certain number of distance computations across all indexes.

Table 2. First round in the URF session where a relevant item for the task was found using the hyperplanes generated for the entire dataset D

Task	Scan	Annoy	eCP	HNSW	IVF
0	**33**	34	34	34	34
1	–	–	–	–	–
2	–	–	–	–	–
3	–	–	–	–	–
4	–	–	–	–	–
5	–	–	–	–	–
6	**12**	12	12	12	12
7	–	–	–	–	–
8	**23**	24	24	24	**23**
9	**2**	2	2	2	**2**
10	–	–	–	–	–
11	**25**	32	32	32	32
12	**19**	20	**19**	20	**19**
13	**32**	–	–	–	–
14	11	11	**4**	**4**	12
15	10	10	10	10	10
16	**36**	37	**36**	37	37
17	–	–	–	–	–
18	**3**	–	–	–	–
19	**23**	–	–	–	–
20	**20**	25	–	–	–
21	–	–	–	–	–
22	–	–	–	–	–
23	**4**	15	15	15	15
Solved	14	11	10	10	10
Best	12	2	5	3	4

Table 3. First round in the URF session where a relevant item for the task is found using the hyperplanes generated from each index

Task	Annoy	eCP	HNSW	IVF
0	–	–	**33**	47
1	**25**	–	–	–
2	–	–	–	–
3	–	**35**	–	–
4	**33**	-	–	43
5	–	–	–	–
6	20	**6**	22	12
7	–	–	–	–
8	9	17	18	**7**
9	**2**	**2**	**2**	**2**
10	–	–	–	–
11	**20**	40	32	43
12	39	–	**17**	20
13	33	27	**24**	–
14	18	15	**11**	17
15	**6**	8	7	10
16	29	11	37	**9**
17	–	–	–	–
18	12	5	8	**3**
19	**9**	18	23	30
20	16	18	17	**15**
21	–	–	–	–
22	–	32	–	–
23	17	8	**4**	14
Solved	15	14	14	14
Best	6	4	6	5

eCP and IVF solve some tasks that Scan could not, which again indicates the larger scope of the full scan encountering noise. We further test the assumption of hyperplanes from an index' own session being best, by running the hyperplanes from one index with another index. These results show that while they solve the same tasks, the rounds for many of the tasks differ, ranging from a few rounds to 20+. HNSW with IVF's hyperplanes show the most similar performance. We have omitted the table for these results to not exceed the article length.

Table 4. Results from similar search scope (~6400)

Task	Annoy	eCP	HNSW	IVF
0	-	-	-	47
1	33	-	-	-
2	-	-	-	-
3	-	35	-	-
4	-	-	-	43
5	-	-	-	-
6	26	6	16	12
7	-	-	-	-
8	11	17	-	7
9	2	2	2	2
10	-	-	47	-
11	23	40	27	43
12	31	-	15	20
13	31	27	34	-
14	18	15	11	17
15	6	8	7	10
16	-	11	46	9
17	-	-	-	-
18	12	5	8	3
19	7	18	12	30
20	15	18	17	15
21	-	-	-	-
22	-	32	-	-
23	-	8	34	14
Solved	12	14	13	14
Best	6	6	4	7

Table 5. Results from reduced search scope (~3200)

Task	Annoy	eCP	HNSW	IVF
0	20	-	-	31
1	29	4	-	-
2	-	-	-	-
3	-	-	-	37
4	42	33	-	-
5	45	-	-	-
6	17	8	48	15
7	-	-	-	-
8	20	7	36	7
9	2	2	2	2
10	-	41	-	-
11	27	15	26	-
12	22	11	30	23
13	27	33	26	-
14	22	15	11	17
15	7	8	6	10
16	7	-	-	9
17	-	-	-	-
18	7	8	7	3
19	10	22	8	23
20	16	24	22	20
21	-	-	-	-
22	-	38	44	-
23	30	5	8	5
Solved	17	16	13	13
Best	4	10	5	4

Table 6. Results from reduced search scope (~1600)

Task	Annoy	eCP	HNSW	IVF
0	-	-	-	-
1	-	46	-	31
2	-	47	-	-
3	-	-	-	-
4	18	31	-	-
5	-	-	-	-
6	40	15	25	14
7	-	-	-	-
8	-	38	-	10
9	2	2	2	2
10	-	-	41	-
11	22	33	10	42
12	21	32	45	20
13	31	31	-	-
14	14	10	11	15
15	6	6	10	10
16	29	16	13	32
17	-	-	-	-
18	8	8	7	3
19	10	27	8	16
20	18	28	16	23
21	-	-	-	-
22	35	30	43	-
23	16	7	8	5
Solved	14	17	13	13
Best	6	4	6	7

Overall, from these results we have shown that all indexes, with their recommended settings, are capable of dealing with URF tasks. There is no clear indication for which index is best; while Annoy solves the most tasks, there are still tasks that are solved faster with the other indexes. A point of interest now is the approximation parameter for each index which determines the number of items each index considers or the number of distance computations taking place. With the recommended settings Annoy has the lowest number of average distance computations with ~4300, while HNSW has the highest with ~9400. The approximation parameter is what introduces the quality/time trade-off and is often set based on the specific use case. The transparency of this parameter is better for some indexes than others. For IVF and eCP it is the b and n_{probe} parameter, which is how many clusters to consider during the search. In eCP b is used for each level in its hierarchy. Annoy uses the $search_k$ parameter, which is the number of binary trees it will search. HNSW uses the ef_S parameter, which is the number of candidates to consider while retrieving the top k items.

In Table 4 we compare the performance using settings for each index that result in around 6400 distance computations. For this HNSW's $ef_S = 700$ and for Annoy's $search_k = 14000$, while IVF and eCP remain the same ($b/n_{probe} = 64$), so the only changes worth noting in the table are for Annoy and HNSW. With these settings they both solve fewer tasks, 2 for Annoy and 1 for HNSW, and some tasks where they were the best require more round, leading to other indexes solving them faster or in the same round. For Annoy and HNSW to have the same distance computations, we had to increase Annoy's search parameter while reducing HNSW's. To investigate the effects of a lower scope further, we run the evaluation protocol again with distance computations roughly around 3200 and 1600 for all indexes, depicted on Tables 5 and 6 respectively. When reducing the scope eCP solves more tasks and is seemingly faster than with the higher scope. Annoy improves in terms of tasks solved with 17 at scope 3200, but at 1600 it solves 14. HNSW solves the same number of tasks for both reduced scopes, but not always the same tasks, which hints that certain tasks are better with a lower scope for HNSW and some are better with a larger one. IVF has similar behavior as HNSW when reducing scope. It should also be noted that the lowest scope runs for IVF and HNSW also lead to tasks finishing before the 50 rounds as all items returned had been seen in previous rounds. Fortunately, the relevant item was found in a previous round for those tasks, but in case this occurs earlier, some notion of incremental retrieval that can expand the search within the index is needed. This feature exists in eCP when used with Exquisitor.

5   Discussion

In this section we discuss the insights gained from the experiments and the role of approximation parameters for each index. In Fig. 2 the distribution of distance computations is depicted for each index, where the average distance computation was around 6400. Annoy with $search_k = 14000$ has the highest variability, fluctuating between 4000 and 8000 distance computations. eCP, HNSW and IVF are more consistent[4] and closer to the average target.

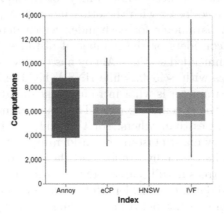

Fig. 2. Distribution of distance computations for the indexes, when the approximation was aimed to be $\sim$6400

---

[4] The 0-valued outliers for HNSW stem from URF sessions stopping early, as everything returned has already been seen, while the actual minimum was around 4700.

Our experiments highlight that each index is able to solve URF tasks on the small LSC dataset, with eCP and Annoy being better than HNSW and IVF. The LSC dataset contains many near-duplicate images, as the dataset depicts the daily life of one person. To get a better picture with a more general dataset and similar tasks, we conducted an experiment on the dataset from Video Browser Showdown 2020 [19] (VBS) consisting of 1 million images. Solving tasks is more difficult in VBS as there are more scenarios to cover, and typically filters are applied to help with the task. With pure URF both Annoy and eCP manage to solve 2–3 tasks out of the 12, while HNSW and IVF solve 1. Similarly to the LSC collection, eCP and Annoy performed best with a lower scope ($\sim$3200). However, given that Annoy still fluctuates between 1000 and 8000 distance computations, eCP remains the better overall choice. HNSW and IVF at similar scope did not manage to solve any task, and HNSW even had multiple sessions stopping early due to all returned items being seen. This is the danger of a small search scope, and is why having an easy to comprehend and reliable approximation parameter is extremely beneficial. With eCP and IVF one can reliably ask for additional b/n_{probe} clusters, knowing the computation time will be roughly the same. However, for Annoy and HNSW this is more difficult.

6 Conclusion

In this paper we investigated the performance of multiple state-of-the-art ANN indexes in user relevance feedback (URF) settings dealing with hyperplane queries. We evaluated 4 indexes, the tree-based approach Annoy, the graph-based approach HNSW, and the cluster-based approaches IVF and eCP. Each of these high-dimensional indexes use some form of approximation that introduces a quality/time trade-off. In interactive URF sessions, fast and reliable response time is crucial. Through our experiments using an automated evaluation protocol simulating URF sessions, we find that each index is able to solve such tasks. We also discovered that a lower setting for the approximation parameter, which reduce the search space, can improve results. However, if it is set too low, the index may not find any new items to present the user. Out of the four, eCP and Annoy perform best overall. We further analyze the approximation parameters of the indexes and find eCP's parameter to be more comprehensible and reliable. The other indexes are still viable for URF on a small scale collection, but it is harder to predict their performance when used at scale. Following up with real user tests and conducting experiments on even larger collections is warranted, to better verify these findings.

Acknowledgments. This work was supported by Icelandic Research Fund grant 239772-051 and by the Innovation Fund Denmark for the project DIREC (9142-00001B).

References

1. Aumüller, M., Bernhardsson, E., Faithfull, A.J.: ANN-benchmarks: a benchmarking tool for approximate nearest neighbor algorithms. Inf. Syst. **87**, 101374 (2020)
2. Bachrach, Y., et al.: Speeding up the Xbox recommender system using a Euclidean transformation for inner-product spaces. In: RecSys, pp. 257–264. ACM (2014)
3. Barraco, M., Cornia, M., Cascianelli, S., Baraldi, L., Cucchiara, R.: The unreasonable effectiveness of CLIP features for image captioning: an experimental analysis. In: 2022 IEEE/CVF Conference on Computer Vision and Pattern Recognition (CVPR), pp. 4662–4670 (2022)
4. Bartolini, I., Ciaccia, P., Waas, F.: FeedbackBypass: a new approach to interactive similarity query processing. In: Very Large Data Bases Conference (VLDB), pp. 201–210 (2001)
5. Bernhardsson, E.: Annoy, github.com/spotify/annoy
6. Dubey, S.R.: A decade survey of content based image retrieval using deep learning. IEEE TCSVT **32**(5), 2687–2704 (2021)
7. Gurrin, C., et al.: Comparing approaches to interactive lifelog search at the lifelog search challenge (LSC2018). ITE TMTA **7**(2), 46–59 (2019)
8. Huang, Q., Lei, Y., Tung, A.K.H.: Point-to-hyperplane nearest neighbor search beyond the unit hypersphere. In: SIGMOD, pp. 777–789. ACM (2021)
9. Indyk, P., Motwani, R.: Approximate nearest neighbors: towards removing the curse of dimensionality. In: Proceedings of the Thirtieth Annual ACM Symposium on Theory of Computing (STOC), pp. 604–613 (1998)
10. Iwasaki, M., Miyazaki, D.: Optimization of Indexing Based on k-Nearest Neighbor Graph for Proximity Search in High-dimensional Data. ArXiv e-prints (2018)
11. Johnson, J., Douze, M., Jégou, H.: Billion-scale similarity search with GPUs. IEEE TBD **7**(3), 535–547 (2021)
12. Khan, O.S., et al.: Interactive learning for multimedia at large. In: Jose, J.M., et al. (eds.) ECIR 2020. LNCS, vol. 12035, pp. 495–510. Springer, Cham (2020). https://doi.org/10.1007/978-3-030-45439-5_33
13. Khan, O.S., Jónsson, B.Þ., Zahálka, J., Rudinac, S., Worring, M.: Impact of interaction strategies on user relevance feedback. In: International Conference on Multimedia Retrieval (ICMR), pp. 590–598. ICMR 2021, ACM (2021)
14. Malkov, Y.A., Yashunin, D.A.: Efficient and robust approximate nearest neighbor search using hierarchical navigable small world graphs. IEEE TPAMI **42**(4), 824–836 (2020)
15. Moise, D., Shestakov, D., Gudmundsson, G., Amsaleg, L.: Indexing and searching 100M images with map-reduce. In: Proceedings of the 3rd ACM Conference on International Conference on Multimedia Retrieval (ICMR), pp. 17–24. ICMR 2013, ACM (2013)
16. Morozov, S., Babenko, A.: Non-metric similarity graphs for maximum inner product search. In: Neural Information Processing Systems (NeurIPS), pp. 4726–4735 (2018)
17. Rui, Y., Huang, T.: Optimizing learning in image retrieval. In: Proceedings IEEE Conference on Computer Vision and Pattern Recognition (CVPR). vol. 1, pp. 236–243 (2000)
18. Rui, Y., Huang, T.S., Ortega, M., Mehrotra, S.: Relevance feedback: a power tool for interactive content-based image retrieval. IEEE TCSVT **8**, 644–655 (1998)
19. Schoeffmann, K.: A user-centric media retrieval competition: the video browser showdown 2012–2014. IEEE Multimedia **21**(4), 8–13 (2014)

20. Shrivastava, A., Li, P.: Asymmetric LSH (ALSH) for Sublinear Time Maximum Inner Product Search (MIPS). In: Advances in Neural Information Processing Systems (NeurIPS), pp. 2321–2329 (2014)
21. Subramanya, S.J., Devvrit, F., Simhadri, H.V., Krishnaswamy, R., Kadekodi, R.: DiskANN: fast accurate billion-point nearest neighbor search on a single node. In: Advances in Neural Information Processing Systems (NeurIPS), pp. 13748–13758 (2019)
22. Zahálka, J., Worring, M.: Towards interactive, intelligent, and integrated multimedia analytics. In: 2014 IEEE Conference on Visual Analytics Science and Technology (IEEE VAST), pp. 3–12. IEEE (2014)
23. Zhou, X., Huang, T.: Relevance feedback in image retrieval: a comprehensive review. Multimedia Syst. **8**, 536–544 (2003)

Accelerating k-Means Clustering
with Cover Trees

Andreas Lang[✉][iD] and Erich Schubert[iD]

TU Dortmund University, Dortmund, Germany
{andreas.lang,erich.schubert}@tu-dortmund.de

Abstract. The k-means clustering algorithm is a popular algorithm
that partitions data into k clusters. There are many improvements to
accelerate the standard algorithm. Most current research employs upper
and lower bounds on point-to-cluster distances and the triangle inequality to reduce the number of distance computations, with only arrays as
underlying data structures. These approaches cannot exploit that nearby
points are likely assigned to the same cluster. We propose a new k-means
algorithm based on the cover tree index, that has relatively low overhead
and performs well, for a wider parameter range, than previous approaches
based on the k-d tree. By combining this with upper and lower bounds,
as in state-of-the-art approaches, we obtain a hybrid algorithm that combines the benefits of tree aggregation and bounds-based filtering.

1 Introduction

One of the most popular clustering algorithms is k-means, often with the standard algorithm taught in textbooks (commonly attributed to Lloyd [11], but
described before by, e.g., Steinhaus [23]). In k-means, the data is approximated
using k centers, which are the arithmetic mean of the partitions, and the goal
is to minimize the sum of squared deviations of all samples and their nearest
centroids. Finding the true optimum is NP-hard [12], and hence we need heuristics such as the standard algorithm. The popularity of k-means and an ever-
increasing amount of data led to many improvements to the standard algorithm.
Most common improvements replicate the convergence of the standard heuristic
exactly, and are hence sometimes called "exact" k-means. The major part of its
runtime is the distance calculations between samples and cluster centers in each
iteration. One way to accelerate k-means is to approximate the data, e.g., by sampling [4] or aggregation, which is used in mini-batch k-means [22], BICO [6], and
BETULA [10], among others. The expected values of the results are very similar
to the standard algorithm, which is not surprising because the means used in
k-means are statistical summaries, too. For "exact" k-means, without lossy data
aggregation, approaches primarily fall into two categories: (1) k-d trees have been

Part of the work on this paper has been supported by Deutsche Forschungsgemeinschaft
(DFG), project number 124020371, within the Collaborative Research Center SFB 876
"Providing Information by Resource-Constrained Analysis", project A2.

used to accelerate k-means [8,14] by assigning subsets of the data to clusters at once, using the distances of the centers to the bounding boxes of the tree nodes, and (2) a large family of approaches which use the triangle inequality to omit unnecessary distance computations, exploiting that many points do not change their cluster after the first few iterations. Philips [15] used the pairwise distances of the centers to identify unnecessary computations, and Elkan [5] additionally uses upper and lower bounds for the distances between points and cluster centers. Hamerly [7] merged the lower bounds to the far cluster centers into a single bound to conserve memory, at the cost of additional distance computations due to looser bounds. State-of-the-art k-means algorithms, like Exponion [13] and Shallot [3], additionally use (hyper-)balls around the closest centers to further reduce computations. None of these methods is always best, but it depends on data dimensionality, data size, the number of clusters, initialization, and the data distribution. As all of these methods compute all distances between points and initial centers in the first round (to obtain the initial bounds), the first iteration is at least as expensive as in the standard algorithm, but it is in the later iterations where these improvements help. Recent proposals, for example, also take the distances to the previous center locations into account [24], or transfer such acceleration techniques to spherical k-means [20] for text clustering by using a similar triangle inequality for cosine similarity [17].

In this paper, we combine the ideas from both of these research directions: we will use an exact data index (a trivial extension of the cover tree [2]), combined with a pruning strategy that uses the triangle inequality, to accelerate the standard k-means algorithm by assigning entire subsets of the data at once. We then use our approach to calculate the upper and lower bounds that are used in Hamerly's [7] algorithm and its derivatives. Which allows us to switch strategies in later iterations when the clusters have become stable and these bounds become effective. We choose the cover tree as it aggregates the data into a hierarchy of ball covers, which allows the direct use of the triangle inequality. This differs from previous approaches using the k-d tree, which used the minimum distance to the bounding boxes [14], respectively hyperplanes implied by these bounding boxes [8]; both of these methods need bounding boxes not used by the original k-d tree. We argue that the ball covers of the cover tree produce more suitable bounds than the bounding boxes used in the k-d tree approaches, and that metric pruning should be superior to the geometric pruning of existing approaches. Furthermore, we hope to achieve improved performance by the way the cover tree controls the data expansion rate (and hence the radius of the nodes). It allows a wider fan-out, whereas the k-d tree is a strict binary tree that will have a higher depth, and more nodes. Last, but not least, a node in the (extended) cover tree is a more compact data structure than the bounding boxes used by the k-d tree approaches by approximately a factor of two: a ball is represented by a center and a radius, whereas the bounding boxes are represented by the midpoint and width in each axis (or alternatively, by a minimum and maximum, but the former is more suitable for k-means). Hence, we expect the new approach to need less memory.

2   Foundations

In this section, we explain the foundations of the cover tree and k-means, using a notation inspired by previous work [13] suitable for both.

2.1   Standard k-Means Algorithm

The standard k-means algorithm is a heuristic to partition the points $s \in X$ into k clusters $\{C_1, ..., C_k\}$. We start with the initial cluster centers $c_1, ..., c_k$ (e.g., sampled from the data) and alternate the following two optimization steps: first, all samples s are assigned to their nearest cluster center by Eq. (1), where $a(s)$ denotes the cluster index, and then the cluster centers c_i are updated by Eq. (2):

$$a(s) \leftarrow \arg\min_{i \in 1,..,k} d(s, c_i), s \in X \tag{1}$$

$$c_i \leftarrow \frac{\sum_{s|a(s)=i} s}{|\{s \mid a(s) = i\}|} \quad i \in 1, ..., k. \tag{2}$$

When no cluster assignment changes, the algorithm stops.[1] While this heuristic does frequently not find the global optimum, this scheme converges to a local fix point when no assignment changes. The selection of the initial cluster centers not only influences the run time of the algorithm but also the quality of the final clustering. Here, the k-means++ [1] initialization has become the most popular choice, as it already provides a (probabilistic) quality guarantee, based on sampling centers from the data proportionally to their expected contribution to the reduction of variance. In this article, we will not further study initialization.

2.2   Triangle Inequality in k-Means

Many accelerated k-means algorithms use the triangle inequality to reduce the number of distance calculations, in particular in Eq. (1) instead of computing the distances between all points and all clusters, as introduced by Phillips [15]. Given sample s and cluster center c_i for which the distance $d(s, c_i)$ is known, and some other c_j with an unknown distance, we can use the triangle inequality

$$d(c_i, c_j) \leq d(s, c_i) + d(s, c_j) \tag{3}$$

$$\Rightarrow \quad d(s, c_j) \geq d(c_i, c_j) - d(s, c_i) \tag{4}$$

to get a lower bound on the distance $d(s, c_j)$, and therefore may be able to avoid computing $d(s, c_j)$. In particular, Eq. (4) yields the implication [15]:

$$d(c_i, c_j) \geq 2d(s, c_i) \quad \Rightarrow \quad d(s, c_j) \geq d(s, c_i) . \tag{5}$$

We can exclude centers from consideration that are far away from the current closest center with this inexpensive filter. By computing all $d(c_i, c_j)$ once at the beginning of each iteration, this is relatively cheap as long as k is not too big.

---

[1] It is also possible to stop early centers' movement is below some threshold.

Another application for the triangle inequality in k-means are per-sample upper bounds on the distance to the assigned cluster and lower bounds on the distances to the other cluster means:

$$u_s \geq d(s, c_{a(s)}) \quad \text{and} \quad l_s \leq d(s, c_j) \quad \forall j \neq a(s).$$

Then $u_s \leq l_s$ implies that $a(s)$ is still the closest cluster by $d(s, c_{a(s)}) \leq u_s \leq l_s \leq d(s, c_j)$. These bounds were introduced by Elkan [5] and later simplified by Hamerly [7] to use only a single lower bound for all other clusters. We increase the upper bound and decrease the lower bound based on the movement of the centers to guarantee their correctness. This becomes more effective once the cluster centers move only very slightly: while cluster centers still move much, we may see $u_s > l_s$ often, and then have to compute the true distances often. For Hamerly's version, a single cluster that moves substantially is enough to require many recomputations, Elkan's version computes fewer distances, but has to update $N \cdot k$ bounds each iteration. The current state-of-the-art algorithms Exponion [13] and Shallot [3] also use the same bounds as Hamerly.

When the distances of a sample s to all cluster means c_i are calculated, the upper bound u_s is set to the distance to the assigned cluster $a(s)$ and the lower bound l_s to the minimum distance to any of the other clusters c_j for $j \neq a(s)$:

$$u_s \leftarrow d(s, c_{a(s)}) \qquad l_s \leftarrow \min_{j \neq a(s)} d(s, c_j).$$

In each iteration, when updating the means, we also have to update the bounds. For each mean c_i, we compute how far it moved from its previous location c_i'. To retain correct bounds, we have to add the distance moved by the nearest mean to the upper bound, and subtract the maximum distance moved by any other cluster mean from the lower bound:

$$u_s \leftarrow u_s + d(c_{a(s)}', c_{a(s)}) \qquad\qquad l_s \leftarrow l_s - \max_{j \neq a(s)} d(c_j', c_j).$$

The various algorithm variants proposed often include additional pruning rules using the triangle inequality, but a thorough discussion is beyond the scope of this paper, see [3,13] for an overview of recent algorithms.

2.3   Cover Tree

The cover tree [2] is a tree-based index structure with linear memory designed to accelerate nearest neighbor and radius search. The key idea of the cover tree is to cover the data with balls of a radius that decreases as we move down the tree; in its theoretical formulation it has an infinite number of levels with radius 2^i; but in practice, there exists a top level where all data is in a single ball and a bottom level where all distinct points have been separated. Levels in-between can be omitted if no changes to the tree structure happen. If the dataset has a finite expansion rate (i.e., the amount of data grows by at most a constant if we double the radius), this index can provide interesting theoretical guarantees for nearest neighbor search. In practice, the cover tree often performs quite well because of its small overhead and as it is inexpensive to build. The ball covers of the cover tree are restricted by the maximum radius in each level, and the tree structure: a sample s that was contained in a node x must remain in a child of x in the next level, and must not move to another ball even if that were closer.

All cover trees obey invariants for their covers N_i at each level $i \in \mathbb{Z}$:

1. (nesting) $N_i \subset N_{i-1}$
2. (cover) $\forall q \in N_{i-1} \exists p \in N_i \; d(p,q) \leq 2^i$ and exactly one p is the parent of q
3. (separation) $\forall p, q \in N_i, d(p,q) \geq 2^i$.

The $p \in N_i$ function as routing objects for searching the tree, and at each level balls of radius 2^i around these objects cover the entire dataset. Instead of storing the N_i at infinitely many levels, we only store levels that differ from the previous. To make the tree navigable from the root, we store for every $p \in N_i$ all $q \in N_{i-1}$ for which p is the parent, and their distance $d(p,q)$; except p itself which always is its own child at distance 0. An interesting side effect of optimization is that since p then is also a routing object in the next level, we can reuse any distance to p that we already computed in all subsequent levels. Two routing objects at the same level have at least the distance $d(p,q) > 2^i$, while descendants are within the radius $d(p,q) \leq 2^i$ to the routing object. While the factor 2^i was used for theoretical results, smaller scaling factors of 1.2–1.3 are typically faster in practice. The scaling factor allows controlling the trade-off between fan-out (width) and depth of the tree. We use the scaling factor of 1.2 in our experiments.

The construction of the cover tree follows the original greedy approach of Beygelzimer et al. [2]. We extend it with a simple bottom-up aggregation afterward to store the sum $S_x = \sum_{p \in x} p$ and the number of samples w_x in each node x. While this increases the memory consumption of the cover tree noticeably (previously, it would store only object references and the distances to the parent), we still have to store only one vector for each node. Not two vectors for the bounding box as in the earlier k-d tree approaches. Because of the higher fan-out, we also have fewer nodes than the k-d tree. Furthermore, we can define a minimum node size, at which we stop building the tree and instead store all remaining points directly with a cover radius of 0. For efficiency, we store all such singleton nodes ($|x| = 1$) more compactly, and omit storing the trivial aggregated values $S_x = x$ and $w_x = 1$, and the radius $r = 0$.

2.4   Bounds Within a Cover Tree

To assign an entire cover tree node that represents a subset of the dataset $x \subseteq X$ to a cluster, all samples $q_x \in x$ have to be closest to the same center. We can bound distances using the node radius $r_x = \max_{q_x \in x} d(p_x, q_x)$:

$$d(p_x, c_i) - r_x \leq d(q_x, c_i) \leq d(p_x, c_i) + r_x . \tag{6}$$

The triangle inequality also yields bounds on the distance of the routing object p_y to a cluster center c_i, given the distance from the parent node to the cluster:

$$d(p_x, c_i) - d(p_x, p_y) \leq d(p_y, c_i) \leq d(p_x, c_i) + d(p_x, p_y) . \tag{7}$$

Combining these yields upper and lower bounds for all samples in a child node, which we can use to prune candidate centers in k-means when $\forall q_y \in y$:

$$d(p_x, c_j) - d(p_x, p_y) - r_y \leq d(q_y, c_j) \leq d(p_x, c_j) + d(p_x, p_y) + r_y \,. \tag{8}$$

3 Cover Tree k-Means

We present a novel algorithm that uses a variation of the cover tree index to filter candidate centers. Using the triangle inequality we calculate bounds, see Section 2.4, to rule out cluster centers and hence reduce the number of distance computations in the assignment phase (i.e., in Eq. (1)) of k-means, using a similar idea as the k-d tree-based approaches [8,14]. However, the new approach uses fewer and smaller nodes for the tree, and as routing objects are reused in the cover tree, needs fewer distance computations. Tree nodes (representing subsets of the data) can be assigned at once, which is particularly beneficial with near-duplicate points. We then further combine this approach with current state-of-the-art stored-bounds k-means algorithms.

3.1 Calculating Distances

During initialization, or when reassigning points, we need the distance from some routing object p_x of cover tree node x to the centers c_i, c_j of a set of candidate clusters $C_i, C_j \in A_x$. For this, we adapt the common lower bound in Eq. (5). But unlike existing algorithms that handle only single samples, we need to ensure that the bounds are true for every sample $q_x \in x$ represented by the node. To avoid computing all distances and to benefit from the tree structure, we use Eq. (6) and the maximum distance r_x of points in the cover tree node. This allows deciding whether we have to calculate the distance $d(q_x, c_j)$ based on the already calculated distance $d(q_x, c_i)$ and the inter-cluster distances $d(c_i, c_j)$:

$$d(c_i, c_j) \geq 2d(p_x, c_i) + 2r_x \geq 2d(q_x, c_i) \tag{9}$$
$$\Rightarrow d(q_x, c_j) \geq d(q_x, c_i) \Rightarrow C_i \notin A_x \,.$$

The inter-cluster distances are computed and stored at the beginning of each k-means iteration, and used many times, as in previous work.

3.2 Assigning Nodes

With all relevant bounds for the routing object p_x computed, node x can be assigned to a cluster $a(x) = C_1$ if $\forall q_x \in x$, the distance to the nearest cluster center c_1 is smaller than the distance to the second-nearest cluster c_2. To decide this, we use the triangle inequality applied to a node, see Eq. (6):

$$d(q_x, c_1) \leq d(p_x, c_1) + r_x \leq d(p_x, c_2) - r_x \leq d(q_x, c_2). \tag{10}$$

Especially for the higher levels of the tree, this inequality often does not hold because nodes have a large radius r_x. Nevertheless, we may be able to eliminate candidates $C_i \in A_x$ for subsequent levels. To rule out that a cluster C_i is the nearest to any sample $q_x \in x$, the distance from its centroid c_i to the routing object $p_x \in x$ has to be larger than the upper bound on the distance to its nearest cluster center, which is a generalization of Eq. (10):

$$d(q_x, c_1) \le d(p_x, c_1) + r_x \le d(p_x, c_i) - r_x \le d(q_x, c_i). \tag{11}$$

As we narrow down the set of candidate cluster centers A_x, eventually only a single center remains, and the entire subtree can be assigned to the same cluster.

If we reassign the node and all its contained points to another cluster, we remove S_y and w_y for all previously assigned nodes $y \subseteq x$ from their old clusters and add S_x and w_x to the new cluster instead. By using the aggregates stored in the tree, this reassignment becomes more efficient. If we cannot assign a node, we process all child nodes recursively.

3.3 Recursion into Child Nodes

When moving to a child node $y \subset x$ in the tree, we can exploit that we store the distances to the parent in the cover tree, and that we know the radius of y. The inequality Eq. (8) combines the known distances to the parent routing object $d(p_x, c_i)$, the distance between parent and child routing objects $d(p_x, p_y)$, and the reduced radius of the child r_y. We now can assign the entire child to the parents' nearest cluster C_1 if $\forall q_y \in y$:

$$d(q_y, c_1) \le d(p_x, c_1) + d(p_x, p_y) + r_y \le d(p_x, c_2) - d(p_x, p_y) - r_y \le d(q_y, c_2). \tag{12}$$

Observe that for $p_x = p_y$ we obtain Eq. (10) with the reduced radius of the next level, but this bound is only tighter if $d(p_x, p_y) \le r_x - r_y$. Since $d(p_x, p_y) > r_y$ for $p_x \ne p_y$ and the radius r_y typically only reduces by the chosen factor 1.2, it frequently will not be sufficient. But as the tree can skip some levels, it occasionally holds and allows skipping some computations. If the above inequality does not hold, we have to recompute at least some distances to assign y. The distance $d(p_y, c_1)$ to the nearest cluster center c_1 of the parent node is most beneficial because it can be used to tighten the inequality in Eq. (12) by eliminating the distance to the parent node on the left-hand side of the equation and allows us to assign y to C_1 if $\forall q_y \in y$:

$$d(q_y, c_1) \le d(p_y, c_1) + r_y \le d(p_x, c_2) - d(p_x, p_y) - r_y \le d(q_y, c_2). \tag{13}$$

The same tightening is applied to Eq. (11) to prune candidate clusters before also recomputing the distances $d(q_y, c_i)$ when moving to the child node. Clusters that do not satisfy the following inequality can be excluded:

$$d(q_y, c_1) \le d(p_y, c_1) + r_y \le d(p_x, c_i) - d(p_x, p_y) - r_y \le d(q_y, c_i). \tag{14}$$

After pruning cluster centers with Eq. (14), the same bounds and calculations are applied recursively to each of the child nodes not yet assigned a cluster.

3.4   Hybrid Cover Tree k-Means

When analyzing the number of distance computations performed per iteration by current state-of-the-art algorithms, like Exponion [13] and Shallot [3], we observe that they correlate with how much the cluster centers move, for the reason explained before. They decrease drastically over the first few iterations as the centers stabilize. Our approach, on the other hand, is only slightly influenced by how far means move, and can already avoid distance computations in the first iteration by reducing the number of candidate cluster centers when traversing the tree. Using the stored aggregates in the tree, we can assign entire subtrees to the same cluster, but the pure tree-based approach benefits little from centers becoming stable after a few iterations. To utilize the best of both worlds, we propose a hybrid algorithm that uses the cover tree only for the first few iterations and then switches to the state-of-the-art Shallot k-means algorithm [3]. Any other algorithm based on Hamerly's [7] bounds could be used instead, as we can efficiently obtain upper and lower bounds from our cover tree, which gives these algorithms an efficient start. Obtaining all $k \cdot N$ bounds for Elkan's algorithm would be more effort. Our approach is not equivalent to initializing with cluster centers obtained with the cover tree, but we prune distance computations when computing the bounds using the tree by filtering candidate means.

For single points, this is trivial (but we still save, as we may have a reduced set of candidates for the second-nearest center). If we assign an entire node x to a cluster C_1, we do not know the exact distance to the nearest two clusters, but we obtain upper and lower bounds by our inequalities above:

$$u_{q \subset x} = d(p_x, c_1) + r_x \tag{15}$$

$$l_{q \in x} = d(p_x, c_2) - r_x. \tag{16}$$

When assigning a subtree, we can also obtain bounds for child nodes $y \subset x$ without additional distance computations:

$$u_{q \in y} = d(p_x, c_1) + d(p_x, p_y) + r_y \tag{17}$$

$$l_{q \in y} = d(p_x, c_2) - d(p_x, p_y) - r_y. \tag{18}$$

While computing $d(p_y, c_1)$ for each child node y would yield tighter bounds, we can simply leave this to the next iteration of the subsequent k-means algorithm. Depeding on the center movements, these will need to be refined anyway, and do not need to be tight. For initializing the Shallot algorithm, we also need to give it the identity of the second-nearest cluster; more precisely, the cluster for which the lower bound was obtained. While it is not guaranteed that the second-nearest cluster to the routing object in the tree is the second-nearest cluster for all points in the node, the Shallot algorithm only requires the bounds to hold. In the regular Shallot algorithm, it can happen that the assumed second-nearest cluster changes unnoticed, this does not affect the correctness of the algorithm.

In summary, when we transition to a stored-bounds algorithm, we can use the bounds used by the Cover-means algorithm to initialize the stored-bounds. These bounds will be less tight, but also much less expensive to compute, as computing the initial bounds is a bottleneck of all stored-bounds approaches.

Table 1. Overview of the datasets used in the experiments

Name	N	D	domain
ALOI	110250	27, 64	color histograms
MNIST	70000	10, ..., 50	autoencoder
CovType	581012	54	remote sensing
Istanbul	346463	2	tweet locations
Traffic	6.2M	2	accident locations
KDD04	145751	74	biology

4   Evaluation

Algorithms: In our experiments, we compare our new cover-tree-based approaches to the *Standard* k-means algorithm, as well as state-of-the-art improvements. The k-d tree filtering variant of *Kanungo* et al. [8] represents current tree-based methods. From the stored-bounds family, we include the popular and easy-to-understand *Hamerly's* [7] and *Elkan's* [5] k-means algorithm, and the state-of-the-art algorithms *Exponion* [13] and *Shallot* [3]. We include two new approaches in our evaluation. First, the base variant, *Cover-means*, uses the cover tree for accelerating k-means. Second, a hybrid of our cover tree variant with Shallot (as explained in Sect. 3.4) denoted as *Hybrid*.

Parameterization: The cover tree brings some additional hyperparameters for the algorithms. We decided not to tune these parameters for specific datasets, but instead we identified a set of default values that will usually work well, and do not vary them in the experiments. Much of these are set to keep the cover tree small, as to not introduce overhead for constructing the tree. The cover tree is built with a scaling factor of 1.2; while larger values may further accelerate clustering, the increased fan-out comes with an increased construction time, in particular for uniform data regions. To further keep the tree small, we stop splitting at a minimum node size of 100 samples. Smaller leaf nodes would allow more pruning, but also increase the construction time. For the hybrid approach, we switch strategies after 7 iterations. This value has likely the most potential for further turning for particular datasets; in particular, for easy synthetic datasets, it may be beneficial to switch later. On the other hand, on too easy datasets or with lucky initialization, k-means may already have converged before. We include the construction time of the cover tree, except for the experiment listing time per iteration and the parameter sweep experiment. We evaluated the same 10 random initializations generated by k-means++ [1] for each algorithm, and usually study medium to large $k = 10 \ldots 1000$.

Implementation: To make the benchmarks more reliable, we keep implementation differences to a minimum and implement all algorithms in a shared codebase, as recommended by Kriegel et al. [9]. We implemented the algorithms in Java using the ELKI framework [18] because it already contains optimized and tested versions of the comparison algorithms. All algorithms are run single-core on an exclusively used AMD EPYC 7302 CPU to reduce the confounding factors.

Datasets: Because the performance of the k-means algorithms strongly depends on the dataset, we evaluate our approach on various real-world datasets, Table 1. For the MNIST dataset which we reduced in dimensionality with an autoencoder, we chose 10, 20, 30, 40 and 50 dimensions. The ALOI data is available in multiple dimensionalities [21], of which we selected 27 and 64 dimensions. The Istanbul and Traffic datasets contain the coordinates of Tweets respectively traffic accidents, and hence are low-dimensional.

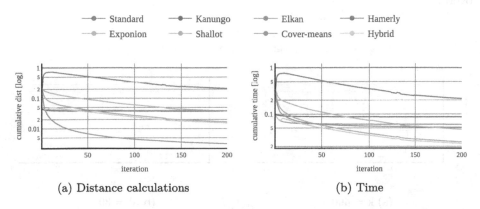

(a) Distance calculations   (b) Time

Fig. 1. Commulative evaluation in relation to the Standard algorithm vs. iterations on the ALOI 64D dataset for $k = 400$

Experimental Results: In Fig. 1, we study the cumulative number of distance computations and the cumulative time on the ALOI 64D dataset for k=400 over the iterations (i.e., convergence) of the algorithms. The construction of the tree is not included here. Both measurements are normalized by the full Standard algorithm to improve readability, and can be interpreted as relative savings.

For the number of distance calculations in Fig. 1a, the algorithms can be categorized into three groups: The Standard algorithm obviously does not skip any distance calculations. The tree-based algorithm of Kanungo and the cover tree, only need a fixed fraction of that (5%-10% on this data). When the cluster centers move only a little after the first few iterations, they exhibit a constant performance. The third group are the stored-bound-based algorithms, which exhibit a decreasing number of distance computations. The performance of Elkan's algorithm is noticeably better than the others on this metric, Hamerly's performs worst, while Exponion and Shallot improve over Hamerly's as expected. Our new Hybrid method combines the early savings of the tree methods with the good late performance of Shallot (here, switching to Shallot later would likely be better).

Figure 1b shows the related run time. The Standard algorithm is again the baseline, and we have similar groups of behavior, but some interesting differences. The most significant difference is that Elkan's algorithm now is considerably worse. While Elkan's can save many distance computations, it has to maintain many more bounds, which introduces a constant overhead per iteration. Exponion and Shallot can save distance computations at little constant overhead, and hence improve over Hamerly's significantly. The pure tree-based approaches do not benefit from convergence much, and also show a constant cost per iteration on the right part. Since the Hybrid version switches to the successful Shallot approach after reaping the benefits of the tree, it uses the least time overall.

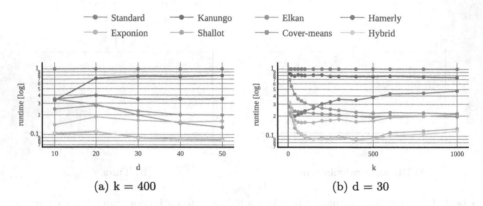

Fig. 2. Runtime in relation to the Standard algorithm vs. k respectively d on the MNIST dataset

Next, we look at the MNIST dataset in Fig. 2 to examine how the different algorithms scale with the dimensionality d and the number of clusters k. Figure 2a shows the runtime in relation to the runtime of the Standard algorithm over multiple different dimensionalities. Most algorithms except Elkan scale about the same as the Standard algorithm. Kanungo's algorithm only has a slight advantage over the standard algorithm and is not really well-suited for high dimensional data. In high-dimensional data, the cost of distance computations increases, and hence the benefits here of Elkan's algorithm over the alternatives become more pronounced, as also observed in other studies.

In Fig. 2b, we can observe that the scaling behavior with k is slightly more interesting. For very low k the benefit of all acceleration techniques is rather low, especially for the tree accelerations, and Kanungo's algorithm does not improve for higher k. Over all k, we can see that Hamerly's algorithm scales worse than the standard algorithm. Exponion and Shallot also show slightly worse scaling properties than Elkan and Cover-means. Here the improvement for our Hybrid approach increases, and switching strategies later could bring further advantages.

In Table 2 we extend our view to the other datasets, and give the number of distance calculations relative to the Standard algorithm. For the stored-bound algorithms, Hamerly always uses the most distance calculations, followed by Exponion and Shallot. Elkan's algorithm has the least of all, as is expected, except for the traffic dataset that has a very high number of samples and is very beneficial for the tree-based methods that can assign many points at once. Our Hybrid approach is usually slightly better than the Shallot algorithm but does not come near Elkan's k-means, while the Cover-means based approach is usually worse than Exponion. We can observe that the performance of the tree-based approaches (Kanungo's algorithm more so than the cover tree) appears to depend more on the dataset. In particular, the low-dimensional Istanbul and traffic datasets are beneficial for the tree-based approaches, whereas Kanungo's algorithm struggles with the complex KDD04 dataset and uses even more distance calculations than the Standard algorithm.

Table 2. Relative number of distance calculations compared to the Standard algorithm for $k = 100$

	CovType	Istanbul	KDD04	Traffic	MNIST10	MNIST30	ALOI27	ALOI64
Kanungo	0.006	**0.002**	1.450	**0.000**	0.149	0.370	0.036	0.048
Elkan	**0.004**	0.002	**0.025**	0.001	**0.007**	**0.009**	**0.005**	**0.006**
Hamerly	0.099	0.078	0.364	0.090	0.198	0.213	0.229	0.253
Exponion	0.016	0.010	0.341	0.009	0.075	0.130	0.060	0.075
Shallot	0.012	0.006	0.311	0.006	0.034	0.061	0.030	0.043
Cover-means	0.012	0.003	0.807	0.001	0.097	0.180	0.044	0.063
Hybrid	0.005	0.003	0.310	0.003	0.031	0.057	0.027	0.038

Table 3. Relative run time compared to the Standard algorithm for $k = 100$

	CovType	Istanbul	KDD04	Traffic	MNIST10	MNIST30	ALOI27	ALOI64
Kanungo	0.068	0.123	4.035	0.182	0.470	0.798	0.133	0.130
Elkan	0.114	0.520	**0.193**	0.652	0.454	0.226	0.180	0.104
Hamerly	0.139	0.171	0.383	0.173	0.262	0.238	0.262	0.278
Exponion	0.064	0.132	0.369	0.142	0.150	0.161	0.107	0.109
Shallot	0.062	0.134	0.346	0.145	**0.120**	**0.098**	0.084	0.080
Cover-means	0.072	0.092	1.121	0.135	0.352	0.313	0.138	0.123
Hybrid	**0.051**	**0.084**	0.457	**0.130**	0.133	0.102	**0.082**	**0.076**

As seen before, there may be overheads in the computations overlooked when counting only distance computations, in particular for Elkan's algorithm. Table 3 shows the total run time of all algorithms, including the construction of the trees. The KDD04 dataset shows that for high dimensional data, Elkan's algorithm often is the fastest because saving distance computations is the most beneficial then. While the additional bounds help for larger k (as they are updated individually), the memory overhead becomes an issue quickly. On the other datasets, the Shallot algorithm is usually the fastest state-of-the-art algorithm. Our cover tree approach is most of the time faster than Kanungo's, Hamerly's and Elkan's algorithms, but it cannot compete with Exponion or Shallot. Combining the cover tree with Shallot in our Hybrid approach leads to the overall best results. The results likely could be further improved by tuning the hyperparameters, e.g., changing from the Cover-means to the Shallot algorithm at some "optimal" point, or increasing the leaf size for the larger data sets, but we have not yet developed a heuristic for this and do not want to overfit to this benchmark.

Lastly, Table 4 shows the relative runtime for a complete parameter sweep for the individual datasets, as one would often need to do in practice when the "true" number of clusters k is not known. Here, we measure the time it takes to run the algorithms for 10 different initializations (restarts, because k-means may converge to different local fix points) and 16 different values for k (to find the best number of clusters). Then, the "best" clustering can be chosen by a heuristic such as the "Elbow" method, or any of the better alternatives [19]. In this scenario, the dataset is used multiple times, and we can reuse the cover-tree we built to amortize the construction cost. In this task, we see Elkan again be the fastest on difficult high dimensional data like the KDD04 dataset, but not being able to handle the complete traffic dataset because of memory overhead for storing all bounds. On all other datasets, our Hybrid approach is the fastest except for the 10D MNIST dataset, where the Shallot algorithm is faster.

Table 4. Relative runtime compared to the Standard algorithm with multiple restarts (parameter sweep to choose k)

	CovType	Istanbul	KDD04	Traffic	MNIST10	MNIST30	ALOI27	ALOI64
Kanungo	0.040	0.112	5.090	0.162	0.409	0.903	0.114	0.116
Elkan	0.093	0.609	**0.171**	-	0.351	0.187	0.121	0.065
Hamerly	0.211	0.208	0.453	0.238	0.338	0.347	0.284	0.304
Exponion	0.040	0.145	0.492	0.162	0.154	0.172	0.077	0.077
Shallot	0.037	0.145	0.414	0.154	**0.121**	0.100	0.059	0.050
Cover-means	0.028	0.059	1.015	0.093	0.272	0.248	0.086	0.077
Hybrid	**0.020**	**0.056**	0.463	**0.089**	0.122	**0.095**	**0.055**	**0.047**

5   Conclusion

We show that tree-based k-means algorithms can be beneficial, in particular for huge datasets because they can assign many points at once. Our new Cover-means approach outperforms the earlier approaches based on the k-d tree on most datasets. The use of the triangle inequality in cover tree k-means for pruning the set of candidate clusters makes it easier to combine this with the other approaches than the bounding box-based approach of the k-d tree methods. The cover tree also uses less memory, as storing the ball needs fewer parameters than storing a bounding box, but also because the cover tree has a higher fan-out and lower depth than the k-d tree. Our new method is in particular well-suited for clustering with a large number of clusters k, where scalability becomes more important. For small k, it will often be sufficient to cluster a subsample of the data [4] on a single CPU, and we will usually find a sufficiently good result because of the stability properties of k-means [16], if a *stable* result exists.

We also show how a hybrid of index-tree-based and stored-bound-based approaches combines the benefits of both worlds and improves over the performance of state-of-the-art approaches for many scenarios. While our hybrid approach that combines the cover tree with the Shallot algorithm is very basic, there are new challenges when fully integrating both approaches in future work: when the current hybrid switches to the Shallot strategy after a fixed number of iterations, it no longer exploits redundancy in the dataset, but also uses individual bounds for each point. The results on the large Traffic dataset suggest that we may want to be able to stick longer to tree-based aggregation for performance for huge datasets, and we have not yet developed a heuristic for this.

The source code of this algorithm will be made available in the ELKI clustering toolkit [18], and we hope that it lays the ground for future research on further accelerating k-means by combining the strength of tree-based and stored-bounds-based algorithms.

References

1. Arthur, D., Vassilvitskii, S.: k-means++: the advantages of careful seeding. In: ACM-SIAM Symposium on Discrete Algorithms, SODA, pp. 1027–1035 (2007)
2. Beygelzimer, A., Kakade, S.M., Langford, J.: Cover trees for nearest neighbor. In: International Conference on Machine Learning, ICML, pp. 97–104 (2006). https:// doi.org/10.1145/1143844.1143857
3. Borgelt, C.: Even faster exact k-means clustering. In: International Symposium on Intelligent Data Analysis, IDA, pp. 93–105 (2020). https://doi.org/10.1007/978-3-030-44584-3_8
4. Bradley, P.S., Fayyad, U.M., Reina, C.: Scaling clustering algorithms to large databases. In: KDD
5. Elkan, C.: Using the triangle inequality to accelerate k-means. In: International Conference on Machine Learning, ICML, pp. 147–153 (2003)
6. Fichtenberger, H., Gillé, M., Schmidt, M., Schwiegelshohn, C., Sohler, C.: BICO: BIRCH meets coresets for k-means clustering. In: ESA, pp. 481–492 (2013). https://doi.org/10.1007/978-3-642-40450-4_41
7. Hamerly, G.: Making k-means even faster. In: SIAM Data Mining, SDM, pp. 130–140 (2010). https://doi.org/10.1137/1.9781611972801.12
8. Kanungo, T., Mount, D.M., Netanyahu, N.S., Piatko, C.D., Silverman, R., Wu, A.Y.: An efficient k-means clustering algorithm: analysis and implementation. IEEE Trans. Pattern Anal. Mach. Intell. **24**(7), 881–892 (2002). https://doi.org/10.1109/TPAMI.2002.1017616
9. Kriegel, H., Schubert, E., Zimek, A.: The (black) art of runtime evaluation: are we comparing algorithms or implementations? Knowl. Inf. Syst. **52**(2), 341–378 (2017). https://doi.org/10.1007/s10115-016-1004-2
10. Lang, A., Schubert, E.: BETULA: fast clustering of large data with improved BIRCH CF-trees. Inf. Syst. **108**, 101918 (2022). https://doi.org/10.1016/j.is.2021.101918

11. Lloyd, S.P.: Least squares quantization in PCM. IEEE Trans. Inf. Theory **28**(2), 129–136 (1982). https://doi.org/10.1109/TIT.1982.1056489

12. Mahajan, M., Nimbhorkar, P., Varadarajan, K.R.: The planar k-means problem is NP-hard. In: WALCOM: Algorithms and Computation, pp. 274–285 (2009). https://doi.org/10.1007/978-3-642-00202-1_24

13. Newling, J., Fleuret, F.: Fast k-means with accurate bounds. In: International Conference on Machine Learning, vol. 48, pp. 936–944 (2016)

14. Pelleg, D., Moore, A.W.: Accelerating exact k-means algorithms with geometric reasoning. In: KDD, pp. 277–281 (1999) https://doi.org/10.1145/312129.312248

15. Phillips, S.J.: Acceleration of k-means and related clustering algorithms. In: International Workshop on Algorithm Engineering and Experiments, ALENEX, pp. 166–177 (2002). https://doi.org/10.1007/3-540-45643-0_13

16. Pollard, D.: Strong consistency of k-means clustering. Ann. Stat. **9**(1), 135–140 (1981)

17. Schubert, E.: A triangle inequality for cosine similarity. In: Similarity Search and Applications, SISAP, pp. 32–44 (2021). https://doi.org/10.1007/978-3-030-89657-7_3

18. Schubert, E.: Automatic indexing for similarity search in ELKI. In: Similarity Search and Applications, SISAP (2022). https://doi.org/10.1007/978-3-031-17849-8_16

19. Schubert, E.: Stop using the elbow criterion for k-means and how to choose the number of clusters instead. SIGKDD Explor. **25**(1), 36–42 (2023). https://doi.org/10.1145/3606274.3606278

20. Schubert, E., Lang, A., Feher, G.: Accelerating spherical k-means. In: Similarity Search and Applications, SISAP, pp. 217–231 (2021). https://doi.org/10.1007/978-3-030-89657-7_17

21. Schubert, E., Zimek, A.: ELKI Multi-View Clustering Data Sets Based on the Amsterdam Library of Object Images (ALOI). https://doi.org/10.5281/zenodo.6355684. Zenodo, June 2010

22. Sculley, D.: Web-scale k-means clustering. In: World Wide Web, WWW, pp. 1177–1178 (2010). https://doi.org/10.1145/1772690.1772862

23. Steinhaus, H.: Sur la division des corp materiels en parties. Bull. Acad. Polon. Sci **1**, 801–804 (1956)

24. Yu, Q., Chen, K., Chen, J.: Using a set of triangle inequalities to accelerate k-means clustering. In: Similarity Search and Applications, SISAP, pp. 297–311 (2020). https://doi.org/10.1007/978-3-030-60936-8_23

Is Quantized ANN Search Cursed? Case Study of Quantifying Search and Index Quality

Gylfi Þór Guðmundsson[(✉)] and Björn Þór Jónsson

Reykjavik University, Menntavegur 1, 102 Reykjavik, Iceland
{gylfig,bjorn}@ru.is

Abstract. Traditional evaluation of an approximate high-dimensional index typically consists of running a benchmark with known ground truth, analyzing the performance in terms of traditional result quality and latency measures, and then comparing those measures to competing index structures. Such analysis can give an overall indication of the suitability of the index for the application that the benchmark represents. When the index inevitably fails to return the sought items for some queries, however, this methodology does not help to explain why the index fails in those cases. Furthermore, when considering many different parameter settings, the process of repeatedly indexing the entire collection is prohibitively time-consuming. In this paper, we define three causes for failures in hierarchical quantized search. We show that the two failure cases that relate to the index can be evaluated and quantified using only the index structure and ground-truth data. In our evaluation, we use eCP, a lightweight algorithm that builds the index hierarchy top-down a priori without any costly segmentation of the dataset, and show that significant insight can be gained into the quality of the index structure, or lack thereof.

Keywords: High-dimensionsional indexing · Hierarchical vectorial quantization · Evaluation methodology

1 Introduction

Approximate near-neighbour search techniques (ANN) are used to accomplish efficient similarity search over a large volume of high-dimensional data. To evaluate such a search engine, we need data and queries where we know the "correct answers," also known as the *ground-truth*. The evaluation process can then be best described as "proof by doing" as we first instantiate the search engine by indexing the evaluation data and then use the queries in the ground-truth to run ANN search and calculate metrics such as *precision*, *recall* and *mAP*. This "full-scale" approach is not only time consuming but also inadequate, as when the search fails those metrics neither explain why nor quantify by how much.

Today's state-of-the-art large scale ANN algorithms are based on proximity graph algorithms [7], which are very costly to construct. Compression-based

© The Author(s), under exclusive license to Springer Nature Switzerland AG 2023
O. Pedreira and V. Estivill-Castro (Eds.): SISAP 2023, LNCS 14289, pp. 163–170, 2023.
https://doi.org/10.1007/978-3-031-46994-7_14

techniques [6] are also popular, but they struggle with billions of features. In comparison, older tree-like hierarchical vectorial quantizers (HVQ) can be constructed quickly, have a lower memory footprint, and have been used at Web-Scale, but they give lower quality results. It is therefore of interest to study where HVQ approaches fail, to understand whether they can be improved to be more competitive.

The goal of the quantizer is to partition the data collection C into small segments. Let us define r as the representative for each such segment and R as the set of all representatives. At *indexing time*, segmentation is based on finding the most similar $r \in R$ for each vector v in the dataset. Then, at *search time*, one (or a few) most similar $r \in R$ are identified and only those segments make up the search result. Let us define the truly most similar representative $r \in R$ as r^t and a ranked list result as L^R. Finding r^t requires a scan over R, but as R is often large, an ANN search using a tree-like hierarchy is used instead. Let us denote the most similar representative found via the index as r^i and a ranked list result as L^i.

For a query q, let us define N_q as the set of its true neighbours. N_q can be said to form a *neighbourhood* and the density of data in this region, e.g. whether the near-neighbour relation is reciprocated or not, is a good indicator of how difficult it is to match q with its N_q. A "good" neighbourhood is when q and its N_q are dense and well represented, i.e., N_q falls into one (or a few) segment(s) similar to q. A "bad" neighbourhood is either a dense grouping that has no good representation, or a neighbourhood so sparse that some of N_q fall into segments that are far down the list of representatives most similar to q.

We can now identify three possible causes for HVQ search failures:

1. **Representation Failure (RF)** occurs when no representative $r \in R$ is a clear best option for a given neighbourhood, i.e., for both q and its N_q. In such a case, the vectors are likely to be fragmented over many segments, causing the ANN search to fail despite correctly indexing the vectors, i.e., for each $n \in N_q$, $r_q^t \neq r_n^t$.
2. **Index Hierarchy Failure (IHF)** occurs when the index is not assigning vectors correctly. That is, for a given vector v the most similar segment identified by the index, r_v^i, does not match the most similar representative r_v^t found by scanning all segment representatives R. Note that IHF can occur both when indexing the data and when using the index to identify what segment(s) should be used in the ANN search.
3. **Segment Search Failure (SSF)** occurs when the ANN search fails to find q's most similar $v \in C$ despite looking in the right segment. This can happen in systems that compress, aggregate, or otherwise approximate the search process inside segments.

Evaluating RF is all about looking at the neighbourhood. Using the ground truth for query q we obtain the ranked list of most similar segment representatives for q, denoted by L_q^R. For each neighbour n in N_q, from the ground truth, we then get the most similar segment r_n^t. We can now quantify RF by examining how far down the list L_q^R we find r_n^t. Note that getting the truly most similar segments is costly as it requires a full scan of all index segments $r \in R$. To evaluate IHF, we need to know a) the index-assigned segment r_v^i and b) the ranked list of most similar segments L_v^R for the vector v. Quantifying IHF is then based on how far down the list L_v^R we find r_v^i. If neither RF nor IHF occurred, but the ground truth indicates that the result is not correct, then SSF has occurred.

In this paper, we report on a case study of the eCP algorithm, a cluster-based HVQ. Since eCP reads whole clusters, SSF does not occur. Our results indicate that RF is much more prevalent with eCP than IHF. Furthermore, we demonstrate that while enhancing the representation with k-Means clustering can be done 50x faster by using only the index, the impact on RF is minimal.

The remainder of this paper is structured as follows. In Sect. 2 we explain our evaluation methods and metrics. Section 3 then present the results of our experiments, and finally we conclude the paper in Sect. 4.

2 Evaluating Indexed ANN Search Failures

Our aim is to understand and quantify how and why the ANN search, using HVQ, fails. The algorithm we choose to evaluate in this initial case study is called Extended Cluster Pruning (eCP) [2]. eCP is a simple yet versatile HVQ, which was initially developed for content-based image retrieval using SIFT features [4] and has been extensively evaluated at large scale [3]. The index construction is very efficient as the index is built a priori in a single top-down pass using randomly sampled vectors from the dataset. This is a great advantage as that means we can evaluate the performance of eCP using just the index and a ground-truth benchmark, without doing any of the costly segmenting. To be clear, eCP makes no effort to improve on the randomly chosen segment representatives. To compensate for potentially poor representation, however, it does support *search-expansion* both inside the index hierarchy as well as for the number of segments to retrieve and scan. In this case study, we consider search expansion at query time, processing SE clusters to find near neighbours. We now explain how we evaluate and quantify the possible causes for HVQ search failures. Since eCP is not susceptible to SSF, we can also estimate recall using only the index structure.

2.1 Evaluating Representation Failure (RF)

To evaluate the RF we need a neighbourhood (q and N_q, both obtained from the ground truth) and eCP's set of representatives R. By scanning R we derive L_q^R, the ranked list of q's most similar representatives, as well as the optimal assignment r_n^t for each $n \in N_q$. If r_n^t is at the top of L_q^R, we know that q and n are in the same segment. How far down the list we find each r_n^t tells us how far the ANN search will need to expand such that it finds that neighbour. Remember that q has many neighbours, so we choose to average this rank, creating a metric we call *Average Rank* (AR). Furthermore, we also add a second metric that is based on counting how many neighbours have a rank lower than some value X. We call his metric *Optimal Recall* (OR) as if we set X equal to the SE parameter used, it tells us exactly what portion of N_q we can hope to find in an ANN search.

2.2   Evaluating Index Hierarchy Failure (IHF)

To evaluate and quantify IHF, we consider a set of vectors v and, as before, we scan R to derive L_v^R. Then, using the index structure, we derive the assigned representative r_v^i, for a given SE setting. The metrics we use to quantify IHF is based on looking at where in the ranked list L_v^R we find the index-assigned r_v^i. If $r_v^i = r_v^t$, where r_v^t is the segment at the top of L_v^R, then the index assignment is optimal. Otherwise, we can use the rank of r_v^i in L_v^R to measure how far off the index assignment is. In our experiments, we report a) how many optimal assignments we have and b) how many are within SE of the optimal. This gives a clear indication of how well the index is doing.

2.3   Evaluating Recall in Absence of Segment Search Failure (SSF)

eCP is not susceptible to SSF, as the ANN search identifies and then scans whole segment(s) to create the final k-NN result. This allows us to calculate what the recall of eCP's ANN search would be using only the index and ground-truth data. For each query q, we obtain L_q^i, the ranked list of the SE most similar representatives for segments that should be scanned. By looking up each $n \in N_q$ from the ground truth to get r_n^i, we can compute recall by counting how many of the r_n^i assignments are anywhere in L_q^i.

3   Evaluation

3.1   Setup

Dataset and Ground-Truth: We use BIGANN [1] in our evaluation. The full set has 1B 128-dimensional SIFT features, but we use the 100M subset as this is sufficient to build a large eCP index. The original ground-truth consists of 10k queries with 1k NNs for each, but to make the computational load manageable we use a subset that consists of the first 50 queries along with all of their 1k near neighbours, for 50k vectors in total. We should note that BIGANN is a difficult dataset. The baseline recall given for track 1 of the *BigANN benchmark* using the BIGANN data is 63.5% recall and the best competitors got 71.4% [7].

Indices and Search Settings: Guidelines exist regarding picking the "right" number of clusters to build an eCP index. As we do not intend to build the full search engine, however, we build four different 3-level deep indices using $R=$ 40k, 80k, 160k and 320k vectors to study the impact of index size. The R vectors are randomly sampled from the 100M SIFT subset. In all experiments we perform search expansion, with maximum $SE = 20$, but in the analysis we consider the impact of varying the SE parameter from 1 to 20..

Hardware and Software: Experiments are all run on a single machine with an Intel i9-7900X CPU, 64 GiB of RAM, and a 1 TB Samsung 960 Pro SSD disk. The OS is Ubuntu 18_04 and we use Spark 2.4.5 with Java–openJDK version 11.0.17 and Scala version 2.11.12. Note that the original SIFT features are 128 dimensions of unsigned 8 bits (0–255), but since Java does not support unsigned data, we scale the values to be *Byte* (-128 to 127).

Table 1. Evaluation of Representation Failure (RF). Average Rank (AR) is the average location of r_n^t in L_q^R across all 50k queries. Optimal Recall (OR) counts how many of the 1k neighbours have a rank lower than $SE = 20$, averaged across 50 queries

Index	AR	OR
40k	55.76	602.54
80k	88.78	547.42
160k	132.12	489.96
320k	222.35	408.82

3.2 Experiment 1: Representation Failure (RF)

In this experiment we use the 50 queries, each having 1k NNs, and evaluate on all 4 index sizes. In Sect. 2.1 we defined the two metrics, AR and OR, that we report. The results are presented in Table 1.

Here, AR is the average rank of the N_q–1k vectors, averaged over all 50 queries. The 40k index has AR of ~56 while 320k has AR of ~222. This implies that adding more clusters seems to give the "bad" neighbourhoods more options, spreading the NNs even further, making it even harder to match q with its N_q.

The OR metric is even more interesting as it essentially indicates the best-case recall for a given search expansion. Here, we average the OR metric over the 50 queries, using $SE = 20$. The 40k index scores ~603 out of 1k queries, while the 320k scores ~409, which can be read as "optimal recall" of 60.3% and 40.9% respectively.

While interesting in itself, the true value of the OR is that it allows us to put the ANN search results of later experiments into context. The main conclusion we can draw from this experiment, however, is that the RF is significant.

3.3 Experiment 2: Index Hierarchy Failure (IHF)

Here we focus on evaluating whether the index is able to assign vectors correctly. From the index, we retrieve for each v of the 50k NN vectors r_v^i and we also scan R to get L_v^R, each vector's ranked list of most similar representatives. The results are presented in Fig. 1.

The x-axis indicates the search expansions used, while the y-axis shows the average across the 50k queries. As was stated in Sect. 2.2 we report two metrics. The first counts how many of the index assignments are optimal, i.e., $r_v^i = r_v^t$, which is shown with the solid lines. The second metric, reported with the dotted lines, counts how many index assignments are within SE of the optimal

Our first observation is that the results are almost identical despite the largest index (320k) having 8 times more segments than the smallest (40k). Second, we observe that without any search expansion ($SE = 1$), the indices are only correctly assigning about 25–30% of vectors. As we expand the search, this ratio grows and at $SE = 5$, the indices already retrieve over 50%. The reason why the

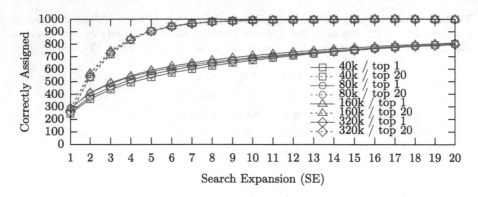

Fig. 1. Evaluation of Index Hierarchy Failures (IHF). The x-axis shows the expansion setting, SE, while the y-axis shows the correctly assigned NNs. Solid lines indicate when $r_v^i = r_v^t$ while the dotted lines indicate that r_v^i is in the top SE clusters of L_q^i

top item of the ranked list improves, is that the search expansion is applied at all levels of the index, reducing the branching errors at the upper levels.

Turning to the dotted lines, which show the neighbours correctly located in the top SE clusters of the ranked list, we observe that they grow even faster. For $SE = 1$, they are identical, but at $SE = 5$ more than 90% of assignments are found. This means that at $SE = 5$, 50% of the neighbours are correctly assigned and another 40% is within 5 of optimal assignment. What remains to be seen is whether this is good enough for the ANN search.

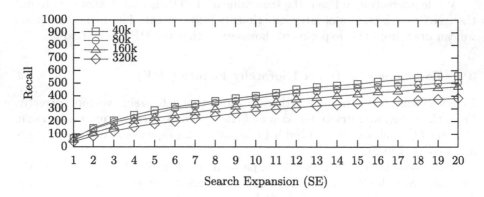

Fig. 2. Evaluation of Recall. The x-axis shows the search expansion, SE, while the y-axis shows the recall, averaged across the 50 queries

3.4   Experiment 3: Estimating Recall

Having investigated RF, producing a baseline for optimal recall, and investigated IHF, we can now check how well the eCP indices actually do in an ANN search.

As said in Sect. 2, this is done by searching for both q and N_q and checking whether the $n \in N_q$ assigned representative r_n^i is in the ranked list of similar representatives for the original query, L_q^i, using search expansion parameter SE. The results are plotted in Fig. 2.

We observe that, as expected, the recall degrades with index size and at first glance the results are not impressive. The 40k index peaks at only 555 matches (~56% recall) and the 320k at 382 (~38%). That is well below the 63.5% baseline from [7]. But if consider the OR metric from the previous experiment (at our maximum expansion), we observe that 40k index is in fact scoring ~56% out of a maximum of ~60% and the 320k index at ~38% out of a maximum of ~40.9%. From this we can assert that despite eCP's indexing hierarchy being very simplistic, it is only responsible for a small fraction of the search failures.

Table 2. Evaluation of whether k-Means can improve **RF** and ANN recall. Results shown are AR, OR, time the k-Means took and search results for 40k indices after the given number of steps of k-Means

Steps	AR	OR	Running time	ANN Recall
KM0 (original)	55.76	602.54	n/a	555.26
KM1 (full)	50.30	657.76	~50h	606.82
KM1 (index)	53.04	648.32	48m	602.20
KM2 (index)	51.14	654.26	~2h	614.34
KM10 (index)	45.91	677.00	~10h	606.80

3.5   Experiment 4: Can K-Means Fix RF?

A maximum recall of 60% for the 40k index cannot be called a great result. A common proposal to addressing representation is to run k-Means clustering. In this section, we evaluate the impact of this using AR and OR. The results are presented in Table 2.

The first line of the table repeats the data from previous experiments for reference, showing values for the original eCP index. The two following lines, with KM1, represent one iteration of k-Means, assigning the entire 100m dataset using the representatives only or using the index. Comparing to all representatives is more precise but requires more than 4 trillion distance calculations, which took about 50 h. In comparison, using the index to speed up the k-Means assignments is more then 10x faster, taking only about 50 min. We observe that while considering all representatives yields better results, a multi-step clustering process is infeasible with that approach even at this moderate scale.

The final two lines show the results from running 2 and 10 steps of the k-Means process, respectively. We observe that while both AR and OR improve slightly, recall eventually decreases, meaning that the positive impact of k-Means is moderate and RF remains a major issue of the eCP index.

4 Conclusions

In this paper we have investigated ANN search using a hierarchical vectorial quantizer. We have shown that we can evaluate the quality of a) the segment representatives, b) the index hierarchy and c) the ANN search for both successful (recall) and failed queries. This we can do using only the index structure and the evaluation *ground-truth* data. Using algorithms such as eCP, that build their index without any actual segmenting/clustering of the full dataset, means we can perform index evaluation with minimal effort.

When we observe poor ANN search results, such as 58% recall, the inclination is to blame the indexing hierarchy, especially when it is as simple and naive as that of the eCP algorithm. By measuring the representation failure, however, we could put the index assignments into perspective. As it turns out, the eCP index itself is only responsible for a fraction of the failed queries.

We also evaluated whether k-Means clustering could alleviate the representation issues but the results show that clustering at large scale a) is prohibitively expensive, even when using the index to speed it up 10-fold, and b) only of a marginal benefit.

What makes multi-layer similarity graph algorithms, such as HNSW [5], obtain such high-quality results is that they address the representation problem by being highly selective when picking representatives at each layer in their indexing/graph structure and by not limiting the possible search/scan to a fixed region once hierarchy is traversed. However, they pay for this ability with the added complexity, footprint, and high construction time.

References

1. Amsaleg, L., Jégou, H.: BIGANN: abillion-sized evaluation dataset, corpus-texmex.irisa.fr. Accessed 2 June 2023
2. Gudmundsson, G.Þ., Jónsson, B.Þ., Amsaleg, L.: A large-scale performance study of cluster-based high-dimensional indexing. In: Proceedings of the international workshop on Very-Large-Scale Multimedia Corpus, Mining and Retrieval (VLS-MCMR), pp. 31–36 (2010)
3. Gudmundsson, G.Þ, Jónsson, B.Þ, Amsaleg, L., Franklin, M.J.: Prototyping a web-scale multimedia retrieval service using spark. ACM Trans. Multimed. Comput. Commun. Appl. (TOMM) 14(3s), 1–24 (2018)
4. Lowe, D.G.: Distinctive image features from scale-invariant keypoints. Int. J. Comput. Vision (IJCV) 60, 91–110 (2004)
5. Malkov, Y., Ponomarenko, A., Logvinov, A., Krylov, V.: Approximate nearest neighbor algorithm based on navigable small world graphs. Inf. Syst. 45, 61–68 (2014)
6. Matsui, Y., Uchida, Y., Jégou, H., Satoh, S.: A survey of product quantization. ITE Trans. Media Technol. Appl. (MTA) 6(1), 2–10 (2018)
7. Simhadri, H.V., et al.: Results of the NeurIPS 2021 challenge on billion-scale approximate nearest neighbor search. In: NeurIPS 2021 Competitions and Demonstrations Track, pp. 177–189. PMLR (2022)

Minwise-Independent Permutations
with Insertion and Deletion of Features

Rameshwar Pratap[1]([✉])([iD]) and Raghav Kulkarni[2]

[1] IIT Hyderabad, Telangana, India
rameshwar@cse.iith.ac.in
[2] Chennai Mathematical Institute, Chennai, Tamil Nadu, India
kulraghav@gmail.com

Abstract. The seminal work of Broder *et al.* [5] introduces the minHash algorithm that computes a low-dimensional sketch of high-dimensional binary data that closely approximates pairwise Jaccard similarity. Since its invention, minHash has been commonly used by practitioners in various big data applications. In many real-life scenarios, the data is dynamic and their feature sets evolve over time. We consider the case when features are dynamically inserted and deleted in the dataset. A naive solution to this problem is to repeatedly recompute minHash with respect to the updated dimension. However, this is an expensive task as it requires generating fresh random permutations. To the best of our knowledge, no systematic study of minHash is recorded in the context of dynamic insertion and deletion of features. In this work, we initiate this study and suggest algorithms that make the minHash sketches adaptable to the dynamic insertion and deletion of features. We show a rigorous theoretical analysis of our algorithms and complement it with supporting experiments on several real-world datasets. Empirically we observe a significant speed-up in the running time while simultaneously offering comparable performance with respect to running minHash from scratch. Our proposal is efficient, accurate, and easy to implement in practice.

Keywords: Sketching algorithms · Jaccard similarity estimation · Streaming algorithms · Locality sensitive hashing (LSH)

1 Introduction

Sets are one of the popular ways to embed data points, and their pairwise similarities are captured using Jaccard similarity. For a pair of sets $U, V \subseteq [d]$, their Jaccard similarity is defined as $|U \cap V|/|U \cup V|$. The seminal work of Broder *et al.* [5] suggests the minHash algorithm that computes a low-dimensional representation (or *sketch*) of the high-dimensional binary data that closely approximates the underlying pairwise Jaccard similarity. We discuss it as follows:[1]

---

[1] We note that binary vectors and sets give two equivalent representations of the same data object. Let the data elements be a subset of a fixed universe. In the

O. Pedreira and V. Estivill-Castro (Eds.): SISAP 2023, LNCS 14289, pp. 171–184, 2023.
https://doi.org/10.1007/978-3-031-46994-7_15

Definition 1 (Minwise Independent Permutations [5]). Let S_d be the set of all permutations on $[d]$. We say that $F \subseteq S_d$ (the symmetric group) is min-wise independent if for any set $U \subseteq [d]$ and any $u \in U$, when π is chosen at random in F, we have

$$\Pr[\min\{\pi(U)\} = \pi(u)] = 1/|U|. \tag{1}$$

For a permutation $\pi \in F$ chosen at random and a set $U \subseteq [d]$, Broder *et al.* [5] define minHash as follows $\text{minHash}_\pi(U) = \arg\min_u \pi(u)$ for $u \in U$. For a pair of points, $U, V \subseteq [d]$, and π is chosen at random in F, we have the following

$$\Pr[\text{minHash}_\pi(U) = \text{minHash}_\pi(V)] = |U \cap V|/|U \cup V|. \tag{2}$$

The above characteristic demonstrates the locality-sensitive nature (LSH) [13] of minHash, and as a consequence, it can be effectively used for the approximate nearest neighbour search problem. minHash is successfully applied in several real-life applications such as computing document similarity [3], item-set mining [2], faster de-duplication [4], all-pair similarity search [1], document clustering [6], building recommendation engine [11], near-duplicate image detection [9], web-crawling [12,18].

This work considers the scenario where features are dynamically inserted and/or deleted from the input. We emphasize that this natural setting may arise in many applications. Consider the *"Bag-of-Word" (BoW)* representation of text, where first, a dictionary is created using the important words present in the corpus such that each word present in the dictionary corresponds to a feature in the representation. Consequently, the embedding of each document is generated using this dictionary based on the frequency of the words present. Consider the downstream applications where the task is to compute pairwise Jaccard similarities between these documents, and the dimensionality of the *BoW* representation is high due to the large dictionary size. We can use minHash to compute the low-dimensional sketch of input documents. It is natural to assume that the dictionary is evolving; new words are inserted, and unused words are deleted. One evident approach to handle such a dynamic scenario is to run the minHash from scratch on the updated dictionary, which is expensive since it involves generating fresh min-wise independent (random) permutations. Note that during the insertion/deletion of features in the dataset, we consider inserting/deleting the same features in all the data points. To clarify this further, let $\mathcal{D} = \{X_i\}_{i=1}^n$ be our dataset, where $X_i \in \{0, 1\}^d$. Considering the addition/removal of the j-th feature, the j-th feature gets inserted/deleted in the point X_i. Similarly, the corresponding j-th feature is inserted/deleted in all the remaining points in $\mathcal{D}$. Note that we don't consider the case when data points are dynamically inserted or deleted in the dataset.

---

corresponding binary representation, we generate a vector whose dimension is the size of the universe, where for each possible element of the universe, a feature position is designated. To represent a set into a binary vector, we label each element's location with 1 if it is present in the set and 0 otherwise.

Problem statement: minHash for dynamic insertion and deletion of features: In this work, we focus on making minHash adaptable to dynamic feature insertions and deletions of features. We note that the insertion/deletion of features dynamically leads to the expansion/shrinkage of the data dimension.

We note that in practice a d dimensional permutation required for minHash is generated via the universal hash function $h_d(i) = ((ai + b) \mod p) \mod d$, where p is a large prime number and a, b are randomly sampled from $\{0, 1, \ldots p-1\}$; typically $((ai + b) \mod p) > d^2$. This hash function generates permutations via mapping each index $i \in [d]$ to another index $[d]$ that can be used to compute the minHash sketch. We note that in the case of dynamic insertions/deletion of features, even using universal hash functions to compute the minHash sketch doesn't give an efficient solution. We illustrate it as follows. Suppose we have a minHash sketch of data points using the hash function $h_d(.)$. Consider the case of feature insertion, where the dimension d increases to $d+1$, and therefore, we require a hash function $h_{d+1}(.)$ to generate a $(d+1)$-dimensional permutation. Note that the permutation generated via $h_{d+1}(.)$ can potentially be different on several values of $i \in [d+1]$. Therefore, just computing $h_{d+1}(d+1)$, taking the corresponding input feature, and taking the minimum of this quantity with the previous minHash would not suffice to compute minHash after feature insertion. If implemented naively, this re-computation step takes $O(d)$ in the worst case. A similar argument also holds in the case of feature deletion.

1.1 Our Contribution

In this work, we consider the problem of making minHash adaptable to dynamic insertions and deletions of features. We focus on cases where features are inserted/deleted at randomly chosen positions from 1 to d. We argue that this is a natural assumption that commonly occurs in practice. For example, in the context of BoW, a word's position in the dictionary is determined via a random hash function that randomly maps it to a position from 1 to d. Therefore, when a new word is added to the dictionary, its final position in the representation appears as a random position (from 1 to d). A similar argument is also applicable to feature deletion. We summarize our contributions as follows:

⋄ **Contribution 1:** We present algorithms (Sect. 2) that makes minHash sketch adaptable to single/multiple feature insertions. Our algorithm takes the current permutation and the corresponding minHash sketch; values and positions of the inserted features as input and outputs the minHash sketch corresponding to the updated dimension.

⋄ **Contribution 2:** We also suggest algorithms (deferred to the full version of the paper [21] due to the space limit, discussed in Section 4 of [21]) that makes minHash sketch adaptable for single/multiple feature deletions. It takes the data points, current sketch, and permutations used to generate the same; positions of the deleted features and outputs the minHash sketch corresponding to the updated dimension.

---

[2] These hash functions are called universal hash functions (see Chapter 11 of [10].

Our work leaves the possibility of some interesting open questions: to propose algorithms when features are inserted or deleted adversarially (rather than uniformly at random from 1 to d, as considered in this work). We hope that our techniques can be extended to handle this situation.

Our Techniques and Their Advantages: A major benefit of our results is that they do not require generating fresh random permutations corresponding to the updated dimension (after feature insertions/deletions) to compute the updated sketch. We implicitly generate a new permutation (required to compute the sketch after feature insertion/deletion) using the old d-dimensional permutation and also show that it satisfies the min-wise independence property (Definition 1). We further give simple and efficient update rules that take the value and position of inserted/deleted features, and output the updated minHash sketch. To show the correctness of our result, we prove that the sketch obtained via our update rule is the same as obtained via computing minHash from scratch using the implicitly generated permutation mentioned above. For both insertions and deletion cases, our algorithms give significant speedups in dimensionality reduction time while offering almost comparable accuracy with respect to running minHash from scratch. We validate this by running extensive experiments on several real-world datasets (Sect. 3 and Table 3). We want to emphasize that our algorithms can also be easily implemented when permutations are generated via random hash functions.

Applicability of our Result in Other Sketching Algorithms for Jaccard Similarity: We note that there are several improved variants of minHash are known such as one-permutation hashing [15, 22], b-bit minwise hashing [14, 16], oddsketch [20] that offer space/time efficient sketches. We would like to highlight that our algorithms can be easily adapt to these improved variants of minHash, in case of dynamic insertion and deletion of features. One permutation hashing divides the permuted columns evenly into k bins. For each data point, the sketch is computed by picking the smallest nonzero feature location in each bin. In the case of dynamic settings, our algorithms can be applied in the bin where features are getting inserted/deleted. Both b−bit minwise hashing [14] and oddsketch [20] are two-step sketching algorithms. In their first step, the minHash sketch of the data points is computed. In the second step of b-bit minwise hashing, the last b-bits (in the binary representation) of each minHash signature is computed. In contrast, in the second step of oddsketch, one bit of each minHash sketch is computed using their proposed hashing algorithm. As both results compute the minHash sketch in their first step, we can apply our algorithms to compute the minHash sketch in case of feature insertion/deletion. This will make their algorithms adaptable to dynamic feature insertions and deletions.

Recently, some hashing algorithms have been proposed that closely estimate the pairwise Jaccard similarity [7, 8, 19] without computing their minHash sketch. However, to the best of our knowledge, their dynamic versions (that can handle dynamic insertions/deletions of features) are unknown. Several improvements of the LSH algorithm [23] have been proposed that are adaptable to the dynamic/streaming framework. However, a significant difference is in the underlying problem statement. These results aim to handle dynamic insertion and deletions of data points, whereas we focus on dynamic insertions and deletions of the features (Table 1).

2   Algorithm for Feature Insertion

Table 1. Notations

Data dimension	d	Input data point $\{0,1\}^d$ or input set	X
Set $\{1,\ldots,d\}$	$[d]$	Data point after feature insertion $\{0,1\}^{d+1}$	X'
Position of the inserted feature	m	Original d-dim. permutation $(a_1,\ldots,a_d)$ s.t. $a_i \in [d]$	π
Value of the inserted feature	b	Lifted $(d+1)$-dim. permutation $(a'_1,\ldots a'_{d+1})$ s.t. $a'_i \in [d+1]$	π'_m
No. of $1's$ in X	$|X|$	Set of non-zero indices of X, i.e., $\{i|x_i = 1\}$	J
Size of the set J	$|J|$	minHash of X with π, i.e., minHash$_\pi(X)$	h_{old}

We first give our algorithm for a single feature insertion.

2.1   One Feature Insertion at a Time – liftHash

The liftHash (Algorithm 2) is our main algorithm for updating the sketch of data points consisting of binary features. It takes a d dimensional permutation π and the corresponding minHash sketch h_{old} π as input. In addition, it takes an index m and a bit value b, corresponding to the position and the value of the binary feature, to be inserted, respectively, and outputs updated hash value h_{new}. We show that h_{new} corresponds to a minHash sketch of the updated feature vector. To show this, we use liftPerm (Algorithm 1), which extends the original permutation π to a $(d+1)$ dimensional min-wise independent permutation. Note that the liftPerm algorithm is used solely for the proof and not required in the liftHash algorithm.

The main intuition of our algorithm is that we can (implicitly) generate a new $(d+1)$-dimensional permutation by reusing the old d-dimensional permutation (Algorithm 1), and can update the corresponding minHash w.r.t. the new $(d+1)$-dimensional permutation via a simple update rule (Algorithm 2). Consider a d dimensional input vector $X = (x_1, x_2, \ldots, x_d)$. A permutation π of $\{1, 2, \ldots, d\}$ can be thought of as imposing the following ordering on the indices of X: $\pi(1), \pi(2), \ldots, \pi(d)$. After feature insertion, we want the (implicit) liftPerm algorithm to generate a new permutation π' of $\{1, 2, \ldots, d+1\}$ that still maintains the ordering that was imposed by π. We show that such an extension is achievable with high probability assuming (i) feature insertion is happening at a random position and (ii) our binary feature vector is sparse. This helps us guarantee (with high probability) that π' is min-wise independent if π is min-wise independent (see Theorem 2). Finally, we show that the sketch obtained by the liftHash algorithm is the same one produced by applying the minHash with respect to the output π' of the liftPerm algorithm (see Theorem 3).

Algorithm 1: liftPerm(π, r).

1 **Input:** d-dim permutation π, a number r.
2 **Output:** $(d+1)$-dim. permutation π'.
3 **for** $i \in \{1, \ldots, d+1\}$ **do**
4   **if** $i \leq r$ **then**
5         $\pi'(i) = \pi(i)$
6   **else**
7         $\pi'(i) = \pi(i-1)$
8   **end**
9 **end**
10 **for** $i \in \{1, \ldots, d+1\}/\{r\}$ **do**
11   **if** $\pi'(i) \geq \pi'(r)$ **then**
12         $\pi'(i) = \pi'(i) + 1$
13   **end**
14 **end**
15 **return** π'

Algorithm 2: liftHash(π, m, b, h_{old}).

1 **Input:** $h_{old} := \text{minHash}_\pi(X)$, π, $m \in [d]$, $b \in \{0, 1\}$.
2 **Output:** $h_{new} := \text{liftHash}(\pi, m, b, h_{old})$.
3 Denote $a_m = \pi(m)$.
    /* m is the position of the inserted feature */
4 **if** $h_{old} < a_m$ **then**
5      $h_{new} = h_{old}$
6 **else**
7   **if** $b = 1$ **then**
8         $h_{new} = a_m$
9   **end**
10   **if** $b = 0$ **then**
11         $h_{new} = h_{old} + 1$
12   **end**
13 **end**
14 **return** h_{new}

We illustrate our algorithm with the following example and then state its proof of correctness.

Example 1 We illustrate our Algorithms using the following example. We assume that the index count starts with 1. Let $X = [1, 0, 0, 1, 0, 1, 0]$ be the data point, and $\pi = [6, 3, 1, 7, 2, 5, 4]$ be the original permutation. Then $\text{minHash}_\pi(X)$ is 5. Further, let us assume that we insert the value $b = 1$ at the index $m = 2$. Therefore $a_m = \pi(m) = 3$. The updated value $X' = [1, 1, 0, 0, 1, 0, 1, 0]$ and due to Algorithm 1 by setting $r = m = 2$, we obtain $\pi'_m = [7, 3, 4, 1, 8, 2, 6, 5]$. We calculate the value of h_{new} outputted by Algorithm 2: as $h_{old} = 5 > a_m = 3$ and $b = 1$, then we have $h_{new} = \text{liftHash}(\pi, m, b, h_{old}) = a_m = 3$. Further, $\text{minHash}_{\pi'_m}(X') = 3$. Therefore, we have $h_{new} = \text{minHash}_{\pi'_m}(X')$.

[3]The following theorem gives proof of the correctness of Algorithm 1, and shows that the permutation π' outputted by the algorithms satisfies the minwise independent property (Definition 1), with high probability. At a high-level proof of Theorem 2 relies on showing the bijection between the ordering on the indices of X by the original d-dimensional permutation π, and $(d+1)$-dimensional permutation π'. We show that this bijection holds with probability 1 when inserted feature value $b = 0$, and holds with a high probability when $b = 1$.

Theorem 2 *Let $\pi = (a_1 \ldots, a_d)$ be a minwise independent permutation, where $a_i \in [d]$, and r be a random number from $[d]$. Let π and r be the input to Algorithm 1. Then for any $X \in \{0,1\}^d$ with $|X| \leq k$, the permutation $\pi' = (a'_1 \ldots, a'_{d+1})$, where $a'_i \in [d+1]$, obtained from Algorithm 1 satisfies the condition stated in Equation (1) of Definition 1, with probability at least $1 - O(k/d)$.*

Theorem 3 gives proof of the correctness of Algorithm 2. We show that the sketch outputted by Algorithm 2 is the same as obtained by running minHash using the $(d+1)$-dimensional permutation obtained by Algorithm 1 on the updated data point after one feature insertion.

Theorem 3 *Let π'_m be the $(d+1)$-dimensional permutation outputted by Algorithm 1 by setting $r = m$. Then, the sketch obtained from Algorithm 2 is exactly the same as the sketch obtained with the permutation π'_m on X', that is, $h_{new} := \text{liftHash}(\pi, m, b, h_{old}) = \text{minHash}_{\pi'_m}(X')$.*

Remark 1 We remark that in order to compute the minHash sketch of X', Algorithm 2 requires only h_{old}, b, m, the value of $\pi(m)$. Whereas *vanilla* minHash requires a fresh $(d+1)$ dimensional permutation to compute the same.

Remark 2 We can extend our results for multiple feature insertion by repeatedly applying Theorem 2, and Theorem 3 along with the probability union bound. However, the time complexity of the algorithm obtained by sequentially inserting n features will grow linearly in n as observed in the empirical results (Fig. 1, Sect. 3). In the next subsection, we present an algorithm that performs multiple insertions in parallel, which helps us achieve much better speedups.

2.2 Algorithm for Multiple Feature Insertions – multipleLiftHash

Results presented in this subsection are extensions to that of Subsect. 2.1. The intuition of our proposal is that we can (implicitly) generate a new $(d+n)$-dimensional permutation (n is the number of inserted features), using the old

---

[3] We defer the proofs of Theorems 2, 3, 5, 6, to the full version of this paper [21] due to space limit.

Table 2. Notations

No. of inserted features	n	Position of inserted features $\{m_i\}_{i=1}^n$, $m_i \in$ $[d+1]$	M
X after n features insertion $\{0,1\}^{d+n}$	X'	Set of inserted bits $\{b_1, \ldots, b_n\}$ with $b_i \in$ $\{0,1\}$	B
multipleLiftHash(M, π, B, h_{old})	h_{new}	Lifted $(d+n)$-dim. permutation	π'_M

d-dimensional permutation. By exploiting the sparsity of input and the fact that inserted bits are random positions, we show that the updated permutation satisfies the min-wise independent property with high probability. Further, we suggest a simple update rule aggregating the existing minHash sketch and the minHash restricted to inserted position and outputs the updated sketch.

Algorithm 3: partialMinHash(π, M, B)

---

1 **Input:** Permutation π, a sorted set of indices $M = \{m_1, \ldots m_n\}$, and set of inserted bits $B = \{b_1, \ldots, b_n\}$
2 **Output:** The min value of π (with appropriate shift) restricted to only those indices m_i of M that correspond to non-zero b_i.
3 $\pi_{M,B} = \{\pi(m_i) \mid i \in \{1, \ldots, n\}$ and $b_i = 1\}$
4 **return** $\min\{\pi_{M,B}(k) + 1\}$

Algorithm 4: multipleLiftPerm(π, R).

---

1 **Input:** Permutation π; R with $|R| = n$.
2 **Output:** $(d+n)$-dim. permutation π'.
3 $R \leftarrow \text{sorted}(R)$ /* sorting array R in ascending order */
4 **for** $i \in \{1, 2, \ldots n\}$ **do**
5 | $R[i] = R[i] + i - 1$
6 **end**
7 $\pi' = \pi$ /* Initialization */
8 **for** $i \in \{1, \ldots n\}$ **do**
9 | $\pi' = \text{liftPerm}(\pi', R[i])$ /* Calling Algorithm 1 with $\pi = \pi'$ and $r = R[i]$ */
10 **end**
11 **return** π'

Algorithm 5: multipleLiftHash(M, π, B, h_{old}).

---

1 **Input:** $h_{old} := \text{minHash}_\pi(X)$, permutation π, M and B.
2 **Output:** $h_{new} := \text{multipleLiftHash}(M, \pi, B, h_{old})$.
3 Let $\pi_M := \{\pi(m) : m \in M\}$.
4 $a_M = \text{partialMinHash}(\pi, M, B)$
5 $h_{new} = \min(h_{old} + |\{x \mid x \in \pi_M \text{ and } x \leq h_{old}\}|, a_M)$ /* Picking the minimum between partialMinHash and shifted value of h_{old}. */
6 **return** h_{new}

Algorithm 5 takes h_{old}, M, B, and π as input, and outputs the updated sketch h_{new}. Algorithm 5 uses Algorithm 3 to obtain the value of partialMinHash – minimum π value restricted to the inserted indices only with inserted bit value 1, from which it obtains multipleLiftHash for the updated input. Algorithm 4 is implicit and is used to prove the correctness of Algorithm 5. Algorithm 4 takes the permutation π and M as input, and outputs a $(d + n)$-dimensional permutation π'_M which satisfies the condition stated in Equation (1) for X, with $|X| \le k$. We show this in Theorem 5. Then in Theorem 6, we show that $h_{new} = \text{minHash}_{\pi'_M}(X')$. As π'_M satisfies, the condition stated in Equation (1) for sparse X, then due to Equation (2) and [5] the sketch of data points obtained from Algorithm 5 approximates the Jaccard similarity.

Example 4 Suppose $X = [1, 0, 0, 1, 0, 1, 0]$ and $\pi = [6, 3, 1, 7, 2, 5, 4]$ are input point and original permutation, respectively. Then the value of h_{old} is 5. Let $M = [2, 4]$ and $B = [0, 1]$. Thus, in this case $\pi'_M = [7, 3, 4, 1, 8, 9, 2, 6, 5]$ and $X' = [1, 0, 0, 0, 1, 1, 0, 1, 0]$. Consequently we have, partialMinHash$(\pi, M, B) = 2 < h_{old} + |\{x \mid x \in \pi_M \text{ and } x \le h_{old}\}| = 5 + 1 = 6$. Therefore, minHash$_{\pi'_M}(X') = 2$.

We have the following theorems for the correctness of the algorithms presented in this subsection. A proof of the Theorem 5 follows similarly to the proof of Theorem 2 along with the probability union bound, and the proof of Theorem 6 is a generalization of proof of Theorem 3.

Theorem 5 *Let π be a minwise independent permutation. Let $M = \{m_1, \ldots, m_n\}$ such that m_i is chosen uniformly at random from $\{1, \ldots, d\}$. Then for any $X \in \{0, 1\}^d$ with $|X| \le k$, the permutation π'_M obtained from Algorithm 4 satisfies the condition stated in Equation (1) of Definition 1, with probability $1 - O(kn/d)$.*

Theorem 6 *Let π'_M be the $(d+n)$-dimensional permutation outputted by Algorithm 4, if we set $R = M$. Then, the sketch obtained from Algorithm 5 is exactly the same as the sketch obtained with the permutation π'_M on X', that is,* multipleLiftHash$(\pi, M, B, h_{old}) = \text{minHash}_{\pi'_M}(X')$.

Along similar lines, we give algorithms for single and multiple-feature deletions. Due to space limit, we defer it to Section 4 of the full version of this paper [21].

3   Experiments

Hardware Description: CPU model name: Intel(R) Xeon(R) CPU @ 2.20 GHz; RAM:12.72 GB; Model name: Google Colab.

Datasets and Baselines: We perform our experiments on *"Bag-of-Words"* representations of text documents [17]. We use the following datasets: NYTimes news articles (number of points = 300000, dimension = 102660), Enron emails (number of points = 39861, dimension= 28102), and KOS blog entries (number of points = 3430, dimension = 6960).

We consider the binary version of the data, where we focus on the presence/absence of a word in the document. For our experiments, we considered a random sample of 500 points from the NYTimes and 2000 points for Enron and KOS.

We compare the performance of our algorithms multipleLiftHash and multipleDropHash with respect to running minHash from scratch on the updated dimension, and we refer to it as vanilla minHash. We also note the performance of sequential versions of single feature insertion/deletion algorithms – liftHash and dropHash, respectively. We give implementation details of the baseline algorithms as the following link https://tinyurl.com/y98yh6k3.

Table 3. Speedup of our algorithms $w.r.t$ their vanilla minHash version

Experiment	Method	NYTimes		Enron		KOS	
		Max.	Avg.	Max.	Avg.	Max.	Avg.
Feature Insertions	multipleLiftHash	54.91×	51.96×	9.61×	9.17×	24.4×	23.11×
	liftHash	91.23×	87.38×	13.96×	12.66×	35.00×	35.50×
Feature Deletions	multipleDropHash	109.5×	105.31×	18.6×	17.01×	46.02×	43.94×
	dropHash	78.34×	72.79×	15.95×	14.89×	38.24×	35.71×

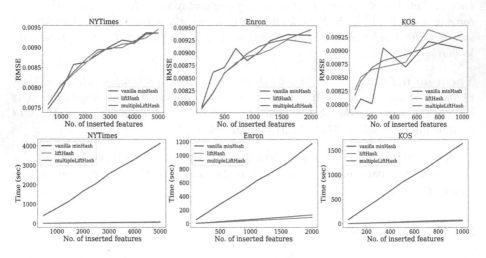

Fig. 1. Comparison among liftHash, multipleLiftHash, and vanilla minHash on the task of feature insertions. Vanilla minHash corresponds to computing minHash on the updated dimension. We iteratively run liftHash n times, where n is the number of inserted features

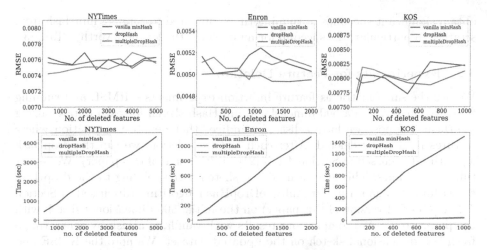

Fig. 2. Comparison among dropHash, multipleDropHash, and vanilla minHash on feature deletions. We iteratively run dropHash n times, where n is the number of deleted features

3.1   Experiments for Feature Insertions

We use two metrics for evaluation: a) RMSE: to examine the quality of the sketch, and b) running time: to measure the efficiency. We first create a 500 dimensional minHash sketch for each dataset using 500 independently generated permutations. Consider that we have a set of n random indices representing the locations where features need to be inserted. For each position, we insert the bit 1 with probability 0.1 and 0 with probability 0.9. We then run the liftHash algorithm (Algorithm 2) after each feature insertion; we repeat this step until n feature insertions are done. This gives a minHash sketch corresponding to the liftHash algorithm. We again run our multipleLiftHash algorithm (Algorithm 5) on the initial 500 dimensional sketch with the parameter n. We compare our methods with vanilla minHash by generating a 500 dimensional sketch corresponding to the updated datasets after feature insertions.

For computing the RMSE, our ground truth is the pairwise Jaccard similarity on the original full-dimensional data. We measure it by computing the square root of the mean (over all pairs of sketches) of the square of the difference between the pairwise ground truth similarity and the corresponding similarity estimated from the sketch. A lower RMSE is an indication of better performance. We compare the RMSE of our methods with that of vanilla minHash by generating a fresh 500 dimensional sketch. We summarise our results in Fig. 1.

Insights: Both of our algorithms offer comparable performance (under RMSE) with respect to running minHash from scratch on the updated dimension. That is, our estimate of the Jaccard similarity is as accurate as the one obtained by computing minHash from scratch on the updated dimension. Simultaneously, we obtain significant speedups in running time compared to running minHash from scratch. In particular, the speedup for multipleLiftHash is noteworthy (Table 3).

3.2 Experiments for Feature Deletion

We use the same metric as feature insertion experiments – RMSE and running time. We first create a 500 dimensional minHash sketch for each dataset using minHash. Suppose we have a list of n indices that denote the position where features need to be deleted. After each feature deletion, we run the dropHash algorithm (discussed in Section 4.1 of [21] - full version of this paper). We repeat this step n times. This gives a minHash sketch corresponding to the dropHash algorithm. We again run our multipleDropHash algorithm (discussed in Section 4.2 of [21] - full version of this paper) on the initial 500 dimensional sketch with the parameter n. We compare our results with vanilla minHash by generating a fresh 500 dimensional sketch on the updated dataset. We note the RMSE and running time as above. We summarise our results in Fig. 2.

Insights: Again, both our algorithms offer comparable performance (under RMSE) with respect to running minHash from scratch. Similar to the previous case, we obtained a significant speedup in running time *w.r.t.* computing minHash from scratch. In particular, the speedup obtained in multipleDropHash is quite prominent. We summarise a numerical speedup in Table 3.

Remark 3 Our current implementation of multipleLiftHash makes multiple passes over indices to be inserted, whereas multipleDropHash makes only one pass over the deleted indices. This is reflected in higher speedup values for multipleDropHash in Table 3. We believe an optimized implementation for multipleLiftHash would further improve the speedup.

4 Conclusion and Open Questions

We present algorithms that make minHash adaptable to dynamic feature insertions and deletions of features. Our proposals' advantage is that they do not require generating fresh permutations to compute the updated sketch. Our algorithms take the current permutation (or its representation using universal hash function [10]), minHash sketch, position, and the corresponding values of inserted/deleted features and output updated sketch. The running time of our algorithms remains linear in the number of inserted/deleted features. We comprehensively analyse our proposals and complement them with supporting experiments on several real-world datasets. Our algorithms are simple, efficient, and accurately estimate the underlying pairwise Jaccard similarity. Our work leaves the possibility of several interesting open questions: (i) extending our results for dense datasets in the case of feature insertions; (ii) extending our algorithms for the case when features are inserted/deleted adversely; (iii) improving our algorithms when we have prior information about the distribution of features; for example, features distribution follows *Zipf's law* etc.; (iv) improving theoretical guarantees and obtaining further speedups by optimizing our algorithms.

Acknowledgments. We sincerely thank Biswadeep Sen for providing their valuable input on the initial draft of the paper.

References

1. Bayardo, R.J., Ma, Y., Srikant, R.: Scaling up all pairs similarity search. In: Proceedings of the 16th International Conference on World Wide Web, WWW 2007, pp. 131–140. Association for Computing Machinery , New York, NY, USA (2007)
2. Bera, D., Pratap, R.: Frequent-itemset mining using locality-sensitive hashing. In: Dinh, T.N., Thai, M.T. (eds.) COCOON 2016. LNCS, vol. 9797, pp. 143–155. Springer, Cham (2016). https://doi.org/10.1007/978-3-319-42634-1_12
3. Broder, A.Z.: On the resemblance and containment of documents. In: . Proceedings of Compression and Complexity of Sequences 1997, pp. 21–29. IEEE (1997)
4. Broder, A.Z.: Identifying and filtering near-duplicate documents. In: Giancarlo, R., Sankoff, D. (eds.) CPM 2000. LNCS, vol. 1848, pp. 1–10. Springer, Heidelberg (2000). https://doi.org/10.1007/3-540-45123-4_1
5. Broder, A.Z., Charikar, M., Frieze, A.M., Mitzenmacher, M.: Min-wise independent permutations (extended abstract). In: Proceedings of the Thirtieth Annual ACM Symposium on Theory of Computing, STOC 1998, pp. 327–336. Association for Computing Machinery, New York, NY, USA (1998)
6. Broder, A.Z., Glassman, S.C., Nelson, C.G., Manasse, M.S., Zweig, G.G.: Method for clustering closely resembling data objects, September 12 2000. US Patent 6,119,124
7. Christiani, T., Pagh, R.: Set similarity search beyond minhash. In: Proceedings of the 49th Annual ACM SIGACT Symposium on Theory of Computing, STOC 2017, pp. 1094–1107. Association for Computing Machinery, New York, NY, USA, (2017)
8. Christiani, T., Pagh, R., Sivertsen, J.: Scalable and robust set similarity join. In: 34th IEEE International Conference on Data Engineering, ICDE 2018, Paris, France, April 16–19, 2018, pp. 1240–1243. IEEE Computer Society (2018)
9. Chum, O., Philbin, J., Zisserman, A.: Near duplicate image detection: min-hash and TF-IDF weighting. In: Everingham, M., Needham, C.J., Fraile, R. (Eds.), Proceedings of the British Machine Vision Conference 2008, Leeds, UK, September 2008, pp. 1–10. British Machine Vision Association (2008)
10. Cormen, T.H., Leiserson, C.E., Rivest, R.L., Stein, C.: Introduction to Algorithms, 3rd edn. MIT Press, Cambridge (2009)
11. Das, A.S., Datar, M., Garg, A., Rajaram, S.: Google news personalization: scalable online collaborative filtering. In WWW 2007: Proceedings of the 16th international conference on World Wide Web, pp. 271–280. ACM, New York, NY, USA (2007)
12. Henzinger, M.: Finding near-duplicate web pages: a large-scale evaluation of algorithms. In: Proceedings of the 29th Annual International ACM SIGIR Conference on Research and Development in Information Retrieval, SIGIR 2006, pp. 284–291. Association for Computing Machinery, New York, NY, USA (2006)
13. Indyk, P., Motwani, R.: Approximate nearest neighbors: towards removing the curse of dimensionality. In: Proceedings of the Thirtieth Annual ACM Symposium on the Theory of Computing, Dallas, Texas, USA, May 23–26, 1998, pp. 604–613 (1998)
14. Li, P., König, A.C.: Theory and applications of b-bit minwise hashing. Commun. ACM **54**(8), 101–109 (2011)

184   R. Pratap and R. Kulkarni

15. Li, P., Owen, A.B., Zhang, C.-H.: One permutation hashing. In: Bartlett, P.L., Pereira, F.C.N., Burges, Léon Bottou, C.J.C., Weinberger, K.Q., (Eds.), Advances in Neural Information Processing Systems 25: 26th Annual Conference on Neural Information Processing Systems 2012. Proceedings of a meeting held December 3–6, 2012, Lake Tahoe, Nevada, United States, pp. 3122–3130 (2012)
16. Li, P., Shrivastava, A., König, A.C.: B-bit minwise hashing in practice. In: Proceedings of the 5th Asia-Pacific Symposium on Internetware, Internetware 2013, New York, NY, USA. Association for Computing Machinery (2013)
17. Lichman, M.: UCI machine learning repository (2013)
18. Singh Manku, G., Jain, A., Sarma, A.D.: Detecting near-duplicates for web crawling. In: Proceedings of the 16th International Conference on World Wide Web, WWW 2007, pp. 141–150. Association for Computing Machinery, New York, NY, USA (2007)
19. McCauley, S., Mikkelsen, J.W., Pagh, R.: Set similarity search for skewed data. In Proceedings of the 37th ACM SIGMOD-SIGACT-SIGAI Symposium on Principles of Database Systems, SIGMOD/PODS 2018, pap.63–74, New York, NY, USA, 2018. Association for Computing Machinery (2018)
20. Mitzenmacher, M., Pagh, R. Pham, ,N.: Efficient estimation for high similarities using odd sketches. In: Proceedings of the 23rd International Conference on World Wide Web, WWW 2014, p–118. Association for Computing Machinery, New York, NY, USA, 2014
21. Pratap,R ., Kulkarni, R.: Minwise-independent permutations with insertion and deletion of features. arxiv.org/abs/2308.11240 (2023)
22. Shrivastava, A., Li, P.: Improved densification of one permutation hashing. In: Proceedings of the Thirtieth Conference On Uncertainty In Artificial Intelligence, UAI 2014, pp. 732–741. AUAI Press, Arlington, Virginia, USA, (2014)
23. Sundaram, N., et al.: Streaming similarity search over one billion tweets using parallel locality-sensitive hashing. Proc. VLDB Endow. 6(14), 1930–1941 (2013)

SDOclust: Clustering with Sparse Data Observers

Félix Iglesias[1]([✉]) [iD], Tanja Zseby[1] [iD], Alexander Hartl[1] [iD],
and Arthur Zimek[2]([✉]) [iD]

[1] TU Wien, Gusshausstraße 25/E389, 1040 Vienna, Austria
{felix.iglesias,tanja.zseby,alexander.hartl}@tuwien.ac.at
[2] SDU, Campusvej 55, 5230 Odense, DK, Denmark
zimek@imada.sdu.dk

Abstract. Sparse Data Observers (SDO) is an unsupervised learning
approach developed to cover the need for fast, highly interpretable and
intuitively parameterizable anomaly detection. We present SDOclust, an
extension that performs clustering while preserving the simplicity and
applicability of the original approach. In a nutshell, SDOclust considers *observers* as graph nodes and applies local thresholding to divide
the obtained graph into clusters; later on, observers' labels are propagated to data points following the *observation* principle. We tested SDO-
clust with multiple datasets for clustering evaluation by using no input
parameters (default or self-tuned) and nevertheless obtaining outstanding performances. SDOclust is a powerful option when statistical estimates are representative and feature spaces conform distance-based analysis. Its main characteristics are: lightweight, intuitive, self-adjusted, noise-resistant, able to extract non-convex clusters, and built on robust parameters and interpretable models. Feasibility and rapid integration into real-world applications are the core goals behind the design of SDOclust.

Keywords: clustering · graphs · unsupervised learning · anomalies

1 Introduction

Sparse Data Observers (SDO) is a recent algorithm for anomaly detection [15].
Although it is general purpose, it was originally conceived for network traffic
analysis, a field that demands the fast processing of high data volumes. Beyond
complexity requirements, core goals in SDO are: (a) robust and intuitive parameterization, (b) the use of explainable models, and (c) the capability to effectively identify "novelties" as anomalies, even when they are dense and collective.
SDOstream [12] is an extension of SDO for data streams. Since its publication
in 2018, SDO has been used in diverse applications; e.g., advanced multi-fault
diagnosis of batteries [25], anomaly detection in sensor networks [5].

In this work we propose SDOclust, which is the extension of SDO for clustering, therefore covering the two main branches of unsupervised learning. In the
related literature we find an extensive collection of methods for clustering, each
of them showing pros and cons, and being suitable for specific environments.

© The Author(s), under exclusive license to Springer Nature Switzerland AG 2023
O. Pedreira and V. Estivill-Castro (Eds.): SISAP 2023, LNCS 14289, pp. 185–199, 2023.
https://doi.org/10.1007/978-3-031-46994-7_16

Suffice it to mention the survey by Xu and Tian, in which up to 71 algorithms are recalled and compared [29]. It is well accepted in the field that there is no "best" algorithm, that they must be evaluated based on application and goals, and that finding better evaluation procedures is even more pressing [18]. However, regardless of the algorithm used, experts still run into concerns related to whether the resulting clustering is reliable and reflects the real structure of the data, or whether the parameters used in the configuration are adequate [26]. Further discussing practical common disadvantages, Böhm et al. state that "they all [clustering algorithms] suffer from one or more of the following drawbacks: they focus on spherical or Gaussian clusters, and/or they are sensitive to outliers, and/or they need user-defined thresholds and parameters" [2].

To a large extent, SDOclust overcomes these common issues, fact that underlines its relevance among well-established methods. In another popular survey about clustering [30], Xu and Wunsch emphasize nine characteristics that are desirable in new generation algorithms. We analyze SDOclust in this light:

1. *Arbitrary shapes.* SDOclust is not confined to particular shapes and is able to capture non-convex and even nested clusters.
2. *Large volumes and high-dimensionality.* Based on statistical sampling, SDOclust benefits the larger the data volume both in terms of accuracy and complexity. On the other hand, SDOclust is a distance-based method that operates directly on the input feature space, so it is affected by the curse of dimensionality similarly to equivalent approaches. That said, empirical tests show excellent results up to 1024 dimensions (Sect. 4).
3. *Outlier/noise detection and removal.* SDOclust nests SDO, therefore inherently generating outlierness scores. On the other hand, clustering formation is rarely affected by outliers/noise in SDOclust. Finally, SDOclust does not remove or set outliers in a binary way, leaving the "oulier thresholding" task to be externally tackled according to application requirements.
4. *Low dependency on parameters.* SDOclust solves most scenarios with default parameters, which are also robust, intuitive, and self-adjustable. Challenging cases might require fine parameterization.
5. *Upgradeable models.* SDOclust can process data in chunks, meaning that it updates models with new data without retraining from scratch.
6. *Immune to data-order.* For SDOclust it makes no difference whether patterns are entered sequentially or jumbled.
7. *Guessing the number of clusters.* SDOclust does not require the number of expected clusters as a parameter externally imputed.
8. *Enriched outputs.* In addition to cluster labels, SDOclust outputs outlierness scores and cluster memberships per point, which can be easily converted into purity estimations[1]. Additionally, SDOclust generates low-density models of the data shape by means of *observers*, which are representatives that preserve data geometry and relative density. Overall, outputs in SDOclust provide comprehensive information for visualization, description and post-analysis.

---

[1] We term a data point as *impure* when it lies in unclear zones between clusters.

9. *Mixed data types*. SDOclust is a numerical method and does not natively work on categorical data. In the current implementation, mixed data types require being adapted during preprocessing.

Note that SDOclust adjusts itself under the assumption that data statistical estimates are representative. Also clusters that are overlapping or show very strong differences in density might be difficult to solve and require fine parameterization. For the evaluation, we test SDOclust on a large number of datasets with a wide variety of shapes and challenges. Experiments are divided into: *two-dimensional data*, to provide a visual assessment, and *multi-dimensional data*, to cover more practical clustering application environments. In both cases internal and external validation metrics are shown. As competitors we use HDBSCAN [3] and k-means-- [4]. HDBSCAN is one of the most notable general-purpose clustering algorithm, as it also meets the three key requirements: (a) operating with default parameters in a wide assortment of cases, (b) detecting outliers, and (c) capturing non-convex patterns. k-means-- is selected as it is a most traditional clustering approach in its implementation with outlier detection.

The rest of the paper is organized as follows: Sect. 2 delves into SDO/SDOclust algorithms and parameters. In Sect. 3 we present evaluation data and metrics. Results are shown and discussed in Sect. 4, and two real application examples are explored in Sect. 5. We conclude with main remarks in Sect. 6.

2   Clustering Based on Sparse Data Observers

SDOclust is based on SDO. In this section we briefly explain the basic principles of SDO and then expand in more detail on SDOclust. Configurable parameters are highlighted in **bold** to be later explained in Sect. 2.3, where we discuss default values, automatic adjustment and the implications of their variation.

2.1   SDO

SDO [15] consists of two phases, here referred to as LEARNING and PREDICTION. During the LEARNING phase, the algorithm performs the following steps:

L1 *Sampling*. S being a set of m input data points, O is a subset of k_0 random data points from S ($k_0 \ll m$). Elements in O are called *observers*.

L2 *Observation*. With D as the distance² matrix between each pair of data points of S and O, an observation matrix I is derived by storing the x-closest observers to each data point in S. Intuitively, each data point is "observed" only by its x-closest observers.

L3 *Removal of idle observers*. From the observation matrix I, we can compute P, an array that contains the occurrences of each observer in I, or, in other words, the number of data points that each observer "observes". If an

---

² With *distance*—or the $d(.)$ function—we refer to *Euclidean* distance, but the method is not restricted to it.

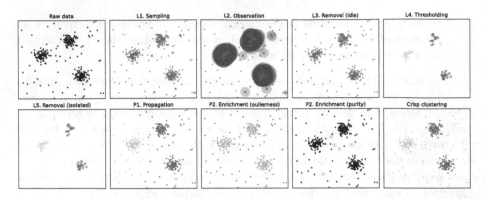

Fig. 1. Description of LEARNING and PREDICTION. L1 samples input data to create the observers set (red 'x's). Observers "observe" input samples in L2 (shadow size represents observations). *Idle* observers are removed in L3. Remaining observers are connected in clusters in L4 by considering cut-off thresholds. In L5 *isolated* observers are removed. Observers' labels are propagated to data points in P1. Outlierness, membership and purity estimations are bound to data points in P2 (*impure* data points cast a shadow). The last plot (right-bottom) shows the derived crisp clustering with outlier detection (outliers in gray color). (Color figure online)

observer does not observe at least q data points, it is considered *idle* and removed from O. This minimizes the chance of selecting outliers as observers. Thus, the final O is a low-density *model* of S formed by k *active* observers ($k < k_0$).

During the PREDICTION phase:

P1 *Observation*. Data points in S, or new objects from a consistent dataset (we keep calling them S for consistency), are evaluated by O. This generates new D and I matrices (but note that now O only contains active observers).

P2 *Outlierness estimation*. For each data point s_i in S, an outlier score y_i is calculated as the average distance from s_i to its x-closest observers.

2.2   SDOclust

SDOclust also implements LEARNING and PREDICTION (Fig. 1). Additionally, it includes an UPDATE phase that is only called in batch mode, i.e., SDOclust processes new data while keeping and updating an already trained model.

The LEARNING phase of SDOclust follows these steps:

L1, L2, L3 *Sampling*, *Observation* and *Removal of idle observers* are exactly the same as in SDO (Sect. 2.1).

   L4 *Thresholding for graph-edge cutting*. Here, observers in O are seen as nodes of an undirected graph to be clustered. We need to create and adjacency matrix A that will ultimately group or separate observers.

For this, instead of using a unique external cutoff threshold, each observer estimates its own locally. Therefore, the cutoff threshold h_i of a given observer o_i is:

$$h_i = d(o_i, o_{i \leftarrow \chi}) \tag{1}$$

i.e., the distance to its χth-closest observer. Thus, the element $a_{i,j}$ of A is:

$$a_{i,j} = \begin{cases} 1 & \text{if } d(o_i, o_j) < h_i \text{ and } d(o_i, o_j) < h_j \\ 0 & \text{otherwise} \end{cases} \tag{2}$$

meaning that observers o_i and o_j will be connected if the distance between them is below h_i and h_j. The purpose of using per-node (or local) cutoff thresholds is to allow solutions with clusters of different densities. On the other hand, this is prone to connect clusters lying very close to each other and create clusters in noise. To compensate this, a ζ coefficient weights the local contribution of h_i with a global contribution $\overline{h}$ obtained as the average of $\{h_1, h_2, ..., h_k\}$. Therefore, the final h_i' remains:

$$h_i' = \zeta h_i + (1 - \zeta)\overline{h} \tag{3}$$

and A is constructed with $\{h_1', h_2', ..., h_k'\}$ instead. From each isolated subgraph of connected observers in A we extract an unique label c_i (any given observer is at least connected to itself), generating the set of unique labels $C = \{c_1, c_2, ..., c_z\}$. Since connected observers share the same label of C, this results in the array of labels $L = \{l_1, l_2, ..., l_k\}$, where $l_i \in C$.

L5 *Removal of isolated observers*: L4 might generate clusters with isolated observers (commonly located in noisy areas). To obtain clean clusters, observers forming subgraphs of less than e nodes are removed (from O, P and L).

During the PREDICTION phase:

P1 *Propagation*. This step runs an *Observation* phase and additionally propagates observers' labels to data points. Therefore, each data point in S, or new data points from a consistent dataset (we continue calling them S), inherits the labels of its x-closests observers in addition to the outlierness score. Unlike with the outlierness score, which is computed as an average distance, observers' labels are categories, making that each data point in S can be multi-labeled. Therefore, if we imagine a space where discovered observers' clusters are orthogonal coordinates $\hat{c}_1, \hat{c}_2, ..., \hat{c}_z$, a data point s_i will obtain a vector label $\overrightarrow{l_i} = (l_{i,1}, l_{i,2}, ..., l_{i,z})$, where $l_{i,j}$ is the sum of the x-closest observers to s_i that belong to cluster c_j.

P2 *Enrichment.* This phase acknowledges that any point s_i obtains an outlierness score y_i and a label vector $\vec{l_i}$. The label vector can be turned into a crisp label: $l_i|_{\text{crisp}}$ marking the *dominant* coordinate[3] of $\vec{l_i}$, into a *membership* vector: $\vec{l_i}|_{\text{memb}} = (\frac{l_{i,1}}{||l_i||_1}, \frac{l_{i,2}}{||l_i||_1}, ..., \frac{l_{i,z}}{||l_i||_1})$, or into a *purity* estimation: $l_i|_{\text{pur}} = ||\vec{l_i}|_{\text{memb}}||_\infty$.

Finally, in the UPDATING phase both LEARNING and PREDICTION are repeated with minor variations. The new subset of data points is sampled for observers while keeping the proportion of one observer per m/k data points in the batch (minimum 1). New observers are added to O, new observations are added to P and, finally, the less-active observers are removed to keep k invariant.

2.3   Parameters: Interpretation, Default Values and Self-tuning

k_0, x and q (or ρ)[4] are intrinsic to SDO and widely discussed in the original paper [15]. We comment on them briefly. If not given externally, the initial number of observers—here termed k_0—is set based on statistical sampling of finite populations. For this estimation, the most variant dimension of the input data after PCA (Principal Component Analysis) is used (Eq. 4):

$$k_0 = \frac{mZ^2\sigma^2}{(m-1)\epsilon^2 + Z^2\sigma^2} \qquad (4)$$

with σ as the variance and CI = 95% (Z = 1.96), error $\epsilon = 0.1\sigma$. This ensures good scalability regardless of the number of data points and dimensions of the input data. For instance, in a n-dimensional dataset S with one thousand data points generated at random with normal distributions ($\mu = \sigma = 1$), SDO/SDOclust ($x = 5$, $\rho = 0.3$) estimates $k \simeq 191$; whereas for the same dataset with one million data points $k \simeq 266$. On the other hand, x expresses neighborhood similarly to k in the k-nearest neighbor algorithm, but without being sensitive to density [31]. Empirical tests confirm the robustness of x and ρ, with $x = 5$ and $\rho = 0.3$ as suitable default values.

Parameters exclusive of SDOclust are:

- χ defines the χth-closest observer of any given observer. It is used to establish the local threshold for cutting-off graph edges (Eq. 1). Small χ is appropriate when solutions with many clusters are expected, whereas high χ should be used when only few clusters are foreseen.
- ζ sets a trade-off between locality and globality in thresholds for cutting-off graph edges, $\zeta = 1$ for purely local and $\zeta = 0$ for purely global (Eq. 3). Local thresholding allows clusters of considerable density difference; however, it tends to merge nearby clusters and form clusters in noise. Global thresholding avoids such merger, but might divide legit clusters.

---

[3] i.e., the coordinate with the highest value. In the absence of a dominant coordinate, the algorithm forces it randomly among the highest candidates.

[4] q is commonly obtained as $q = Q(\rho, P)$, where $Q(.)$ is the *quantile* function.

– e sets the minimum number of observers that a cluster can have. This parameter commonly takes small values and is used to avoid clustering noise.

Empirical tests show that the previous parameters tolerate variations around default values: $\chi = \min(8, \frac{k}{20})$, $\zeta = 0.6$ and $e = 3$. The reason why such parameters work properly in most data scenarios is because, regardless of the number of samples and dimensions, SDOclust summarizes data in a model composed by a number of observers that falls always in a similar order of magnitude.

Finally, since the selection of observers is natively performed at random, SDOclust is not free from stochastic problems (more noticeable in small datasets).

2.4 Complexity

The increase in complexity of SDOclust with respect to SDO is not significant, since the additional operations carried out by SDOclust happen on the observer set which, as seen in Sect. 2.3, if not imposed as an external parameter, converges asymptotically as a function of m. Therefore, the added algorithmic load of SDOclust happens in the thresholding and clustering of the graph during the LEARNING phase. SDOclust works at all times with matrices of size $m \times k$, $k \times k$, $k \times x$, and does not involve iterative processes. Therefore, as in [15], in Big-O notation the complexity of SDOclust is described as $O(mk)|_{m\to\infty} = O(m)$.

3 Evaluation

Implementations of SDOclust, experimental tests and sensitivity analysis for main parameters are freely available in our repository[5], and through a stable DOI-citable version for Reproducible Research in [16]. Datasets can be downloaded from their original sources or—in case that they are further processed or generated by tools—they are also included in our repository. In all experiments, HDBSCAN (as provided in [20]) and SDOclust run with no parameters (default or self-tuned)[6], whereas k-means-- is always imputed with the right number of clusters and outliers to discover[7], extracted from the ground truth (GT).

---

[5] https://github.com/CN-TU/pysdoclust.

[6] HDBSCAN   parameters: $min\_cluster\_size = 5$, $cluster\_selection\_epsilon = 0.0$, $approx\_min\_span\_tree = \text{True}$, $allow\_single\_cluster = \text{False}$, $min\_samples = \text{None}$, $algorithm = \text{'best'}$, $p = \text{None}$, $alpha = 1.0$, $metric = \text{'euclidean'}$, $leaf\_size = 40$, $memory = \text{Memory(location = None)}$, $cluster\_selection\_method = \text{'eom'}$, $gen\_$ $min\_span\_tree = \text{False}$, $core\_dist\_n\_jobs = 4$, $prediction\_data = \text{False}$, $match\_$ ref-$erence\_implementation = \text{False}$;
SDOclust parameters: $x = 5$, $qv = 0.3$, $zeta = 0.6$, $chi\_min = 8$, $chi\_prop = 0.05$, $e = 3$, $chi = \text{None}$, $xc = \text{None}$, $k = \text{None}$, $q = \text{None}$.

[7] k-means-- is tuned with $maximum\_iterations = 1000$ and $tol = 0.0001$, where tol is the convergence criterion for centroid displacement.

3.1   Datasets

Evaluation experiments are in two sets: (a) 15 *two-dimensional datasets* to visually assess performances; (b) 138 *multi-dimensional datasets*. Dataset collections are taken from different sources: (i) The Clustering Basic Benchmark of the University of Eastern Finland [10], which is one of the most known data collections for clustering evaluation[8]. This includes two-dimensional datasets: *s1* [8], *r15* [27], *aggregation* [11], *skewed* [22]; also the low-to-high dimensional set termed *d*, which combines two collections [9,17] and shows datasets from 3 to 1024 dimensions. (ii) Datasets generated or processed with the sklearn.datasets package[9], due to its popularity and widespread use [21], including the two-dimensional datasets named: *rings* and *moons*. Later on, *iris* (Iris Flower), *mallcust* (Mall Customers Segmentation) and *pima* (Pima Indians Diabetes) are popular real datasets here projected into two-dimensional spaces by using t-distributed stochastic neighbor embedding (tSNE) [19]. (iii) Datasets generated with MDCgen [14], a tool to create multi-dimensional data for testing, evaluating, and benchmarking unsupervised classification algorithms. It includes the two-dimensional datasets: *close, separated, complex, high-noise* and *low-noise*; and 6 groups of multi-dimensional datasets, each group addressing a different data challenge, namely: *c* (close clusters, reduced space), *p* (separated clusters, large space), *n* (between 5% and 15% noise aprox.), *h* (between 15% and 30% noise aprox.), *f* (clusters with high density differences), and *x* (complex setups by combining previous challenges).

3.2   Metrics

Experiments are validated externally and internally. For external validation we use the Adjusted Rand Index (ARI) [13], which approaches '1' the better the discovered partition matches the GT, and gives '-0.5' for completely discordant partitions. For internal validation we use the Silhouette index (Sil) [23], which scores '1' when intra-cluster cohesion and inter-cluster separation are maximized. To make Sil consistent, we first remove the top 'n' outliers, where 'n' is given by the GT. We also show the difference between the number of clusters discovered by the algorithm by excess (+) or deficiency (−) compared with the GT.

4   Results and Discussion

Table 1 shows two-dimensional experiments results. SDOclust guesses the number of clusters considerably better than HDBSCAN, which tends to cluster noise and divide bigger structures into micro-clusters. SDOclust average performances are 0.59 ± 0.22 (Sil) and 0.86 ± 0.10 (ARI), vs 0.53 ± 0.27 (Sil) and 0.81 ± 0.19 (ARI) in HDBSCAN and 0.57 ± 0.10 (Sil) and 0.67 ± 0.23 (ARI) in *k*-means--.

---

[8] https://cs.joensuu.fi/sipu/datasets/.
[9] https://scikit-learn.org/stable/datasets.html.

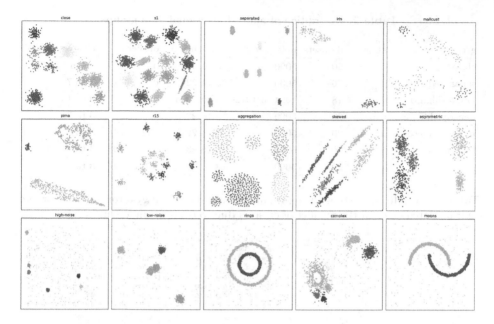

Fig. 2. Clustering of SDOclust on two-dimensional data. Clustered data points take colors based on their cluster, whereas outliers are in gray color (Color figure online)

Table 1. Results of tests with two-dimensional datasets

dataset	inlier samples	dimensions	outliers	clusters GT	clusters HDBSCAN	clusters k-means-	clusters SDOclust	Sil HDBSCAN	Sil k-means-	Sil SDOclust	ARI HDBSCAN	ARI k-means-	ARI SDOclust
close	2000	2	0	11	=1	=	9	0.60	0.50	0.65	0.87	0.81	0.87
s1	5000	2	0	15	+38	=	=	0.24	0.58	0.71	0.54	0.83	1.00
separated	2000	2	0	7	=	=	=	0.91	0.58	0.91	1.00	0.72	1.00
iris	150	2	0	3	=1	=	=	0.89	0.72	0.72	0.57	0.90	0.90
mallcust	200	2	0	4	=	=	+1	0.57	0.62	0.52	0.96	0.89	0.85
pima	768	2	0	3	=	=	+1	0.60	0.61	0.40	1.00	0.72	0.82
r15	600	2	0	15	=	=	-1	0.72	0.67	0.70	0.94	0.90	0.92
aggregation	788	2	0	7	-2	=	-1	0.37	0.44	0.46	0.81	0.63	0.91
skewed	1000	2	0	6	+5	=	-1	0.25	0.44	0.37	0.89	0.60	0.81
asymmetric	1000	2	0	5	-2	=	=	0.68	0.53	0.64	0.47	0.70	0.98
high-noise	2000	2	345	6	+10	=	+1	0.74	0.72	0.90	0.92	0.83	0.83
low-noise	2000	2	111	5	-1	=	-1	0.82	0.68	0.85	0.70	0.61	0.67
rings	3000	2	200	2	+4	=	=	0.08	0.38	0.12	0.97	0.07	0.88
complex	2000	2	155	6	+17	=	-1	0.38	0.51	0.56	0.61	0.62	0.65
moons	3000	2	200	2	+6	=	=	0.13	0.49	0.34	0.96	0.29	0.88

Both internal and external validation highlights SDOclust over its competitors. Figure 2 shows SDOclust clustering. Most noticeable inaccuracies are the tendency of SDOclust to merge very close clusters, more so if they show some overlap or are bridged by intermediate data points. There are different strategies to alleviate this, e.g., reducing ζ and χ while increasing ρ, but such modifications might cause arbitrary divisions in some cases.

Table 2 shows main properties of multi-dimensional dataset collections, while Table 3 provides clustering performances. Again, HDBSCAN and SDOclust run with default parameters, whereas k-means-- is imputed with the expected number of clusters and outliers. Summarizing results, we observe that discrepancies in the number of clusters between GT and SDOclust partitions is minimal, while it is larger for HDBSCAN, prone to overclustering in the presence of outliers. Even

Table 2. Description of multi-dimensional data groups

group	abbrev.	datasets	inlier samples	dimensions	outliers	clusters in GT
close	c	20	5000	3 to 23	0	10 to 14
separated	p	20	5000	5 to 22	0	3 to 7
low-to-high dim.	d	18	1024 to 10126	3 to 1024	0	9 to 16
density diff.	f	20	5000	3 to 22	0	3 to 7
medium noise	n	20	5000	3 to 22	252 to 748	3 to 7
high noise	h	20	5000	3 to 23	858 to 1981	4 to 7
complex	x	20	5000	3 to 22	272 to 747	3 to 7

Table 3. Results of tests with multi-dimensional data. "NCA-error" stands for the "accumulated error in the number of clusters due to over- or underclustering"

group	NCA-error HDBSCAN	NCA-error k-means-	NCA-error SDOclust	Sil HDBSCAN	Sil k-means-	Sil SDOclust	ARI HDBSCAN	ARI k-means-	ARI SDOclust
c	1	35	2	0.83 ± 0.05	0.53 ± 0.10	0.83 ± 0.04	1.00 ± 0.00	0.66 ± 0.39	1.00 ± 0.00
p	0	0	0	0.95 ± 0.01	0.70 ± 0.19	0.95 ± 0.01	1.00 ± 0.00	0.82 ± 0.16	1.00 ± 0.00
d	0	0	0	0.94 ± 0.03	0.64 ± 0.17	0.94 ± 0.03	1.00 ± 0.00	0.77 ± 0.14	1.00 ± 0.00
f	8	0	9	0.70 ± 0.18	0.64 ± 0.14	0.71 ± 0.12	0.95 ± 0.09	0.89 ± 0.11	0.97 ± 0.03
n	67	11	0	0.85 ± 0.04	0.81 ± 0.30	0.97 ± 0.01	0.83 ± 0.06	0.85 ± 0.29	1.00 ± 0.00
h	203	5	0	0.64 ± 0.09	0.81 ± 0.22	0.99 ± 0.00	0.48 ± 0.11	0.80 ± 0.18	1.00 ± 0.00
x	43	0	2	0.74 ± 0.06	0.73 ± 0.16	0.84 ± 0.09	0.82 ± 0.07	0.84 ± 0.17	0.98 ± 0.02

k-means-- shows higher drifts due to global failures when analyzing some scenarios. As for validation indices, SDOclust scores 0.89 ± 0.11 (Sil) and 0.99 ± 0.02 (ARI) in average, vs 0.80 ± 0.14 (Sil) and 0.87 ± 0.19 (ARI) in HDBSCAN and 0.70 ± 0.22 (Sil) and 0.81 ± 0.21 (ARI) in k-means--.

In Fig. 3 we show *critical difference diagrams* from Wilcoxon signed-rank tests over all 153 experiments [7]. The diagrams confirm the statistical differences when comparing results, SDOclust standing out with the best performance.

Fig. 3. Critical difference diagrams comparing algorithm results with Wilcoxon signed-rank tests [7]. Best methods are placed on the right side

The reason behind SDOclust ability to self-tune and obtain suitable clustering lies in leveraging statistical estimations to create simplified models—always with a number of elements (i.e., observers) in a similar order of magnitude—that capture shapes and relative densities independently of the number of input data points and dimensions. This makes possible to calculate and establish neighborhoods, coefficients and thresholds that satisfy a large number of cases.

5   Examples with Real Data

We show two examples of SDOclust for discovering clusters and patterns in real applications. We have selected cases with recent public data and where clustering makes practical sense beyond algorithm testing. We use HDBSCAN as benchmark. As before, both algorithms apply default parameters and are not imputed with the number of clusters to discover. SDOclust establishes outliers per cluster as any data point with an outlierness score beyond the mean plus the Median Absolute Deviation (MAD) over the standard deviation [1].

5.1   Clustering Network Traffic

Communication network traffic is hard to analyze: data spaces are noisy, classes heterogeneous and multiform, and features commonly show non-normal distributions and collinearity. We explore a traffic capture (pcap) from the MAWI samplepoint-F for July 31, 2022[10]. The MAWI Working Group daily publishes 15 min of backbone traffic publicly for research purposes [6]. We extract bidirectional flows between IP addresses (srcIP, dstIP) that exchange information for a maximal duration of 60 s. Since modern network traffic nowadays is mostly encrypted, we select features that are available regardless of encryption[11]: the flow duration (dur) and statistics related the number of packets sent (pkt), the length of packets (len) and the inter-arrival time between packets (iat), both forward (F) and backward (B). The final vector that expresses a sample is:

T:srcIP:dstIP → dur, pktF, Md(lenF), Mn(iatF), pktB, Md(lenB), Mn(iatB)

where T is the *timestamp* and, together with the srcIP and dstIP, forms a unique ID to identify flows. Md() and Mn() stand for the statistical *Mode* and *Mean*. The capture contains about 9 million flows in the given format, of which we randomly sample a thousandth part (8630 flows) for our exploration and normalize them with quantile normalization (the most consistent based on feature distributions). Domain knowledge suggests that higher layer protocols, yet heterogeneously, may influence traffic shapes. We use this information as a tentative benchmark for external validation (i.e., ARI). Analyzed flows account for: 82.5% TCP, 15.1% UDP, 2.1% ICMP, and 0.3% of others protocols.

Results reveal very different performances in terms of the number of clusters between HDBSCAN and SDOclust, although similar in validation scores:

HDBSCAN: 117 clusters, 322 outliers, Sil = 0.89, Sil(inliers) = 0.96, ARI = 0.32
SDOclust: 8 clusters,   53 outliers, Sil = 0.89, Sil(inliers) = 0.89, ARI = 0.31

In Fig. 4 we use tSNE to visualize solutions given by HDBSCAN, SDOclust and protocol distribution. tSNE has proven excellent at capturing the structure of high dimensional data, it performs projections based on local similarity that are consistent with cluster shapes. From this perspective, SDOclust appears to align more naturally with tSNE and the protocol labeling and number of clusters than HDBSCAN. Nevertheless, by removing outliers and due to the high number of clusters found, HDBSCAN shows a remarkable improvement in the internal validation, this indicating its ability to find micro-clusters of high purity.

---

[10] https://mawi.wide.ad.jp/mawi/samplepoint-F/2022/202207311400.html.
[11] Extracted with Go-flows [28].

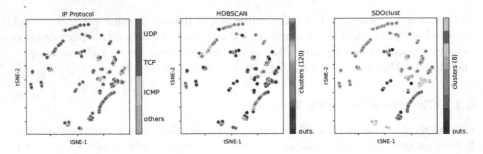

Fig. 4. tSNE proj. (colors are protocol), HDBSCAN and SDOclust clustering. (Color figure online)

5.2   Energy Building Profiles

This second example applies clustering to reveal characteristic profiles in building energy performance. We use real electricity consumption data of the "CN Rectorat" of the Polytechnic University of Catalonia (UPC) during 2022. This data is openly published by the Sirena Project [24][12]. We rearrange hourly consumption data to process 365 vectors (days) of 24 features (hours). Note that each vector is a time series, i.e., a daily energy consumption curve. Since raw data is magnitude consistent and in order not to break dependencies between features, we do not apply any normalization. Euclidean distances estimate similarity as usual since we want to group samples considering both curve shape and volume of electricity. SDOclust and HDBSCAN obtain the following results:

$$\text{HDBSCAN:} \quad 8 \text{ clusters}, \quad 77 \text{ outliers}, \quad \text{Sil} = 0.28, \quad \text{Sil(inliers)} = 0.48$$
$$\text{SDOclust:} \quad 7 \text{ clusters}, \quad 16 \text{ outliers}, \quad \text{Sil} = 0.43, \quad \text{Sil(inliers)} = 0.48$$

Again, differences in validation between SDOclust and HDBSCAN are small, this time also with regard to discovered clusters (both in number and shapes, Fig. 5). The similarity between both clusterings is ARI = 0.76. The deterioration of HDBSCAN in the complete internal assessment (inliers and outliers) is due to its tendency to find many outliers (21% in this case). As for the application, it is tempting to think that discovered profiles correspond to the days of the week; however, the close analysis reveals that they rather show a seasonal character. Taking SDOclust clusters as a reference: (a) P2 is dominant throughout the year; (b) P3 (winter-spring) and P6 (summer-autumn) are almost exclusively for Mondays; (c) P4 occurs mainly in winter and early spring; (d) P1 mostly in April; (e) P5 mainly from late spring to autumn, but excluding the summer holiday period; (f) and P0 mainly during summer.

---

[12] https://upcsirena.app.dexma.com/.

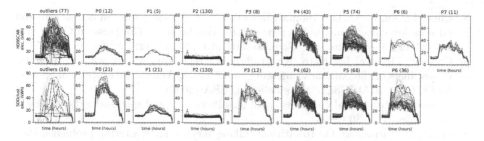

Fig. 5. Clustered profiles disclosed by HDBSCAN (top) and SDOclust (bottom)

These two examples with real data show how HDBSCAN is a method that prioritizes cluster purity (then with a marked risk of overclustering), while SDO-clust prioritizes generalization (therefore with a risk of underclustering).

6   Conclusions

In this paper we have presented SDOclust, a clustering algorithm based on the unsupervised analysis approach pioneered by its predecessor SDO. We have tested SDOclust with default parameters on 155 datasets comprising a wide variety of sizes, dimensions and specific data challenges, yet obtaining excellent results and outperforming the popular HDBSCAN and k-means-- algorithms. Considering default settings, compared to HDBSCAN, which is arguably the most efficient and autonomous alternative, SDOclust tends to find the main top-level spatial partitions and, therefore, is significantly less prone to generate micro-clusters. Key features of SDOclust are: linear complexity, scalability, parameters that are robust, intuitive and self-tuned, resistance to outliers and noise, ability to discover non-convex clusters, use of simple and updateable models, and the generation of rich outputs for detailed post-analysis. Next planned steps mainly involve an incremental version of SDOclust for streaming data.

References

1. Archana, N., Pawar, S.: Periodicity detection of outlier sequences using constraint based pattern tree with mad. Int. J. Adv. Stud. Comput. Sci. Eng. **4**(6), 34 (2015)
2. Böhm, C., Faloutsos, C., Pan, J.Y., Plant, C.: Robust information-theoretic clustering. In: 12th ACM SIGKDD International Conference on Knowledge Discovery and Data Mining, KDD 2006, pp. 65–75. Association or Computer Machine, New York (2006)
3. Campello, R.J.G.B., Moulavi, D., Zimek, A., Sander, J.: Hierarchical density estimates for data clustering, visualization, and outlier detection. ACM Trans. Knowl. Discov. Data **10**(1), 5:1–5:51 (2015)
4. Chawla, S., Gionis, A.: k-means-: a unified approach to clustering and outlier detection. In: 2013 SIAM International Conference on Data Mining, pp. 189–197. SIAM (2013)

5. Chen, L., Xu, L., Li, G.: Anomaly detection using spatio-temporal correlation and information entropy in wireless sensor networks. In: IEEE Congress on Cybermatics: iThings, GreenCom, CPSCom, SmartData, pp. 121–128 (2020)
6. Cho, K., Mitsuya, K., Kato, A.: Traffic data repository at the wide project. In: Proceedings of the Annual Conference on USENIX Annual Technical Conference, ATEC 2000, p. 51. USENIX Association, USA (2000)
7. Demšar, J.: Statistical comparisons of classifiers over multiple data sets. J. Mach. Learn. Res. **7**, 1–30 (2006)
8. Fränti, P., Virmajoki, O.: Iterative shrinking method for clustering problems. Pattern Recogn. **39**(5), 761–765 (2006)
9. Fränti, P., Virmajoki, O., Hautamäki, V.: Fast agglomerative clustering using a k-nearest neighbor graph. IEEE Trans. Pattern Anal. Mach. Intell. **28**(11), 1875–1881 (2006)
10. Fränti, P., Sieranoja, S.: K-means properties on six clustering benchmark datasets. Appl. Intell. **48**(12), 4743–4759 (2018)
11. Gionis, A., Mannila, H., Tsaparas, P.: Clustering aggregation. ACM Trans. Knowl. Disc. Data (TKDD) **1**(1), 4-es (2007)
12. Hartl, A., Iglesias, F., Zseby, T.: Sdostream: low-density models for streaming outlier detection. In: 28th ESANN, Bruges, Belgium, 2–4 October 2020, pp. 661–666 (2020)
13. Hubert, L., Arabie, P.: Comparing partitions. J. Classif. **2**(1), 193–218 (1985)
14. Iglesias, F., Zseby, T., Ferreira, D., Zimek, A.: Mdcgen: multidimensional dataset generator for clustering. J. Classif. **36**(3), 599–618 (2019)
15. Iglesias, F., Zseby, T., Zimek, A.: Outlier detection based on low density models. In: IEEE International Conference on Data Mining Workshops (ICDMW), pp. 970–979 (2018)
16. Iglesias Vázquez, F.: SDOclust Evaluation Tests (2023). https://doi.org/10.48436/3q7jp-mg161
17. Kärkkäinen, I., Fränti, P.: Gradual model generator for single-pass clustering. Pattern Recogn. **40**(3), 784–795 (2007)
18. von Luxburg, U., Williamson, R.C., Guyon, I.: Clustering: science or art? In: Proceedings of ICML Workshop on Unsupervised and Transfer Learning, vol. 27, pp. 65–79. PMLR (2012)
19. Van der Maaten, L., Hinton, G.: Visualizing high-dimensional data using t-sne. J. Mach. Learn. Res. **9**(2579–2605), 9 (2008)
20. McInnes, L., Healy, J., Astels, S.: hdbscan: Hierarchical density based clustering. J. Open Source Softw. **2** (2017). https://hdbscan.readthedocs.io/en/latest/
21. Pedregosa, F., et al.: Scikit-learn: machine learning in Python. J. Mach. Learn. Res. **12**, 2825–2830 (2011)
22. Rezaei, M., Fränti, P.: Can the number of clusters be determined by external indices? IEEE Access **8**, 89239–89257 (2020)
23. Rousseeuw, P.J.: Silhouettes: a graphical aid to the interpretation and validation of cluster analysis. J. Comput. Appl. Math. **20**, 53–65 (1987)
24. Ruiz Martorell, G., López Plazas, F., Cuchí Burgos, A.: Sistema d'informació del consum d'energia i d'aigua de la UPC (Sirena). 1r Congrés UPC Sostenible (2007)
25. Sun, J., Qiu, Y., Shang, Y., Lu, G.: A multi-fault advanced diagnosis method based on sparse data observers for lithium-ion batteries. J. Energy Storage **50**, 104694 (2022)

26. Tsai, C.-Y., Chiu, C.-C.: A clustering-oriented star coordinate translation method for reliable clustering parameterization. In: Washio, T., Suzuki, E., Ting, K.M., Inokuchi, A. (eds.) PAKDD 2008. LNCS (LNAI), vol. 5012, pp. 749–758. Springer, Heidelberg (2008). https://doi.org/10.1007/978-3-540-68125-0_72

27. Veenman, C.J., Reinders, M.J.T., Backer, E.: A maximum variance cluster algorithm. IEEE Trans. Pattern Analy. Mach. Intell. **24**(9), 1273–1280 (2002)

28. Vormayr, G., Fabini, J., Zseby, T.: Why are my flows different? a tutorial on flow exporters. IEEE Commun. Surv. Tutor. **22**(3), 2064–2103 (2020)

29. Xu, D., Tian, Y.: A comprehensive survey of clustering algorithms. Ann. Data Sci. **2** (2015)

30. Xu, R., Wunsch, D.: Survey of clustering algorithms. IEEE Tran. Neural Netw. **16**(3), 645–678 (2005)

31. Zimek, A., Gaudet, M., Campello, R.J., Sander, J.: Subsampling for efficient and effective unsupervised outlier detection ensembles. In: Proceedings of the 19th ACM SIGKDD International Conference on Knowledge Discovery and Data Mining, pp. 428–436 (2013)

Solving k-Closest Pairs
in High-Dimensional Data

Martin Aumüller[1](✉)(iD) and Matteo Ceccarello[2](✉)(iD)

[1] IT University of Copenhagen, Copenhagen, Denmark
maau@itu.dk
[2] University of Padova, Padua, Italy
matteo.ceccarello@unipd.it

Abstract. We investigate the k-closest pair problem in high dimensions, that is finding the $k \geq 1$ closest pairs of points in a set $S \subseteq \mathcal{X}$ in a metric space $(\mathcal{X}, \text{dist})$. This is a fundamental problem in computational geometry with a wide variety of applications, including network science, data mining, databases, and recommender systems. We propose an *exact* algorithm with a controllable *failure probability*, thus allowing the user to specify the desired *recall*. Our algorithm has expected *subquadratic* running time under mild assumption on the distance distribution, relying only on the existence of a Locality Sensitive Hash family for the metric at hand. We complement our theoretical analysis with an experimental evaluation, showing that our approach can provide solutions orders of magnitude faster than current state-of-the-art data structures designed for specific metrics.

1 Introduction

In this paper we study the *k-closest pair problem*: Given a set $S \subseteq \mathcal{X}$ from a metric space $(\mathcal{X}, \text{dist})$, the task is to identify k pairs of distinct points in $S \times S$ that are closest to each other. Solving this problem has numerous applications, for example in network science [29], data mining [22], and databases [21].

A naïve solution is to employ an all-to-all comparison between the points in S. This will result in $O(|S|^2)$ distance comparisons; if S has n elements, the running time will thus become *quadratic*. For metrics such as *Manhattan* distance or *Euclidean* distance, there exist approaches for solving the closest pair problem in d dimensions in time $2^{O(d)} n \log^{O(d)} n$ [8,13], which lead to subquadratic running times if the dimensionality d is small. However, these data structures suffer from the well-known curse of dimensionality because they have an *exponential dependence* on d.

To design *scalable* closest pair algorithms with *subquadratic running time guarantees*, research settled on allowing the results to be *approximate*. In a strong theoretical sense, this means that if the closest pair is at distance r, then an algorithm guarantees to return a pair at distance at most cr for some approximation factor $c > 1$ (with some small failure probability). As we will

O. Pedreira and V. Estivill-Castro (Eds.): SISAP 2023, LNCS 14289, pp. 200–214, 2023.
https://doi.org/10.1007/978-3-031-46994-7_17

discuss in the related work section, many industry-strength solutions use the word more loosely to refer to the inexactness of results. The generally accepted measure of the quality of such approaches is their *recall*, i.e., the fraction of correct pairs identified by the algorithm. These solutions usually do not give strong guarantees on this quality measure. In this work, we propose a Locality-Sensitive Hashing [18] (LSH) based solution with theoretical guarantees on the expected achieved recall. In practice, this means that users only set two parameters: The amount of memory available for the index, and the recall guarantee.

In this work, we propose an extension of the LSH-forest approach by Bawa et al. [5] in the variant described by Aumüller et al. in [4]. In the latter work, the authors describe a query algorithm that carries out a bottom-up traversal of the LSH tries employed by the LSH forest with an adaptive stopping criterion. In Sect. 3, we will describe a traversal approach to solve the k-closest pairs problem. Intuitively, we first build build an LSH forest, which consists of L tries indexing the dataset vectors according to their LSH hash codes. The closest pairs are found by merging nodes in a bottom-up traversal of the trie, keeping track of the best candidate pairs found so far. We prove that our algorithms adapt well to the data distribution: Theorem 2 shows that up to some small additional cost factors, the running time of the proposed algorithm is *asymptotically equivalent* to an LSH-based *clairvoyant* algorithm that queries the part of the LSH forest that minimizes the expected work by knowing *the exact distance distribution*.

In Sect. 4 we describe implementation choices surrounding the proposed approach. For example, we store the trie as a flat array sorted by hash code to support efficient merging of subtrees in the trie, make use of sketches to save the expensive evaluation of distance computations, and discuss details of the parallelization strategy both for index building and the bottom-up traversal. The experimental evaluation is presented in Sect. 5 and compares our approach to several industry-standard baselines such as Meta Research' popular *FAISS* [19] library. We show the competitiveness of our approach to these approaches, in particular under the light that we compare the results to an intensive grid search for best parameters for the other approaches, whereas our approach just takes the available space and a recall guarantee as parameters.

Related Work

LSH Approaches. Locality-sensitive hashing [18] is the de-facto standard for providing theoretically sound algorithms for the approximate near neighbor problem. Popular LSH functions include E2LSH [12] for Euclidean space, SimHash [9] and Crosspolytope LSH [2] for inner product similarity (or cosine similarity) on the unit sphere, and MinHash [6] and 1-bit MinHash [20] for set similarity under Jaccard similarity. Traditionally, the LSH framework aims to solve the (c, r)-near neighbor problem that requires to return a point at distance at most cr to a query point if there exists a point at distance at most r (with some constant probability). The k-NN problem can either be solved using a reduction to different (c, r)-near neighbor problem instances [16], or via direct approaches such as the LSH forest [5] and its variant [4] that we base our work on. In the database community, other directions to LSH-based indexing became very popular. These approaches use locality-sensitive hash functions to project the data points to a lower-dimensional space and index them using I/O-efficient tree data structures. For example, `LSBTree` [24] projects the points to a lower dimension using LSH, employs the Z-order to map points to a single value and indexes these values using B-Trees. A more recent approach called `PM-LSH` [28] indexes the projected points directly using the PM-Tree [23] without applying the Z-order.

Closest Pair Algorithms. One of the seminal papers on efficient solutions to the closest pair problem in high-dimensional data is by Xiao et al. [26]. It mainly focuses on Jaccard similarity but also discusses Cosine, Dice, and Overlap similarity. At a high level, it maintains the input sets in a priority queue ordered by an upper bound on the similarity it can attain with any other set, based on prefix filtering. Sets are extracted in decreasing order of such an upper bound and their similarity with other sets is computed by means of an inverted index on the tokens. Several optimizations to this approach were introduced in the recent paper [25]. Further improvements for sets under the *overlap* similarity are discussed in [27]. In particular, they propose a variant of [26] that evaluates more than one token for each point that is popped from the priority queue. While this approach improves the performance in the case of the overlap similarity, the authors discuss that for the Jaccard similarity it provides little benefit over [26].

For high-dimensional data using Euclidean distance, closest pairs can be found both by the `LSBTree` [24] and by the more recent `PM-LSH` [28]. In particular, `LSBTree` maintains a guess on the smallest k-th distance, and generates candidate pairs from the points whose difference of Z-values is below a threshold derived from the current guess. Another approach for the Euclidean distance was presented in [7]. Using random projections, points are mapped on the real line, where candidate pairs are generated from intervals of consecutive projections.

In general metric spaces, [15] provides a solution used on the *count M-tree* index, a variant of the classical M-tree data structure [11].

2   Preliminaries

Consider a metric space $(\mathcal{X}, \text{dist})$, and let $k > 0$ be an input parameter. Let $S^2 := \{(s, s') \in S \times S \mid s \neq s'\}$ be the set of distinct pairs in S.

Definition 1. The k-*closest pairs* in a set $S \subseteq \mathcal{X}$ are a sequence of distinct pairs $(r_1, s_1), \ldots, (r_k, s_k) \in S^2$ such that: For all other pairs $(r, s) \in S^2$ and for all $i \in \{1, \ldots, k\}$, $\text{dist}(r_i, s_i) \leq \text{dist}(r, s)$.

Informally, the task is to find a set of k closest pairs of points in S. For $k = 1$, this problem is called the *closest pair problem*.

Naïvely, the problem can be solved by enumerating all pairs of points in S^2, leading to $O(|S|^2)$ distance comparisons.

In this paper we present randomized algorithms with probabilistic guarantees. This means that our algorithm receives two input parameters k and $\delta \in (0, 1)$. If a pair $(r, s) \in S^2$ is a k-closest pair, then it is output by the algorithm with probability at least $1 - \delta$. If the quality of the solution is measured using *recall*, the fraction of correct pairs reported by the algorithm, we *expect* a recall of $1 - \delta$.

Definition 2 (Locality-Sensitive Hashing [18]). Let $(\mathcal{X}, \mathrm{dist})$ be a metric space, let T be a set, and let $\mathcal{H}$ be a family of functions $h \colon \mathcal{X} \to T$. For positive reals r_1, r_2, q_1, q_2, with $q_1 > q_2$, $\mathcal{H}$ is (r_1, r_2, q_1, q_2)-sensitive if for $x, y \in \mathcal{X}$ and h sampled uniformly at random from $\mathcal{H}$ we have that:

- $\mathrm{dist}(x, y) \leq r_1 \Rightarrow \Pr[h(x) = h(y)] \geq q_1$
- $\mathrm{dist}(x, y) \geq r_2 \Rightarrow \Pr[h(x) = h(y)] \leq q_2$

As a technical detail, we assume that the LSH family is *monotonic*, i.e., its collision probability function is decreasing with the distance. Moreover, we assume that we can evaluate the probability of collision at a certain distance.[1] We denote the collision probability function with $p \colon \mathbb{R} \to [0, 1]$ and for ease of notation use $p(x, y) := p(\mathrm{dist}(x, y))$ for $x, y \in \mathcal{X}$. Most popular LSH families have this property, such as Euclidean LSH [12] for Euclidean space, random hyperplane hashing [9] and Cross-Polytope hashing [2] for the d-dimensional unit sphere under inner product similarity (or cosine similarity), or 1-bit MinHash described by Li and König in [20] for set similarity under the Jaccard similarity. Since we use LSH functions as a black-box, our results hold for all LSH families that have this property and are not restricted to special cases.

In [4], Aumüller et al. introduced PUFFINN, a highly-optimized implementation of an LSH-based k-nearest neighbor search algorithm. Their work builds upon the LSH forest data structure of Bawa et al. [5] and the adaptive search mechanism described by Dong et al. in [14]. Since our work extends their data structure, we provide a recap of how their data structure works next. See [4] for more details.

Given a set $S \subseteq \mathcal{X}$, two parameters $L, K \geq 1$, and access to an LSH family $\mathcal{H}$, the data structure consists of a collection of L LSH tries of max depth K. The LSH tries are indexed by $j = 1, \ldots, L$. The jth LSH trie is built from the set of strings

$$\{(h_{1,j}(x), \ldots, h_{K,j}(x)) \mid x \in S\}. \tag{1}$$

where $h_{i,j} \sim \mathcal{H}$. The trie is constructed by recursively splitting S on the next (ith) character until $|S| \leq i$ or $i = K + 1$ at which point we create a leaf node in the trie that stores references to the points in S.

Example 1. Figure 1 gives an example with a small set of points in the Euclidean plane, reporting the solution of a top-5 global join. The right hand side of the figure gives an example for an LSH forest with L tries initialized

---

[1] It will be evident from the analysis of the algorithms that an *estimate* on the collision probability function will be sufficient in practice.

with the dataset. Each trie has depth K. Paths from the root to the leaves are labelled with the hash values of the corresponding points. For instance, in the first trie point d has hash value $(0, 1)$, while point f has hash value $(2, 2)$.

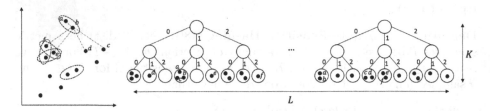

Fig. 1. Left: A set of 12 points in the plane with ellipses marking the 5 closest pairs. Right: L LSH tries of depth K with example distribution of points

Given a query point $q \in \mathcal{X}$ and a failure probability δ, PUFFINN traverses each trie j to find the leaf corresponding to the string $(h_{1,j}(q), \ldots, h_{K,j}(q))$. Starting from that, it traverses the tries in a bottom-up fashion and keeps track of the current kth closest point x'_k. Let p be the probability of a collision under random choice of the LSH of two points at distance $\mathrm{dist}(q, x'_k)$. If the current depth in the tries is i, and $\ln(1/\delta)/p^i$ is smaller than the current index of the trie that is inspected, the algorithm terminates and returns the closest k points as the answer to the query. [4, Lemma 3] shows that the stopping criterion guarantees that each point of the k nearest neighbors of q is found with probability at least $1 - \delta$. Their Lemma 4 states that the algorithm asymptotically inspects $O(\mathrm{OPT})$ candidate points in expectation, where OPT is the expected number of candidate points of a "clairvoyant" LSH-based algorithm that knows the distance distribution.

3   Algorithm, Analysis, and Problem Difficulty

In the following, we describe our algorithm to compute k closest pairs, extending the single query algorithm proposed in [4]. We first introduce some general concepts.

The algorithm makes use of a priority queue to keep track of the current k closest candidate pairs. This priority queue is implemented as a MaxHeap which associates a pair of points $(x, y) \in \mathcal{X}^2$ with their distance $\mathrm{dist}(x, y)$. The number of elements in each priority queue will be at most k, i.e., if the priority queue has k elements and we insert an element, the element with maximum priority is removed. To make this assumption explicit, we call it a k-priority queue. Each operation in such a priority queue can be implemented to run in time $O(\log k)$, for example using a binary heap.

Fix a set $S \subseteq \mathcal{X}$ and two parameters $L, K \geq 1$. First we build L LSH tries of depth at most K as discussed in the previous section. For $x, y \in S$, let $x \sim_{i,j} y$

Algorithm 1: CLOSEST-PAIRS(k, δ)

---

1 PQ $\leftarrow$ empty k-priority queue of unique (pair of points, dist)
2 **for** $i \leftarrow K, K-1, \ldots, 0$ **do**
3   **for** $j \leftarrow 1, 2, \ldots, L$ **do**
4   **foreach** $F \in S_{i,j}$ **do**
5   **foreach** *unchecked* $(x, y) \in F$ **do**
6   **if** $PQ.max() \geq \mathrm{dist}(x, y)$ **then**
7                      $PQ.\mathrm{insert}((x, y), \mathrm{dist}(x, y))$

8   **if** $i == 0 \lor (PQ.\mathrm{size}() == k \land j \geq \ln(1/\delta)/p(PQ.max())^i)$ **then**
9   **return** PQ

---

if $(h_{1,j}(x), \ldots, h_{i,j}(x)) \in T^i$ equals $(h_{1,j}(y), \ldots, h_{i,j}(y)) \in T^i$, i.e., x and y have the same length-i prefix in the jth LSH trie. Let $S_{i,j}$ denote the partition of points in S under the equivalence relation $\sim_{i,j}$. Since a trie can be built in linear time in the concatenated length of the input strings, cf. (1), we summarize the properties of building the trie data structure as follows.[2]

Fact 1. *Given K and L, building an LSH trie for n keys carries out $O(nKL)$ hash function evaluations to build the input strings, and takes time $O(nKL)$ and uses $O(nL)$ words of space to build the L tries representing these strings.*

Algorithm 1 describes the k-closest pair algorithm on a set $S \subseteq \mathcal{X}$ carried out on an LSH trie data structure with parameters $K, L \geq 1$. Given $k \geq 1$ and $\delta \in (0, 1)$, the algorithm initializes an empty k-priority queue. Using this data structure, the algorithm keeps track of the k closest pairs of points so far. We assume that S admits a total order (e.g., the indices of the keys in the set) and identify two points $x, y \in \mathcal{X}$ as the pair (x, y) with $x < y$.

For $i \leftarrow K$, the algorithm carries out all-to-all comparisons in each of the leaves, over all L tries. Next, the trie is traversed in a bottom-up fashion. For each node n on level i in trie j, i.e., the node that represents one set F in $S_{i,j}$, we carry out an all-to-all comparison between those $(x, y) \in F$ that have not been in the same subset in $S_{i+1,j}$. After trie j has been explored, we check the stopping criterion.

Example 2. Restricting Fig. 1 to the first trie presented there, the algorithm carries out $3 + 3 + 1 + 1 = 8$ distance computation in the leaves of that trie. On the level higher up, it carries out $3 + 3 + 6 + 2 = 14$ distance comparisons. We observe that $S_{2,0}$ (the first trie at largest depth) has points $\{a, b, c\}$ and $\{e\}$ as part of the partition, while $S_{1,0}$ has the set $\{a, b, c, e\}$ and $S_{0,0}$ contains all twelve points in a single set.

---

[2] Note here that the keys are the hash codes of the points in S. In many cases, a hash function can be evaluated in time $O(d)$, but many other scenarios exist. For example for set similarity, the MinHash value can be computed independently of the universe size in the size of the set at hand.

To implement Algorithm 1, the leaves, i.e., all sets in $S_{K,j}$, are stored as sets using hashing. When traversing the trie, all-to-all comparisons are carried out between all pairs of child nodes and the sets are merged together. In this way, each distance computation gives rise to two potential priority queue operations taking $O(\log k)$ time. We charge the cost of merging the child nodes to these all-to-all comparisons. Over all iterations of the nested loop, checking the termination criterion takes time $O(KL)$. We summarize this discussion in the following corollary. We separate the potential expensive distance computations from all other operations to make the statements more precise.

Corollary 1. *Let* $S \subseteq \mathcal{X}$ *with* $|S| = n$, *and let* $K, L, k \geq 1$. *Let* C *be the number of pairs* (x, y) *for which a distance is computed in Line 6 of Algorithm 1 in an LSH forest of depth* K *with* L *tries. The algorithm can be implemented to carry out* $O(C)$ *distance computations and all other operations run in time* $O(C \log k + KL)$.

3.1 Analysis

For a dataset $S \subseteq \mathcal{X}$, identify by the sequence $\text{OPT} = ((x_1, y_1), \ldots, (x_k, y_k))$ a sequence of k closest pairs, and denote the best candidates found by Algorithm 1 as $\text{OUT} = ((x'_1, y'_1), \ldots, (x'_k, y'_k))$.

Theorem 1. *Given* $S \subseteq \mathcal{X}$, $k \geq 1$, *and* $\delta > 0$. *Then*

$$\Pr[(x, y) \in OUT] \geq 1 - \delta \quad \forall (x, y) \in OPT.$$

Proof. Fix a pair (x, y) that is part of the output. There are two ways that the algorithm can return in Line 9. First, it can reach level 0, which means that it carried out a linear scan of all pairs of points. Second, it can return because it inspected the jth trie on level i and $j \geq \ln(1/\delta)/p(x'_k, y'_k)^i$. By the monotonicity of the LSH, $p(x'_k, y'_k) \leq p(x, y)$, because $\text{dist}(x'_k, y'_k) \geq \text{dist}(x, y)$. The probability that y did not collide with x in all j tries is

$$\left(1 - p(x, y)^i\right)^j \leq \left(1 - p(x, y)^i\right)^{\ln(1/\delta)/p(x'_k, y'_k)^i}$$
$$\leq \exp\left(-\ln(1/\delta) \cdot p(x, y)^i / p(x'_k, y'_k)^i\right)$$
$$\leq \exp(-\ln(1/\delta)) \leq \delta,$$

where the second-last inequality follows by the monotonicity of the LSH function, and we also used the inequality $1 - z \leq \exp(-z)$ for $z \geq 0$.

While Corollary 1 tells us that the running time of the algorithm is asymptotically equivalent to the number of pairs that are compared to each other, it is not clear how many such pairs will be inspected. To this end, let us define the work of an optimal, clairvoyant LSH-based closest pair algorithm that knows the distance distribution between all pairs of points. Let (x_k, y_k) be a k closest pair of points of maximum distance. We define

$$\text{OPT}(L, K, k, \delta) = \min \left\{ \frac{\ln(1/\delta)}{p(x_k, y_k)^i} \left(i + \sum_{x,y \in S} p(x, y)^i \right) \,\middle|\, 0 \leq i \leq K, \frac{\ln(1/\delta)}{p(x_k, y_k)^i} \leq L \right\} \quad (2)$$

as the expected cost of the LSH-based algorithm that knows the exact distance distribution. The cost on level i includes that each pair of points in the top-k set is found with probability at least $1 - \delta$ if we inspect $j \geq \ln(1/\delta)/p(x_k, y_k)^i$ tries. The expected cost of searching one LSH trie at depth i is $i + \sum_{x,y \in S} p(x, y)^i$. In our expression for the expected query time we use a unit cost model that counts distance computations. As shown in Corollary 1, counting distance computations is asymptotically equivalent to the running time of the algorithm.

The following theorem relates the running time of Algorithm 1 to the running time of the optimal algorithm that knows the full distance distribution.

Theorem 2. *Given a dataset S and parameters L, K, build the LSH trie data structure for S. Given k and δ such that $\ln(k/\delta) \leq L$, with probability at least $1 - \delta$, Algorithm 1 computes the k closest pairs in S in expected time*

$$O\left(OPT(L, K, k, \delta/k) + L(k + K) + nKL\right).$$

Before proceeding with the proof, we remark that the analysis compares the expected time to the clairvoyant variant in the case that we set the failure probability so low that the result is *exact* (with probability at least $1 - \delta$).

Proof. From Fact 1, building the trie takes time $O(nKL)$. Setting the failure probability to δ/k, Algorithm 1 returns the exact k closest pairs with probability at least $1 - \delta$ using a union bound. Conditioning on this event, the algorithm will stop at the largest i such that $\frac{\ln(k/\delta)}{p(x_k, y_k)^i} \leq L$. Denote this level with i', and let i^* be the level used to minimize the work in (2). The expected running time of Algorithm 1 can be bounded by the term

$$\frac{\ln(k/\delta)}{p(x_k, y_k)^{i'}}\left(i' + \sum_{x,y \in S} p(x, y)^{i'}\right) + \left(L - \frac{\ln(k/\delta)}{p(x_k, y_k)^{i'}}\right)\left(i' + 1 + \sum_{x,y \in S} p(x, y)^{i'+1}\right), \tag{3}$$

where the first term bounds the work done on level i', and the second term bounds the work done on level $i' + 1$ on the tries not inspected on the level above. Let T contain all pairs in S^2 that are not k closest pairs. We start by bounding the first term of the summation and continue as follows:

$$\frac{\ln(k/\delta)}{p(x_k, y_k)^{i'}}\left(i' + \sum_{x,y \in S} p(x, y)^{i'}\right) \leq L(k + K) + \frac{\ln(k/\delta)}{p(x_k, y_k)^{i'}} \sum_{(x,y) \in T} p(x, y)^{i'}$$

$$\overset{(i)}{\leq} L(k + K) + \frac{\ln(k/\delta)}{p(x_k, y_k)^{i^*}} \sum_{(x,y) \in T} p(x, y)^{i^*}$$

$$\leq L(k + K) + OPT(L, K, k, \delta/k),$$

where (i) follows from the monotonicity of the LSH collision probability function. The theorem follows by observing that the second summand in (3) is at most a factor of $1/p(x_k, y_k) = O(1)$ larger than the first term.

We remark that the running time of Theorem 1 can be as high as $O(n^2)$ in the worst case. This is true because of the general nature of the problem (for example, by setting $k = \binom{n}{2}$), or because of the hardness of the data distribution.

3.2   Measuring the Difficulty of Closest Pairs

Ahle et al. [1] defined the expansion around the query as a difficulty measure to bound the running time of an LSH-based adaptive query algorithm. Aumüller and Ceccarello gave empirical evidence in [3] that the expansion predicts the indexing difficulty of datasets well in general. For the closest pair problem, we consider the following definition:

Definition 3. Given $S \subseteq \mathcal{X}$ and $k, k' \geq 1$ with $k < k'$, let $((x_i, y_i))_{(x_i, y_i) \in S^2}$ be a sequence of pairs $(x, y) \in S^2$ ordered by their distance. Then $\mathrm{contrast}_{k|k'}(S) :=$ $\frac{\mathrm{dist}(x_{k'}, y_{k'})}{\mathrm{dist}(x_k, y_k)}$ is the *contrast* of the kth to the k'th closest pair.

We use this definition of contrast to bound the running time of the optimal, *clairvoyant* algorithm. By Theorem 1, this also provides a bound on the expected running time of Algorithm 1.

Lemma 1. *Given $S \subseteq \mathcal{X}$ with $|S| = n$, an LSH family $\mathcal{H}$, integers $K, L, k \geq 1$, and $\delta > 0$, let $c^* = \mathrm{contrast}_{k|2k}(S)$. Let p_1, p_2 be the collision probability of the k and $2k$ closest pair, respectively, for $\mathcal{H}$. Let $\rho = \rho(c^*) = \frac{\log(1/p_1)}{\log(1/p_2)}$ and assume that $L = \Omega\left(n^{2\rho}/k^\rho\right)$. Then $OPT(K, L, k, \delta) = O\left(n^{2\rho} k^{1-\rho} \ln(1/\delta)\right)$.*

Proof. Let S' be the set of all pairs that are not among the $2k$ closest pairs. As discussed before, the expected cost on level i of the clairvoyant algorithm is

$$\frac{\ln(1/\delta)}{p_1^i}\left(i + \sum_{x,y \in S} p(x,y)^i\right) \leq \frac{\ln(1/\delta)}{p_1^i}\left(i + 2k + \sum_{(x,y) \in S'} p(x,y)^i\right)$$

$$\leq \frac{\ln(1/\delta)}{p_1^i}\left(i + 2k + \binom{n}{2}p_2^i\right).$$

Setting $i = \frac{\log(n^2/k)}{\log(1/p_2)}$, $\binom{n}{2}p_2^i \leq k$ and $1/p_1^i = (n^2/k)^{\frac{\log 1/p_1}{\log 1/p_2}} = n^{2\rho}/k^\rho$.

For Euclidean space, $\rho(c) = 1/c^2$, so their exists a level for the clairvoyant algorithm with subquadratic expected running time $O(n^{2/c^2} k^{1-1/c^2} \ln(1/\delta))$. As shown in Theorem 2, Algorithm 1 has the same asymptotic running time up to logarithmic factors. Note that $c \geq \sqrt{2}$ yields *sublinear* running time, because the build time of the trie data structure is disregarded. The expected running time of Algorithm 1 is at least $\tilde{O}(nKL)$ for building the trie.

If the contrast is small, the space requirement on L in Lemma 1 is large. Let c^* be the smallest value of c such that $L \geq n^{2\rho(c^*)}/k^{\rho(c^*)}$, and let d^* be the distance of a k-th closest pair. We can carry out the same analysis as in the proof of Lemma 1. In each trie, we have expected cost k for the k closest pairs, and we expect to see no more than $\binom{n}{2} p(c^* d^*)^i \leq k$ pairs at distance larger than $c^* d^*$ for the choice $i = \lfloor \log(L)/\log(1/p_1)\rfloor$. Each pair with a distance in $[d^*, c^* d^*]$ collides with probability at most p_1^i, so overall all tries we expect to see each pair once. Thus, the expected number of pairs inspected is $O\left((Lk + N_{c^*, d^*}(S)) \ln(1/\delta)\right) = O\left((n^{2\rho(c^*)} k^{1-\rho(c^*)} + N_{c^*, d^*}(S)) \ln(1/\delta)\right)$, where $N_{c^*, d^*}(S)$ is the number of pairs with distance in $[d^*, c^* d^*]$.

4   Implementation Choices

Our algorithms are implemented in the framework provided by PUFFINN [4].

Trie Data Structure. We focus on supporting the Cosine and the Jaccard similarity. For these two similarity functions we choose as hash functions 1-bit Min-
Hash [20] and SimHash [9], respectively. Both these hash functions output single
bits: it is thus very natural to represent the strings of hash values described
in Sect. 2 as bitstrings, packing the bits into machine words. We also support
more complex hash functions such as Crosspolytope-LSH [2]. For this LSH family, we view the output hash code (which is an integer $\{0, \ldots, 2d - 1\}$) as a
length-$\lceil \log d + 1 \rceil$ bitstring and concatenate a small number of hash functions.
For intermediate positions in the trie, i.e., those where we use only part of the
output of a single LSH, we estimate the collision probabilities by sampling. Since
evaluating $O(nKL)$ hash values is time-demanding, PUFFINN supports the ten-
soring and pooling approach described by Christiani in [10].

  By viewing the output of the LSH as a bitstring, we can optimize the trie
implementation. Rather than using a pointer-based implementation, we store
point indices, paired with the corresponding bitstring hash values, in a flat array.
The array is then sorted lexicographically by hash value, leading to a more compact and cache efficient data structure. Furthermore, to speed up index construction we rely on radix-sorting, given that the bitstring hash values can be also
interpreted as integers. In this implementation of the trie, the nodes in the same
subtrie at a given depth i are all the consecutive entries of the array sharing the
same length-i prefix.

Sketching. Finally, to further prune similarity computations we use sketching
with a similar setup as the original PUFFINN paper [4]. Each point is associated
with a different 64-bit sketch in each repetition, computed using either 1-bit
MinHash or SimHash, depending on the similarity function. Consider now two
colliding points x and y, and let s_k be the highest distance of any pair currently
in the k-heap to be possibly updated, if $d(x, y) < s_k$. Let τ be the expected
number of different bits in the sketches of points at distance s_k. Before evaluating
the similarity of x and y, we first check the number of different bits in the
corresponding sketches: if such difference is larger than τ, the similarity between
x and y is not computed at all. This has the effect of reducing the number of
similarities being evaluated, at the cost of slightly reducing the recall of the
algorithm.

  In the following, we refer to the implementation of our algorithms as
PUFFINN-join.

Table 1. Datasets used in the experimental evaluation. The last two columns report the relative contrast at 100 pairs and 10 000 pairs [17]

dataset	n	dimensions	RC @ 100	RC @ 10 000
DeepImage	10 000 000	96	7 615.56	2 343.25
Glove	1 193 514	200	38.04	5.15
DBLP	2 773 660	4 405 478	22.52	7.83
Orkut	2 732 271	8 730 857	20.97	2.99

5   Evaluation

This section reports on the results of our experiments, which are tailored to answer the following questions: (Q1) How does our approach compare with the state of the art? (Q2) How does the amount of available memory influence the performance of our algorithm? (Q3) What is the relationship between intrinsic dimensionality measures and the performance of the algorithm?

Experimental Setup. Experiments were run on 2× 14-core Intel Xeon E5-2690v4 (2.60 GHz) with 512 GB RAM using Ubuntu 16.10 (kernel 4.4.0). The code is available at https://github.com/Cecca/puffinn, along with all the scripts to suitably preprocess the datasets.

We focus our evaluation on two metrics: the running time and the recall.

Datasets. Information about the datasets used in this evaluation is reported in Table 1. In particular, we consider two datasets with cosine similarity (Glove and DeepImage) and two datasets under Jaccard similarity (DBLP and Orkut).

In particular, for all datasets we report a summary of the Relative Contrast [17]—i.e. the ratio between the average distance and the k-th distance—which will be useful in interpreting the results [3]. In particular, we expect DeepImage to be easier than Glove, and both to be easier than the two Jaccard datasets. Furthermore, the relative contrast of DeepImage is extremely high. This means that the top pairs of globally closest points are much closer than the average pair, meaning that this dataset is expected to be considerably easier than the others for the global top-k problem.

Baselines. Under the Jaccard similarity we compare with XiaoEtAl [26], whereas for the cosine similarity[3] we consider the LSB-Tree approach [24, Algorithm CP3] Furthermore, for cosine similarity we consider a baseline that uses the HNSW implementation provided by FAISS [19], querying the k-nearest neighbors of each point and then selecting the k closest among the resulting pairs.

---

[3] We omit PM-LSH [28] as its closest-pair implementation is unavailable and as we were unsuccessful in both implementing it ourselves and in reaching out to the authors.

Parameter Choices. For our approach we set the memory given to the index in the range 256MB to 32GB, by powers of two, corresponding to up to $L \approx 2000$, depending on the dataset, for fixed $K = 24$. As for the target recall, we set it to 0.8, 0.9, and 0.99. For HNSW we test $M \in [32, 48]$, $efConstruction \in [100, 500]$, $efSearch \in [k8, k16]$ for a top-k join. For LSB-Tree we test m up to 8, whereas XiaoEtAl takes no parameters. We remark that this is one of the most relevant differences between our approach and the state of the art: while we can specify a desired target recall, all other approaches require to experiment with several combinations of parameters before finding a configuration suitable for the desired quality level.

Table 2. Running times. Missing values are for runs that timed out after 8 h. The last column reports the time for the index construction (not applicable to XiaoEtAl), which is also included in the total time reported in the other columns

dataset	algorithm	Total time (s) for different k					indexing (s)
		1	10	100	1 000	10 000	
Glove	faiss-HNSW	68.1	132.8	551.7	–	–	63.8
	LSBTree	18.2	136.7	2028.4	2127.4	959.3	3.1
	PUFFINN	5.0	5.0	5.0	5.1	6.3	4.7
DeepImage	faiss-HNSW	299.7	533.8	2632.9	–	–	255.4
	LSBTree	112.0	93.4	114.6	176.2	368.6	13.6
	PUFFINN	37.2	37.5	37.1	37.4	37.4	18.9
DBLP	XiaoEtAl	9.3	14.0	9.8	12.1	58.3	0.0
	PUFFINN	4.9	4.9	4.9	4.9	5.0	4.2
Orkut	XiaoEtAl	118.0	122.0	142.3	1170.3	–	0.0
	PUFFINN	24.7	24.8	24.7	24.5	73.3	23.9

Comparison with Baselines. In the first set of experiments we measure the running time required by different algorithms to achieve recall at least 0.9. The results are reported in Table 2. On all datasets, PUFFINN-join runs faster than the baselines. Furthermore, observe that for $k \leq 1000$ the running time of our algorithm remains basically unaffected by the number of pairs returned. This is because all of the runtime, in this setup, is spent building the index, which is independent of the value of k. Finally, observe that compared to LSBTree, our approach is orders of magnitude faster.

Space/Time Tradeoffs. In Fig. 2 we report the space/time tradeoff of our algorithm, at recall 0.9 and $k \in \{100, 10\,000\}$. In particular, we report on the total time (Fig. 2a), which is comprised of the time to build the index (Fig. 2b), and the time to run the join (Fig. 2c). The top row of plots reports the results for $k = 100$, the bottom row for $k = 10\,000$.

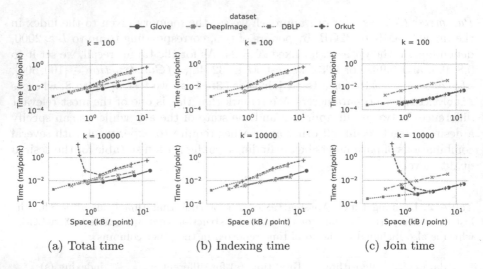

(a) Total time   (b) Indexing time   (c) Join time

Fig. 2. Space/time tradeoff of our algorithm at guaranteed recall 0.9 and $k \in \{100, 10\,000\}$, for the global top-k problem

Observe that the indexing time (Fig. 2b) increases with the space given to the index, as expected. Furthermore, we have the Jaccard-similarity datasets being processed more slowly than the cosine-similarity datasets: this is a consequence of the longer time required to compute MinHash compared to SimHash.

For $k = 100$, on all datasets, the best performance is attained by the configurations using the least space. The reason is in the high relative contrast values of the 100th top pair (see Table 1), which imply that in the few repetitions required to confirm the top pairs there are few other collisions to check. In fact, the index construction time dominates the join time (Fig. 2c) by a large margin.

In contrast with this for $k = 10\,000$ we have that for Orkut increasing the memory usage gives better performance. In fact, the $10\,000$th pair of this dataset has relative contrast of just about 3, meaning that our approach requires either many repetitions or short hash values to confirm it: using more memory allows to use more repetitions on longer hashes, thus reducing the number of collisions to be evaluated. Using too little memory makes the join part of the algorithm dominate on the index construction part, enabling the tradeoff that is observed in the figure.

References

1. Ahle, T.D., Aumüller, M., Pagh, R.: Parameter-free locality sensitive hashing for spherical range reporting. In: SODA, pp. 239–256. SIAM (2017)
2. Andoni, A., Indyk, P., Laarhoven, T., Razenshteyn, I.P., Schmidt, L.: Practical and optimal LSH for angular distance. In: NIPS, pp. 1225–1233 (2015)
3. Aumüller, M., Ceccarello, M.: The role of local dimensionality measures in benchmarking nearest neighbor search. Inf. Syst. **101**, 101807 (2021)

4. Aumüller, M., Christiani, T., Pagh, R., Vesterli, M.: PUFFINN: parameterless and universally fast finding of nearest neighbors. In: ESA, LIPIcs, vol. 144, pp. 10:1–10:16. Schloss Dagstuhl - Leibniz-Zentrum für Informatik (2019)
5. Bawa, M., Condie, T., Ganesan, P.: LSH forest: self-tuning indexes for similarity search. In: WWW, pp. 651–660. ACM (2005)
6. Broder, A.Z.: On the resemblance and containment of documents. In: Compression and Complexity of Sequences 1997, Proceedings, pp. 21–29. IEEE (1997)
7. Cai, X., Rajasekaran, S., Zhang, F.: Efficient approximate algorithms for the closest pair problem in high dimensional spaces. In: Phung, D., et al. (eds.) PAKDD 2018. LNCS (LNAI), vol. 10939, pp. 151–163. Springer, Cham (2018). https://doi.org/10.1007/978-3-319-93040-4_13
8. Chan, T.M.: Orthogonal range searching in moderate dimensions: k-d trees and range trees strike back. In: SoCG, LIPIcs, vol. 77, pp. 27:1–27:15 (2017)
9. Charikar, M.S.: Similarity estimation techniques from rounding algorithms. In: STOC, pp. 380–388. ACM (2002)
10. Christiani, T.: Fast locality-sensitive hashing frameworks for approximate near neighbor search. In: Amato, G., Gennaro, C., Oria, V., Radovanović, M. (eds.) SISAP 2019. LNCS, vol. 11807, pp. 3–17. Springer, Cham (2019). https://doi.org/10.1007/978-3-030-32047-8_1
11. Ciaccia, P., Patella, M., Zezula, P.: M-tree: an efficient access method for similarity search in metric spaces. In: VLDB, pp. 426–435. Morgan Kaufmann (1997)
12. Datar, M., Immorlica, N., Indyk, P., Mirrokni, V.S.: Locality-sensitive hashing scheme based on p-stable distributions. In: SCG, pp. 253–262. ACM (2004)
13. Dietzfelbinger, M., Hagerup, T., Katajainen, J., Penttonen, M.: A reliable randomized algorithm for the closest-pair problem. J. Algor. **25**(1), 19–51 (1997)
14. Dong, W., Wang, Z., Josephson, W., Charikar, M., Li, K.: Modeling LSH for performance tuning. In: CIKM, pp. 669–678. ACM (2008)
15. Gao, Y., Chen, L., Li, X., Yao, B., Chen, G.: Efficient k-closest pair queries in general metric spaces. VLDB J. **24**(3), 415–439 (2015)
16. Har-Peled, S., Indyk, P., Motwani, R.: Approximate nearest neighbor: towards removing the curse of dimensionality. Theory Comput. **8**(1), 321–350 (2012)
17. He, J., Kumar, S., Chang, S.: On the difficulty of nearest neighbor search. In: ICML. icml.cc/Omnipress (2012)
18. Indyk, P., Motwani, R.: Approximate nearest neighbors: towards removing the curse of dimensionality. In: Proceedings of 30th Annual ACM Symposium on the Theory of Computing (STOC), pp. 604–613 (1998)
19. Johnson, J., Douze, M., Jégou, H.: Billion-scale similarity search with GPUS. IEEE Trans. Big Data **7**(3), 535–547 (2021)
20. Li, P., König, C.: B-Bit minwise hashing. In: WWW 2010, pp. 671–680. ACM, New York (2010)
21. Ntarmos, N., Patlakas, I., Triantafillou, P.: Rank join queries in NOSQL databases. Proc. VLDB Endow. **7**(7), 493–504 (2014)
22. Pirbonyeh, M., Rezaie, V., Parvin, H., Nejatian, S., Mehrabi, M.: A linear unsupervised transfer learning by preservation of cluster-and-neighborhood data organization. Pattern Anal. Appl. **22**(3), 1149–1160 (2019)
23. Skopal, T., Pokorný, J., Snášel, V.: Nearest neighbours search using the PM-tree. In: Zhou, L., Ooi, B.C., Meng, X. (eds.) DASFAA 2005. LNCS, vol. 3453, pp. 803–815. Springer, Heidelberg (2005). https://doi.org/10.1007/11408079_73
24. Tao, Y., Yi, K., Sheng, C., Kalnis, P.: Efficient and accurate nearest neighbor and closest pair search in high-dimensional space. ACM Trans. Database Syst. **35**(3), 20:1–20:46 (2010)

25. Wang, H., Yang, L., Xiao, Y.: Setjoin: a novel top-k similarity join algorithm. Soft. Comput. **24**(19), 14577–14592 (2020)
26. Xiao, C., Wang, W., Lin, X., Shang, H.: Top-k set similarity joins. In: ICDE, pp. 916–927. IEEE Computer Society (2009)
27. Yang, Z., Zheng, B., Li, G., Zhao, X., Zhou, X., Jensen, C.S.: Adaptive top-k overlap set similarity joins. In: ICDE, pp. 1081–1092. IEEE (2020)
28. Zheng, B., Zhao, X., Weng, L., Nguyen, Q.V.H., Liu, H., Jensen, C.S.: PM-LSH: a fast and accurate in-memory framework for high-dimensional approximate NN and closest pair search. VLDB J. **31**(6), 1339–1363 (2022)
29. Zhou, X., Wu, B., Jin, Q.: Analysis of user network and correlation for community discovery based on topic-aware similarity and behavioral influence. IEEE Trans. Hum. Mach. Syst. **48**(6), 559–571 (2018)

Vec2Doc: Transforming Dense Vectors into Sparse Representations for Efficient Information Retrieval

Fabio Carrara$^{(\boxtimes)}$ (ID), Claudio Gennaro (ID), Lucia Vadicamo (ID),
and Giuseppe Amato (ID)

ISTI CNR, Pisa, Italy
{fabio.carrara,claudio.gennaro,lucia.vadicamo,
giuseppe.amato}@isti.cnr.it

Abstract. The rapid development of deep learning and artificial intelligence has transformed our approach to solving scientific problems across various domains, including computer vision, natural language processing, and automatic content generation. Information retrieval (IR) has also experienced significant advancements, with natural language understanding and multimodal content analysis enabling accurate information retrieval. However, the widespread adoption of neural networks has also influenced the focus of IR problem-solving, which nowadays predominantly relies on evaluating the similarity of dense vectors derived from the latent spaces of deep neural networks. Nevertheless, the challenges of conducting similarity searches on large-scale databases with billions of vectors persist. Traditional IR approaches use inverted indices and vector space models, which work well with sparse vectors. In this paper, we propose Vec2Doc, a novel method that converts dense vectors into sparse integer vectors, allowing for the use of inverted indices. Preliminary experimental evaluation shows a promising solution for large-scale vector-based IR problems.

Keywords: Inverted Index · Approximate Search · High-Dimensional Indexing · Very Large Databases · Surrogate Text Representation

1 Introduction

Deep learning and artificial intelligence have significantly changed the scientific research landscape, impacting a wide range of fields such as computer vision, natural language processing, and data science. One field that has experienced a profound transformation is Information Retrieval (IR), where the ability to retrieve information accurately and across different modalities has greatly improved. This has been made possible by the advent of neural networks, which have simplified IR problems into searches for vector similarities in latent spaces generated by deep neural networks. These advances have opened up new possibilities for the development of more sophisticated and accurate search systems.

O. Pedreira and V. Estivill-Castro (Eds.): SISAP 2023, LNCS 14289, pp. 215–222, 2023.
https://doi.org/10.1007/978-3-031-46994-7_18

When it comes to indexing billions or trillions of data descriptors, there is limited consensus on which algorithms are the most effective at this scale versus their hardware cost [14]. However, when dealing with large-scale databases containing billions of data objects, such as those found on the web, common approaches of comparing query feature vectors with data feature vectors can become computationally infeasible.

In the context of textual documents, full-text search engines, like Elasticsearch, Solr, or Lucene, have been extensively used and tested in various domains and have already proven to be effective for large-scale indexing and searching. They leverage mature technology and infrastructure, offering scalability, flexibility in querying, and a rich set of features. For these reasons, over the years, we have explored various approaches to enable the searching of non-textual data by transforming it into text format, which can then be indexed and queried using off-the-shelf text search engines. We have called this family of approaches as *Surrogate Text Representation* (STR) [2,3,6].

In this short paper, we propose *Vec2Doc*, an alternative approach that transforms dense vectors into sparse representations that can be efficiently indexed using inverted indices. This technique allows us to take advantage of the well-established performance of inverted indices while also benefiting from the power of deep learning-generated dense vectors. Our method is specifically designed for dense vectors and allows easy indexing and retrieval using standard search engines that employ inverted indices. It extends the Scalar Quantization (SQ) STR approach [2] with a simple but effective idea to expand the codebook used for indexing, leading to improved performance of the inverted index.

The remainder of this paper is structured as follows: Sect. 2 provides an overview of related work, Sect. 3 presents the methodology of Vec2Doc, Sect. 3.1 details the experimental setup and results, and Sect. 4 concludes the paper and outlines directions for future research.

2   Background and Related Work

Most of the approaches for approximate metric search rely on transforming data objects into a different space where the search can be performed more efficiently. This is especially relevant when the original space has a high intrinsic dimensionality, when the dataset being searched is large, or when the actual distance to compare two data objects is computationally expensive. Examples of such techniques include transforming metric objects into binary sketches [7,8], permutations [4,16], and other pivot-based representations [9,15].

In 2010, Gennaro et al. [6] introduced a technique that utilized the distances between data objects and a set of reference objects (pivots) to map the data into a textual representation. Their objective was to convert a global descriptor into a sequence of terms resembling a text document, which could be processed by a text search engine such as Lucene. To achieve this, the mapping should be designed to approximate the original distance function, ensuring that the distance between textual documents and queries reflects the original similarity between data objects and data queries. The main advantage of encoding data objects as text was the ability to leverage off-the-shelf text retrieval engines for performing similarity searches. This approach was initially referred to as the Surrogate Text Representation (STR) approach. However, over the years, various techniques have been proposed to transform descriptors into textual documents, and the term STR has come to encompass the broader family of approaches of this nature [1–3]. Some of these techniques have been designed to work on general metric spaces, while others are specialized for vector spaces.

In this work, our focus is on STR techniques specialized to index and search dense real vectors, such as data descriptors extracted with deep neural networks. These techniques can be mathematically formalized as space transformations of the form $f : \mathbb{R}^d \to \mathbb{N}^m$, where each original vector y is mapped into an integer-valued vector $\overline{y}$. The key idea is to interpret $\overline{y}$ as a term frequency vector based on a synthetic codebook $\mathcal{C} = \{\tau_1, \dots, \tau_m\}$ of m terms. Consequently, the associated text document for vector y is obtained by concatenating the codebook terms with space separators, where each term τ_i is repeated a number of times equal to $\overline{y}_i$. We indicate with $T_{f,\mathcal{C}}(\cdot)$ the overall transformation from the original vectors to the text documents, which depends on both the function f and the used codebook $\mathcal{C}$. For example, if $f(y) = \overline{y} = [3, 1, 0, 2]$ and $\mathcal{C} = \{\text{"A"}, \text{"B"}, \text{"C"}, \text{"D"}\}$, the resulting text document associated with y would be $T_{f,\mathcal{C}}(y) = $ "A A A B D D". The rationale behind this approach is that the abstract transformation f corresponds to the function that precisely generates the vectors used internally by the search engine based on the *vector space model* [12], particularly in the case of a simple TF-weighting scheme.

To ensure compatibility with text retrieval engines and the efficiency of the inverted index, it is important that f generates a sparse vector with non-negative components. Various STR approaches exist, differing in their specific methods for handling negative values, achieving sparsification, and performing the final real-to-integer discretization. For example, in [2,3], the use of the Concatenated Rectified Linear Unit (CReLU) activation function is employed to prevent the presence of negative values in the transformed vectors. In [3] a Voronoi partitioning scheme is employed, where different codebooks are utilized for each partition. This approach aims to increase the sparsity of the transformed data, leading to more efficient indexing and retrieval processes.

3   Vec2Doc

The Vec2Doc is a generalization of the Scalar Quantization STR approach [2]. The SQ transforms a dense vector $y \in \mathbb{R}^d$ into a term frequency vector $\overline{y} \in \mathbb{N}^m$ through four main steps:

1. *Centering and Random Orthogonal Projection:* $y_1 = R(y - \mu) \in \mathbb{R}^d$ where $R \in \mathbb{R}^{d \times d}$ is a random orthogonal matrix and $\mu \in \mathbb{R}^d$ is set to center the data to zero mean. This step is used to uniform the distribution of vector components. In particular, the random rotation helps distribute the information along all the components of the vector in order to limit the presence of

unbalanced posting lists in the final inverted file (important for the efficiency of inverted indices). In [2] the centering operation is applied only to the data object but not to the queries in order to preserve dot-product similarities.

2. *Positivization*: $\boldsymbol{y}_2 = \text{CReLU}(\boldsymbol{y}_1) \in \mathbb{R}^{2d}$, where the Concatenated Rectified Linear Unit transformation [13] is applied to transform the vector into a positive one. The CReLU operation involves concatenating both the original vector and its negation and then setting all negative values to zero, i.e., $\text{CReLU}(\boldsymbol{v}) = \max([\boldsymbol{v}, -\boldsymbol{v}], \boldsymbol{0})$ where the max is applied element-wise.

3. *Sparsification*: $\boldsymbol{y}_3 = g_\gamma(\boldsymbol{y}_2) \in \mathbb{R}^{2d}$ where g_γ is a component-wise thresholding function, i.e., $g_\gamma(x) = x$ if $x > \gamma$, 0 otherwise.

4. *Integer Quantization*: $\boldsymbol{y}_4 = \lfloor s\boldsymbol{y}_3 \rfloor \in \mathbb{N}^{2d}$ where $\lfloor \cdot \rfloor$ denotes the floor function and s is a multiplication factor > 1 that works as a *quantization factor* to transform float components into integer.

The primary drawback of this approach is its limitation in terms of the dimensionality of the resulting term frequency vectors, which is fixed at $2d$, where d represents the dimensionality of the original vector. Consequently, the vocabulary size necessary for indexing the data with inverted files remains constant, posing an inconvenience when dealing with large datasets. To clarify further, if the number of posting lists is fixed, the length of each posting list may become excessive for large datasets, ultimately impacting search efficiency negatively.

To overcome this issue, in [3] we proposed the VP-SQ approach where the data is clustered in Voronoi cells and a different vocabulary can be used for each cell. In this paper, instead, we propose a simple but effective idea to extend the vocabulary size without using centroids. Nevertheless, our new Vec2Doc approach can be used alone or in combination with the Voronoi partitioning strategy to improve performance further.

The Vec2Doc transformation employs a semi-orthogonal transformation to expand the dimensionality of the vectors before the real-to-integer discretization process. Specifically, the initial step of the SQ approach (step 1. described above) is replaced with the following transformation:

$$\boldsymbol{y}_1 = A\boldsymbol{y} \in \mathbb{R}^m, \tag{1}$$

where $A \in \mathbb{R}^{m \times d}$ is a semi-orthogonal matrix (i.e., $A^T A = I$) with $m > d$. This transformation is applied to both data and query vectors without centering the former. The purpose of this transformation is to increase the vector dimensionality while preserving the dot product:

$$< A\boldsymbol{v}, A\boldsymbol{w} >= \boldsymbol{v}^T A^T A\boldsymbol{w} = \boldsymbol{v}^T \boldsymbol{w} =< \boldsymbol{v}, \boldsymbol{w} > .$$

Moreover, since the semi-orthogonal matrix is randomly chosen, it behaves similarly to a random rotation, effectively distributing information across the different dimensional components.

To sample a random semi-orthogonal matrix, we follow [11] and apply the following recurrent formula to a random normally-distributed real-valued matrix

$$A \leftarrow A - \frac{1}{2}A\left(A^T A - I\right) \tag{2}$$

that efficiently converge into a semi-orthogonal matrix in a few iterations.

3.1 Experiments

For evaluation, we adopt two approximate nearest neighbor benchmarks, which are Glove-100 [10] and NYTimes-256 [5], collecting $\sim$ 1M 100-dimensional and $\sim$ 280k 256-dimensional real-valued vectors as search set respectively. Both provide 10k test queries and desired results (100 nearest neighbors). We L_2-normalize all vectors to perform cosine similarities as inner products.

We measure the search effectiveness with the Recall@10 metric. For efficiency, we measure the search cost in an implementation-independent way by counting the number of accessed posts in the posting lists in the inverted index, which also corresponds to the number of multiply-add operations needed to compute scores for all data points. In addition, we measure the index size as the total number of posts in the posting lists.

First, we compare our improved SQ scheme against the original one proposed in [2] in Fig. 1. Each line is obtained by choosing the desired vocabulary size (i.e., the number of rows m of the semi-orthogonal transformation A in Eq. 1) and varying the threshold γ that controls the sparsification level. We set the quantization factor $s = 10^5$ for all experiments. We note that, as m increases, the obtained recall increases when considering a fixed search cost, thus achieving a better recall-speed trade-off on both benchmarks. However, this is paid in terms of space; as m increases, the index size increases to maintain the same recall values. We also observe diminishing returns as m increases.

Next, we test our proposal in combination with the Voronoi partitioning scheme proposed in [3]. In brief, the search set is divided into c partitions via k-means at index time, and at query time, only the n partitions having the closest centroids to the query are accessed. In each partition, whichever STR technique can be adopted as long as each partition has its vocabulary that does not share tokens with other partitions. In Fig. 2, we compare the Voronoi-partitioned version of our proposal with the Voronoi-partitioned SQ method (VP-SQ). Each line is obtained by varying all the method parameters (number of centroids $c \in \{128, 256, 512, 1024, 2048, 4096, 8192\}$, number of partitions accessed at query time $n \in \{1, 2, 4, 8, \ldots, c\}$, plus the parameters of the underline STR technique, i.e., sparsification threshold γ, vocabulary size m) and by keeping only the configurations belonging to the Pareto frontier. Due to the large number of parameter configurations to be tested, we report results only on the smaller NYTimes-256 benchmark. We can see that our proposal provides an improved effectiveness-efficiency trade-off also in the Voronoi-partitioning version.

These preliminary experiments demonstrate that the Vec2Doc approach can be effectively utilized in conjunction with Voronoi-partitioned STRs, offering an easy-to-implement strategy to enhance the vocabulary size within each Voronoi cell. In most cases, this achieves the optimal trade-off between efficiency and effectiveness.

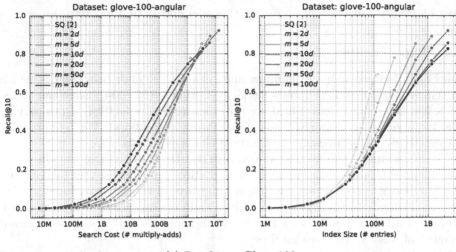

(a) Results on Glove 100.

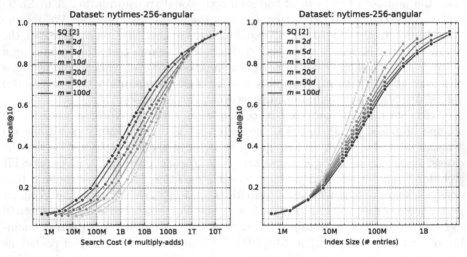

(b) Results on NYTimes 256.

Fig. 1. Effectiveness (Recall@K) vs. Search Cost (# of multiply-add needed to compute scores) and Index Size (# of entries in the index) of our proposal and of the SQ baseline [2]. Each line is obtained by choosing the desired vocabulary size m and varying the sparsification threshold γ.

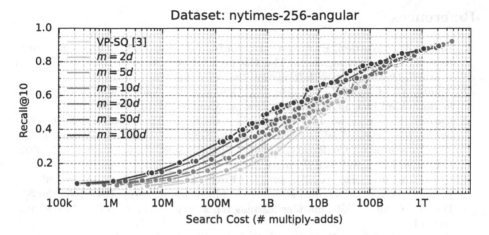

Fig. 2. NYTimes-256. Effectiveness (Recall@K) vs. Search Cost (# of multiply-add needed to compute scores) of the Voronoi-partitioned versions of our proposal and of SQ [3]. Each line is obtained by plotting the Pareto-optimal configurations when varying the number of partitions c, the number of accessed partitions n, the vocabulary size m, and the sparsification threshold γ.

4 Conclusion

In this paper, we have presented Vec2Doc, a novel approach for transforming dense vectors into sparse representations specifically designed to address the challenges of large-scale information retrieval tasks. Our approach enables an expansion of the vocabulary size utilized in STR encoding, thereby positively impacting search efficiency.

However, our approach is not without limitations. One of the challenges we aim to address in future work is the problem of out-of-distribution queries such as those arising from cross-modal features. Currently, Vec2Doc's performance may be hindered when dealing with cross-modal embeddings due to the application of the CReLU activation function. This limitation can lead to poor performance when handling datasets with mixed data modalities, such as text and images.

To overcome this challenge, future research will explore alternative feature transformation techniques that better handle cross-modal features. Additionally, we plan to investigate the scalability of Vec2Doc for increasingly larger datasets and the potential integration with other advanced IR systems to improve its applicability and performance further.

Acknowledgements. This work was partially funded by AI4Media - A European Excellence Centre for Media, Society, and Democracy (EC, H2020 n. 951911), SUN - Social and hUman ceNtered XR (EC, Horizon Europe n. 101092612), and National Centre for HPC, Big Data and Quantum Computing - HPC (CUP B93C22000620006).

References

1. Amato, G., Bolettieri, P., Carrara, F., Falchi, F., Gennaro, C.: Large-scale image retrieval with elasticsearch. In: The 41st International ACM SIGIR Conference on Research Development in Information Retrieval, pp. 925–928 (2018)
2. Amato, G., Carrara, F., Falchi, F., Gennaro, C., Vadicamo, L.: Large-scale instance-level image retrieval. Inf. Process. Manage. **57**(6), 102100 (2020)
3. Carrara, F., Vadicamo, L., Gennaro, C., Amato, G.: Approximate nearest neighbor search on standard search engines. In: Similarity Search and Applications: 15th International Conference, SISAP 2022, Bologna, Italy, October 5–7, 2022, Proceedings, pp. 214–221. Springer (2022)
4. Chávez, E., Figueroa, K., Navarro, G.: Effective proximity retrieval by ordering permutations. IEEE Trans. Pattern Anal. Mach. Intell. **30**(9), 1647–1658 (2008)
5. Dua, D., Graff, C.: UCI machine learning repository (2017). http://archive.ics.uci.edu/ml
6. Gennaro, C., Amato, G., Bolettieri, P., Savino, P.: An Approach to Content-Based Image Retrieval Based on the Lucene Search Engine Library. In: Lalmas, M., Jose, J., Rauber, A., Sebastiani, F., Frommholz, I. (eds.) Research and Advanced Technology for Digital Libraries, pp. 55–66. Springer, Berlin, Heidelberg (2010). https://doi.org/10.1007/978-3-642-15464-5_8
7. Higuchi, N., Imamura, Y., Mic, V., Shinohara, T., Hirata, K., Kuboyama, T.: Nearest-neighbor search from large datasets using narrow sketches. In: ICPRAM, pp. 401–410 (2022)
8. Mic, V., Novak, D., Zezula, P.: Binary sketches for secondary filtering. ACM Trans. Inform. Syst. (TOIS) **37**(1), 1–28 (2018)
9. Novak, D., Zezula, P.: Ppp-codes for large-scale similarity searching. Transactions on Large-Scale Data-and Knowledge-Centered Systems XXIV: Special Issue on Database-and Expert-Systems Applications, pp. 61–87 (2016)
10. Pennington, J., Socher, R., Manning, C.D.: GloVe: global vectors for word representation. In: Empirical Methods in Natural Language Processing (EMNLP), pp. 1532–1543 (2014)
11. Povey, D., et al.: Semi-orthogonal low-rank matrix factorization for deep neural networks. In: Interspeech, pp. 3743–3747 (2018)
12. Salton, G., McGill, M.J.: Introduction to Modern Information Retrieval. McGraw-Hill Inc, New York, NY, USA (1986)
13. Shang, W., Sohn, K., Almeida, D., Lee, H.: Understanding and improving convolutional neural networks via concatenated rectified linear units. In: Proceedings of the 33rd International Conference on Machine Learning. ICML 2016, vol. 48, pp. 2217–2225. JMLR.org (2016)
14. Simhadri, H.V., et al.: Results of the neurips'21 challenge on billion-scale approximate nearest neighbor search. In: NeurIPS 2021 Competitions and Demonstrations Track, pp. 177–189. PMLR (2022)
15. Vadicamo, L., Connor, R., Chávez, E.: Query filtering using two-dimensional local embeddings. Inf. Syst. **101**, 101808 (2021)
16. Vadicamo, L., Gennaro, C., Falchi, F., Chávez, E., Connor, R., Amato, G.: Re-ranking via local embeddings: a use case with permutation-based indexing and the nsimplex projection. Inf. Syst. **95**, 101506 (2021)

Similarity Search with Multiple-Object Queries

Richard Connor[1]([✉])([iD]), Alan Dearle[1]([iD]), David Morrison[1]([iD]),
and Edgar Chávez[2]([iD])

[1] University of St Andrews, St Andrews, Scotland, UK
{rchc,al,dm236}@st-andrews.ac.uk
[2] CICESE, Ensenada, Baja California, Mexico
elchavez@cicese.edu.mx

Abstract. Within the topic of similarity search, all work we know assumes that search is based on a dissimilarity space, where a query comprises a single object in the space.

Here, we examine the possibility of a multiple-object query. There are at least three circumstances where this is useful. First, a user may be seeking results that are more specific than can be captured by a single query object. For example a query image of a yellow hot-air balloon may return other round, yellow objects, and could be specialised by a query using several hot-air balloon images. Secondly, a user may be seeking results that are more general than can be captured by a single query. For example a query image of a Siamese cat may return only other Siamese cats, and could be generalised by a query using several cats of different types. Finally, a user may be seeking objects that are in more than a single class. For example, for a user seeking images containing both hot-air balloons and cats, a query could comprise a set of images each of which contains one or other of these items, in the hope that the results will contain both.

We give an analysis of some different mathematical frameworks which capture the essence of these situations, along with some practical examples in each framework. We report some significant success, but also a number of interesting and unresolved issues. To exemplify the concepts, we restrict our treatment to image embeddings, as they are highly available and the outcomes are visually evident. However the underlying concepts transfer to general search, independent of this domain.

1 Introduction

In recent years, the field of content-based retrieval has witnessed significant advancements in the area of nearest neighbour search, enabling efficient retrieval of similar items from large collections. However, traditional nearest neighbour queries often overlook the inherent relationships between multiple complementary queries, limiting their ability to provide comprehensive results for conjunctive search scenarios. This paper introduces the concept of conjunctive similarity queries, which aims to enhance traditional nearest neighbour search by extending it to handle multiple complementary queries simultaneously. Specifically, we explore the challenges and potential problems encountered in attempting to build answers for conjunctive queries. By addressing the limitations of traditional nearest neighbour search methods and offering a more comprehensive retrieval approach, conjunctive similarity queries have the potential to revolutionise content-based retrieval systems. This paper sets the foundation for future research in this exciting and promising area, aiming to bridge the gap between isolated queries and more holistic retrieval scenarios.

In this article, to exemplify the concepts, we restrict our treatment to image embeddings, as they are highly available and the outcomes are visually evident. However the underlying concepts transfer to general search, independent of this domain. In the context of images, the proposed approach is complementary to simultaneously searching a collection with both text-based and image-based queries; in this paper we use a pure image-based embedding.

To motivate this paper consider the output of the nearest neighbour search shown in Fig. 1a. The query is the top leftmost image - a photograph of an albatross. In total there are 25 images in these results which are of an albatross[1]. By contrast in Fig. 1b we show the output of a *conjunctive* query, where the query subject comprises a set formed by the best few results of the first query. The conjunctive query gives 82 images of albatrosses.

We stress that this is conceptually a single query, rather than the collated output of six different queries, and in fact the technique used here has approximately the same evaluation cost as the single-image query.

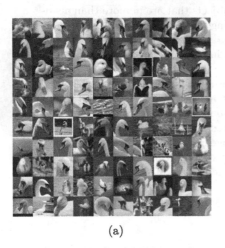

(a)   (b)

Fig. 1. (a) The 100NN results of a traditional single image query of an albatross. Many of the results are images of swans. (b) The 100NN results of a six-image conjunctive query for albatross images. The majority of the results are now albatrosses.

---

[1] We describe the data set and how these categorisations were made later in the paper.

There are (at least) three scenarios where conjunctive query could be a useful mechanism:

query specialisation As in the above example, a user may be using a query image in order to find images of albatrosses, in which case a conjunctive query comprising images of several more specific images gives better results.

query generalisation In this scenario, the user may be seeking images of seabirds in general, in which case the initial results may be more specialised than desired. This is an increasingly prevalent scenario as embeddings improve and collections become larger. A conjunctive query comprising images of several different types of bird can give more suitable results in this case.

subject combination query In this scenario, a user has a number of query objects in different subject domains, and wishes to find objects from a collection that are somewhat similar to all of them. For example, the individual query items may be images of seabirds and boats, where the user wishes to find images which contain both of these subjects.

As far as we know, the notion of addressing these issues through a *conjunctive query* mechanism has not previously been explored. Note that in all of these cases, the result of the single conjunctive query should contain results that would not be found from performing separate queries over each element of the conjunctive query set. The intention is that the conjunctive query mechanism performs a single coherent query over some abstraction of its multiple-object argument.

The rest of this article is structured as follows. In Sect. 2 we outline three different possible formal models for conjunctive query, and in Sect. 3 we give some examples within each of the models. In Sect. 4 we report on the results of experiments we have performed. Finally we summarise the progress we have made, and list a number of open issues we have encountered.

2 Formal Models of Conjunctive Query

Similarity search is normally defined over a dissimilarity space (U, d), where U is some universal domain and d is a dissimilarity function. Search is performed with respect to a large finite space $S \subset U$. The general requirement is to efficiently find members of S which are most similar to an arbitrary member of U given as a query, where the function d gives the only way by which any two objects may be compared. [1,13] summarise a large volume of research in this domain.

In this paper we explore an extension of this concept where, instead of searching using a single element of U given as a query, we search using a set of elements $Q = \{q_1, ..., q_n\}$ where $q_i \in U$.

We seek general models of search in this domain; a little care is required to ensure that investigations are grounded in a useful formal framework.

Normally, for a single query q, the desired result can be defined as $\{s \in S \mid d(q, s) \leq t\}$, for some *threshold* value t which gives a useful size of result set[2]. For our context, we require to extend this. We propose that the following will suffice:

- we maintain the definitions of spaces (U, d) and (S, d) as above
- instead of a query $q \in U$, search is defined in terms of a query $Q = \{q_1, ..., q_n\}$ where $q_i \in U$
- we require some more general numeric dissimilarity function $\theta : \mathcal{P}(U) \times U \to \mathbb{R}^+$ where $\mathcal{P}(U)$ is the powerset of U. θ then defines an ordering on S according to the query Q and each element $s \in S$, so that the result of a query will be $\{s \in S \mid (\theta(Q, s) \leq t\}$ for some suitable threshold t

We have outlined three general techniques for defining θ, as follows.

2.1   Aggregate Measures over Dissimilarity

Simplest, if we assume a similarity space (U, d), any form of aggregation may be used over d to derive θ. For example, $\theta(Q, s)$ can be very simply defined as $\sum_{q \in Q} d(q, s)$.

2.2   Generative Functions

A second general model is that, from a query set Q, a single numeric function is generated, and used to order the finite data set. This requires a generative function $\sigma : \mathcal{P}(U) \to (U \to \mathbb{R})$. The result of the conjunctive query is then $\{s \in S \mid f(s) \leq t\}$, where $f = \sigma(Q)$, for some appropriate t.

One particular case of this is when f re-applies the domain distance function d. The idea is to find a single element $q' \in U$ that best represents the conjunctive query Q in the original dissimilarity domain. In this case a function σ^- maps Q to a value q', and then $f(s) = d(s, q')$, or more formally $\sigma(Q) = \lambda u.d(u, \sigma^-(Q))$.

2.3   Divergence Functions

The term divergence is not, in general, formally defined. Here, we define a *divergence* function δ as a (positive, numeric) dissimilarity measure over a finite subset[3] of a domain U, i.e. $\delta : \mathcal{P}(U) \to \mathbb{R}$. The notion is that the function returns some indicator of a general dissimilarity over all the elements of its argument; for example, a collection of identical objects should return 0, while a collection of objects that have little in common with each other should return a high value.

A normal binary dissimilarity function would thus be a special case of this more general divergence: given a divergence function δ, a dissimilarity function d can be defined as $d(x, y) = \delta(\{x, y\})$.

---

[2] Note that this definition encompasses both *range* and *nearest-neighbour* search.

[3] As for full generality we do not wish to exclude repeated elements, we are really discussing *bags* rather than sets.

In the context of conjunctive query, a divergence function δ can be used by ordering the search space in terms of the divergence of the query Q, with each element of S added in turn: the nearest neighbour to a query Q is thus the database object which gives the smallest divergence when that object is added to the query.

Formally, the solution to a conjunctive query Q over a finite set S is therefore

$$\{s \in S \mid \delta(Q \cup \{s\}) \leq t\}$$

for some appropriate value of t.

2.4   Evaluation Cost

Finally in this section we note a significant difference in the potential cost of mechanisms within the different categories. For any aggregation mechanism, it will be necessary to calculate all distances between every object within Q and every object within S. For any generative mechanism which reverts to a normal distance function, not only is the cost that of a single distance against S, but any mechanism used to optimise a normal similarity search can be re-used for this purpose. Divergence functions may vary between these two extremes, and in fact our examples include one function within each cost category.

3   Conjunctive Search Mechanisms

Here we outline some examples in each of the above categories.

3.1   Aggregation Functions

As mentioned, $\theta(Q, s)$ can be very simply defined as $\sum_{q \in Q} d(q, s)$. Other functions such as harmonic or geometric means are obvious contenders and have been used in agglomerative clustering [9]. A problem with such approaches is that the distances between every element of S and every element of Q require to be evaluated.

Fagin [6] gives a mechanism for extracting database objects with multiple properties, where the database has pre-calculated orderings for objects in some pre-defined categories. Fagin's algorithm gives an optimisation for searching the multiple lists to find the best-ranking solutions, but again requires all distances between S and Q to be pre-calculated.

3.2   Generative Functions

One example of a generative function is the "perfect point" strategy. Suppose for example $U = (\mathbb{R}^n, \ell_2)$. For each element of the query set Q, an effective near-neighbour distance is calculated, i.e. a distance within which similar objects are found in S. From this set of distances, a point $q' \in \mathbb{R}^n$ can be calculated, with the distance from q' to each $q_i \in Q$ being equal to the near-neighbour distance of q_i. The result of a conjunctive search based on this principle is then $\{s \in S \mid \ell_2(q', s) \le t\}$.

We note that the point q' almost undoubtedly will not exist within (S, d), especially if the elements of Q are disparate. However q' is a value which should be similar to any element of U which is equally similar to all elements of Q, and the idea is that the nearest elements of S should therefore inherit this property.

3.3   Divergence Functions

We have experimented with two divergence functions, *MSED* and *nSimplex volume*. MSED can be applied to any *probability* space, that is a vector space where each vector contains only positive or zero values that sum to 1. nSimplex volume can be applied to any space that is finitely embeddable in Hilbert space, which includes Euclidean, Cosine, Jensen-Shannon and Quadratic Form spaces.

MSED. The information-theoretic metric Structural Entropic Distance (SED) was first introduced in [3]. While initially proposed as a metric over labelled tree structures, the core evaluation is over probability vectors, that is any domain $\mathbb{R}^n$ where for each element v, $v_i \ge 0$ and $\sum_i v_i = 1$. This more general metric was evaluated in [11].

SED as a pairwise distance metric at its heart compares the Shannon entropy[4] of two vectors with that of their arithmetic mean. The key observation is that, if the two are equal, then so also is their mean; however the less similar they are, the higher the relative entropy of their mean. This function is normalised by the form

$$SED(v, w) = \frac{C(\frac{v+w}{2})}{\sqrt{C(v) \cdot C(w)}} - 1$$

where $C(x) = e^{H(x)}$, to give an outcome in the range $[0, 1]$. 0 implies the two input vectors are identical, and 1 implies that no individual dimension has a non-zero value in both input vectors, i.e. their dot product is zero and they are therefore orthogonal.

In [11] we observed that this function generalises to a variadic input, rather than just a pair of values. A normalised form of this notion may derived, for a set of n probability vectors V, as:

$$MSED(V) = \frac{1}{n-1} \left(\frac{C\left(\sum_i \frac{V_i}{n}\right)}{\sqrt[n]{\Pi_i C(V_i)}} - 1 \right)$$

Again, an outcome of 0 implies all elements of V are identical, and an outcome of 1 implies that all elements of V are mutually orthogonal. For calculations of the term $MSED(Q \cup \{s\})$ where Q is fixed for many different values of s, the majority of the cost can be amortised given prior knowledge of Q.

---

[4] defined by $H(v) = -\sum_i v_i \ln v_i$.

nSimplex Volume. In [4] it is shown that, in many metric spaces, a finite set of objects can be used to form an *nSimplex*. An nSimplex is a specific mapping from a set of n objects to a simplex[5] in $(n-1)$-dimensional Euclidean space. The observation underlying the divergence mechanism is that, for a set of similar objects, the volume of this simplex will be relatively small.

The specific construction of the nSimplex is iterative over the objects used as its basis, at each stage forming a simplex whose last Euclidean point forms an apex of the previous simplex, in one further dimension. That is, an nSimplex in n dimensions can be formed from an nSimplex σ in $(n-1)$ dimensions and an object u, by creating a new point in n dimensions according to the distances measured between u and each object already represented in σ. Thus, the representation of u in the new simplex is the only point with a non-zero coordinate in the nth dimension, with this final coordinate representing the altitude of that point over the base simplex σ.

In the context of conjunctive search, the divergence function is the volume of this simplex. A base simplex is formed from the elements of Q, and each $s \in S$ is used to construct a new apex. The volume of each resulting simplex is therefore directly proportional to the altitude of this apex, that is its final coordinate, making the ordering of the volumes very simple to extract.

Note that although the cost of constructing the base simplex is incurred only once per conjunctive query, for each object of the database a distance must be calculated to each element of Q, making this mechanism potentially expensive as with aggregate functions.

4 Experiments

In this section we describe some of the experiments we have conducted to explore the concept of conjunctive queries[6] At this stage, our outcomes are primarily qualitative: these are novel semantic queries and we do not at present have a way of constructing a framework for objective measurements[7].

We first describe the experimental setup in general and some of the infrastructure used to qualitatively assess the efficacy of the mechanisms.

The data used for all the experiments consists of one million images from the MIRFLICKR-1M image collection [8,10]. We have encoded these images using two different convolutional neural networks: one to provide a set of feature vectors to be be used in the search process, and another to provide categorical data to be used as ground truth on image similarity.

---

[5] A simplex is an object constructed from a set of points in n-dimensional space, by considering each point as a vertex which is joined to all of the other points. For example, a tetrahedron is a simplex formed from four points in 3D space.

[6] All the code for these experiments can be found on github: https://github.com/ MetricSearch/sisap2023.git.

[7] We note that constructed ground truth for even a single query requires $\binom{n}{2}$ observations.

The feature vectors used in the experiments are encoded using the Dino2 ViT-S/14 network [12]. Dino2 is a state-of-the-art self-supervised pre-training method for computer vision tasks. Cosine distance is the normal metric to use over these embeddings, and we used this in most cases. MSED however is defined over probability vectors. To obtain these we applied the RELU transform to the raw data, followed by ℓ_1 normalisation. We measured the use of SED over this transformed data for normal search to be almost, but not quite, as good as Cosine distance over the raw data.

We used the Resnet18 model [7] to extract categorical labels for the images. Resnet18 is trained over ImageNet [5] data and categories.

For testing all of the conjunctive mechanisms outlined in Sect. 3, we have used the Dino2 embeddings. Many of our observations are based on the relatively subjective judgements we are able to make by repeating many different queries and looking at the outcomes. For generalisation and combination queries, we do not know of a better judgement mechanism. We re-emphasise that the data and code are both publicly available for interested researchers to do their own experiments.

4.1   Measuring Specialisation Queries

For specialisation queries, we have developed an objective test methodology based on the Resnet18 categorisation of the data. Given this, a significant reason for using Dino2 embeddings is that the network is trained independently of the ImageNet categorisation. That is, we only use Dino2 to perform any search task, and we use Resnet18 only to measure the quality of the search.

We take as an assumption that a high-quality classification implies a strong semantic similarity in cases where the categorisation is based on a particularly high *softmax* score. That is, two data items placed in the same category, both with high softmax values according to the classifier, are very likely to be visually similar. Furthermore, we assume that two objects, one with a high softmax score and the other with a low score, are very unlikely to be visually similar. Both of these assumptions are relatively straightforward to test.

The MIRFLICKR collection is large enough that many images are categorised correctly, even although the image set was extracted pseudo-randomly from uncategorised Flickr photographs, independent of subject matter. Over 55,000 images in the collection have a *softmax* value of greater than 0.9. In the experiments we use queries drawn from a subset of the data for which the relevant category contains between 100 and 184 such images; there are 100 such categories. For each individual test, we judge the success of a query by the number of images returned within the 100 nearest neighbours which have the same categorisation as the query, according to the softmax values.

For example, referring back to Figs. 1a and 1b, there are 147 images in the data set for which "albatross" is the highest category. In the results of the single query 13 of them are categorised the same as the query. For the conjunctive query, 82 of the top 100 results are categorised as "albatross".

In each case where we have also visually checked the outcomes; in fact the visual inspection in almost all cases looks rather better than is implied by the categorical scoring technique; however the latter is entirely objective.

4.2   Specialisation Query Experiments

In this section we compare conventional metric search, using query by example, with the conjunctive query approach with multiple query images, using the methodology described in Sect. 4.1.

Using the Dino2 embeddings we performed normal nearest-neighbour queries using Cosine distance to act as a baseline for the evaluation of the different conjunctive techniques.

We also ran four different variants of conjunctive queries: nSimplex, MSED, a simple average, and the perfect point strategy, all as documented in Sect. 3.

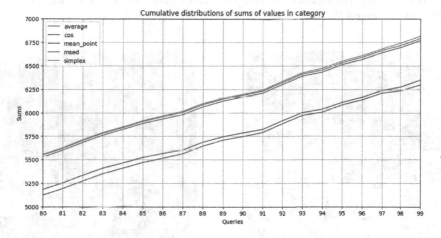

Fig. 2. Cumulative sums of specialisation queries. *cos* denotes the single-query baseline, others are conjunctive query results.

Figure 2 shows the cumulative sums from the 100 experiments conducted as described above. To make the results clearer to read we show the cumulative results for the last 20 queries. The plots are of the cumulative sums for the nearest 100 results which are in the same category as the query. For reference the maximum possible cumulative total from the database is 9744.

From the figure it can be seen that three of the conjunctive queries (aggregate, nSimplex and MSED) are significantly better than the corresponding single-object queries. The perfect point strategy improves only a little on the single-object query. nSimplex performs the best overall, closely matched by aggregate average and MSED, although these differences may not be statistically significant.

It is also worth noting that the computational cost of both nSimplex and aggregate is much higher than perfect point or MSED, both of which cost approximately the same as the single-object query. For nSimplex and aggregate, it is necessary to measure the distance between each point in the conjunctive query to all points in the database.

4.3 Generalisation Query Experiments

To demonstrate how generalisation may be used with conjunctive queries consider Fig. 3a. The first figure shows the results from a traditional nearest neighbour for query using the image in the top left hand corner of the grid of a Siamese cat (category 284). All the results are of Siamese cats. This is an excellent result if the user were indeed looking for Siamese cats. However, they may have been seeking cats in general. The second image in Fig. 3b shows the result of a conjunctive query using images drawn from the following ImageNet categories: tabby cat (281), tiger (282), Persian cat (283), Siamese cat (284), Egyptian cat (285) and leopard (288).

(a) The results of a traditional single image NN query for Siamese cat images

(b) The results of a six image NN conjunctive query for mixed cats

Fig. 3. Results from Siamese cats and general cat searches

The results show cats drawn from all of the categories in the query, and indeed other types of cats which are not in any of the categories. They are subjectively very obviously different from the single-category results shown in Fig. 3.

We do not report quantitatively on the efficacy of the searches from this and the following section as we have not as yet performed any objective measurement. However similar results are obtained for other similar search tasks, for example by grouping different categories of dogs, fish etc. We have noted that MSED appears to function rather better than any of the other techniques we have applied to this task, but have no measurement to justify this observation.

4.4   Combination Query Experiments

Combination queries are perhaps the most obvious use for conjunctive queries in general, and indeed this is where we started work within this domain.

After many attempts, we had had no success with any combination query mechanism. We were very hopeful that both nSimplex volume and perfect point queries would give us results in this category, but neither did. However at that point we tried MSED and the results were, in relative terms, almost unbelievably better. There are still many issues to understand, but at this point we believe MSED is the only mechanism that works well for this task, at least within the domain of image embeddings.

We give a single example of this to illustrate how the technique may be employed. The Resnet18 categories 440 and 441 are "beer bottle" and "beer glass". The 100 images with the highest categorical scores in each of these categories are shown in Figs. 4a and 4b below. Note that Fig. 4 contains 3 cups or glasses and Fig. 4b contains 5 bottles. Figure 5 shows the top 100 nearest neighbours from a conjunctive query formed from the best three images from each category. These images show: 19 images containing just glasses, 41 images containing just bottles (or cans) and 30 containing both categories.

(a) The images with the highest scores in category 440 (beer bottles)

(b) The images with the highest scores in category 441(beer glasses)

Fig. 4. Results from beer-bottles and beer-glasses searches

Fig. 5. Results from the conjunctive query for bottles and glasses

5   Conclusions

Our proposal of querying with multiple objects enables a more comprehensive and refined exploration of an object space than can be achieved by traditional dissimilarity queries. By allowing a set of objects as a query, a system can capture the collective characteristics and features present in the data. This allows for a more holistic representation of the desired concept or theme. As a result, the ranked set of objects in the solution provides a more focused and targeted selection that aligns with the specific attributes present in the query set. At this early stage we have concentrated on results in the single domain of search over image embeddings. Although our techniques are general and not image-specific, it is of course possible that they do not transfer to other semantic domains.

We have identified three specific classes of user task for which conjunctive querying provides solutions: specialisation, where a single-object search gives overly general results; generalisation, where a single object search gives overly specific results; and combination, where results which satisfy more than a single search topic are desired. We have shown successful examples in each of these, where a relevant user task has been satisfied.

We have identified a formal framework which serves to allow investigation into the domain, as well as identifying three different classes of functions. The classification highlights some issues of the cost of conjunctive query; for example aggregate functions require up to nk distance evaluations for n data and k query objects, while generative and some divergence functions require a maximum of only n calculations.

Objective measurement of outcomes is a significant problem. We have outlined a mechanism for measuring query specialisation, but generalisation and combination are more difficult, which makes it impossible to give any very strong and general conclusions as to the different mechanisms tested.

Our early results can be summarised as follows. For query specialisation, the outcomes are in two groups: the aggregate sum, nSimplex volume, and MSED all give notably better results than the baseline similarity query and the "perfect point" strategy. For generalisation and combination, our perception is that MSED gives significantly better results than any of the other mechanisms. This perception is very clear, however we do not know a way of measuring it convincingly. We invite interested researchers to access our code base.

In terms of query efficiency, MSED and "perfect point" both give a relatively tolerable $\mathcal{O}(n)$ calculation cost, compared with the the the $\mathcal{O}(kn)$ incurred by all of the other mechanisms, for n data and a conjunctive query with k objects. Furthermore some of these strategies can be significantly optimised; for example generative functions result in a standard distance-based search, and the inverted index incremental evaluation strategy shown in [2] can be applied to MSED.

6   Further Work

This article represents a first effort to implement and evaluate a number of different mechanisms towards the novel concept of conjunctive search. We have made significant progress, but there are very many open issues.

Confirming initial results In particular for subject combination queries, our results are patchy: for some combinations of queries they work very well, for others not at all. There are many different possible explanations for this, some technical and some based simply on unknown limitations of the data sets we have used so far.

Other semantic domains In particular, we have so far experimented only with images, and in fact a specific set of images. We have used a number of different embeddings and found similar results, so we predict these results will carry forward to other image sets and queries, and also to other spaces represented by embeddings. However this requires to be checked. We can easily do so with word embeddings, high quality sets of which are freely available. It would be very interesting to try language model embeddings, where the three query types have clear and useful parallels.

Objective measurement of outcomes For specialisation queries on images, we have found a reasonable working model of an objective test of different techniques. As yet we are less clear about how to test generalisation or combination queries. Without such objective tests it is difficult to be very confident in comparing mechanisms with each other.

Efficient conjunctive search Almost all of the effort expended in the similarity search domain has been on how to evaluate queries efficiently against very large data sets. Many of these techniques transfer directly to this domain, in particular for techniques which result in a normal similarity search against the database. However it may be that sets of restrictive postulates exist in this domain, as alternatives to e.g. the metric postulates, which allow scaleable or other more efficient search to avoid exhaustive calculation against the database, in which case the more expensive metrics such as nSimplex volume may become relatively more usable.

Potential Application Domains We are currently exploring avenues for exploiting these results. One of these is in data-linkage in which it is advantageous to be able to search for groups of related records; for example the birth records of siblings in a single family. Recommender systems could allow suggestions for future purchases based on a conjunctive query comprising items already purchased within a category, which for example may capture a certain fashion sense. In the domain of drug discovery, a researcher might be interested in finding a peptide that possesses antibacterial and anticancer properties: a conjunctive similarity query might aim to retrieve such peptides. Lastly in the field of histology, it may be possible to generalise over a number of different pathological cell images in order to detect others of a similar type with a different visual presentation.

Acknowledgements. This work is partly supported by ESRC grant ES/W010321/1 "2022-2026 ADR UK Programme".

References

1. Chávez, E., Navarro, G., Baeza-Yates, R., Marroquín, J.L.: Searching in metric spaces. ACM Comput. Surv. **33**(3), 273–321 (2001). https://doi.org/10.1145/502807.502808
2. Connor, R., Cardillo, F.A., Moss, R., Rabitti, F.: Evaluation of jensen-shannon distance over sparse data. In: Brisaboa, N., Pedreira, O., Zezula, P. (eds.) SISAP 2013. LNCS, vol. 8199, pp. 163–168. Springer, Heidelberg (2013). https://doi.org/10.1007/978-3-642-41062-8_16
3. Connor, R., Simeoni, F., Iakovos, M., Moss, R.: A bounded distance metric for comparing tree structure. Inf. Syst. **36**(4), 748–764 (2011)
4. Connor, R., Vadicamo, L., Rabitti, F.: High-dimensional simplexes for supermetric search. In: Similarity Search and Applications: 10th International Conference, SISAP 2017, Munich, Germany, 4–6 October 2017, Proceedings, vol. 10, pp. 96–109. Springer, Heidelberg (2017). https://doi.org/10.1007/978-3-319-68474-1_7
5. Deng, J., Dong, W., Socher, R., Li, L.J., Li, K., Fei-Fei, L.: Imagenet: a large-scale hierarchical image database. In: 2009 IEEE Conference on Computer Vision and Pattern Recognition, pp. 248–255. IEEE (2009)
6. Fagin, R., Lotem, A., Naor, M.: Optimal aggregation algorithms for middleware. In: PODS, PODS 2001, pp. 102–113. Association for Computing Machinery, New York (2001). https://doi.org/10.1145/375551.375567
7. He, K., Zhang, X., Ren, S., Sun, J.: Deep residual learning for image recognition. CoRR abs/1512.03385 (2015). arxiv.org/abs/1512.03385

8. Huiskes, M.J., Lew, M.S.: The mir flickr retrieval evaluation. In: MIR 2008: Proceedings of the 2008 ACM International Conference on Multimedia Information Retrieval. ACM, New York (2008)
9. Jain, A.K., Murty, M.N., Flynn, P.J.: Data clustering: a review. ACM Comput. Surv. **31**(3), 264–323 (1999). https://doi.org/10.1145/331499.331504
10. Mark J. Huiskes, B.T., Lew, M.S.: https://press.liacs.nl/mirflickr//
11. Moss, R., Connor, R.: A multi-way divergence metric for vector spaces. In: Brisaboa, N., Pedreira, O., Zezula, P. (eds.) SISAP 2013. LNCS, vol. 8199, pp. 169–174. Springer, Heidelberg (2013). https://doi.org/10.1007/978-3-642-41062-8_17
12. Oquab, M., Darcet, T., Moutakanni, T., Vo, H., Szafraniec, M., Khalidov, V., et al.: Dinov2: learning robust visual features without supervision (2023). arxiv.org/abs/2304.07193
13. Zezula, P., Amato, G., Dohnal, V., Batko, M.: Similarity Search: The Metric Space Approach, vol. 32. Springer, Heidelberg (2006). https://doi.org/10.1007/0-387-29151-2

Diversity Similarity Join for Big Data

Yasin N. Silva[1]([✉]), Juan Martinez[1], Pedro Castro Cea[1], Humberto Razente[2], and Maria C. Nardini Barioni[2]

[1] Loyola University Chicago, Chicago, USA
{ysilva1,jmartinez29,pcastro}@luc.edu
[2] Universidade Federal de Uberlandia, Uberlândia, Brazil
{humberto.razente,camila.barioni}@ufu.br

Abstract. The Similarity Join (SJ) has become one of the most popular and valuable data processing operators in analyzing large amounts of data. Various types of similarity join operators have been effectively used in multiple scenarios. However, these operators usually generate a large output size and many similar output pairs that represent almost the same information. In previous work, a new operator called Diversity Similarity Join (DSJ) has been proposed to address these issues. DSJ generates a smaller scale output and more meaningful and diverse result pairs. This operator, however, was proposed as a single node operator crucially limiting its scalability properties. In this paper, we propose the Distributed Diversity Similarity Join (D2SJ) operator, an approach that enables SJ diversification on big datasets. We present the design guidelines and implementation details on Apache Spark, a popular big data processing framework. Our experimental results with real-world high-dimensional data show that the proposed operator has excellent performance and scalability properties.

Keywords: Diversity Similarity Join · Big Data · Performance Evaluation · Spark · MapReduce

1 Introduction

Today, big data has unprecedentedly spread to all kinds of industries. Big Data-driven decision-making has become very popular, and many applications produce and process massive amounts of data. While operators with exact-based semantics, such as the Natural Join and grouping/aggregation operators are widely used, many application scenarios, such as social-media platforms, biomedical information processing, and sensor data processing, can significantly benefit from similarity-aware operators (operators that identify and leverage similarities in the data). One of the most useful types of similarity operators is the distance range join (or sometimes referred to simply as the similarity join). This operator finds the pairs of records from two datasets that are separated at most by a distance threshold provided as a parameter (ε) [1]. Multiple similarity join implementation algorithms have been previously proposed. Some of them rely on distributed frameworks and can process massive datasets. However, the similarity join operator can

O. Pedreira and V. Estivill-Castro (Eds.): SISAP 2023, LNCS 14289, pp. 238–252, 2023.
https://doi.org/10.1007/978-3-031-46994-7_20

generate a massive amount of result pairs. Moreover, many of these output pairs can be very similar to others adding little value to the analysis process, and the output size can grow quickly when the distance threshold grows. These characteristics generate the need to diversify the output of this operator.

The idea of output diversification has been studied primarily in the context of other data operators such as range and k-nearest-neighbor search. To the best of our knowledge, the only paper directly addressing the problem of diversifying the output of the similarity join was proposed by Santos et al. [2]. This previous work proposed the Diversity Similarity Join algorithm which integrates two phases. In the first phase, the standard output of the similarity join between two datasets R and S is computed. In the second phase, the records from S that are within the distance threshold from a given record in R are processed to identify a diversified subset. Crucially, however, this previous algorithm was proposed for a single-node scenario and cannot directly scale to multiple nodes to process big datasets. In this paper, we propose the Distributed Diversity Similarity join (D2SJ) operator, a fully distributed approach to diversify the similarity join output that can be used with big datasets and multiple data types and distance metrics. The main contributions of this paper are:

- We introduce the design elements of D2SJ. A distributed operator to diversify the output of the similarity join suitable to process big datasets.
- We present the implementation details for Apache Spark [3], one of the most popular big data frameworks.
- We comprehensively assess the performance and scalability properties of D2SJ and a baseline solution. We study the performance of the operators when key parameters are increased (data size, number of nodes, dimensionality, and similarity distance threshold).
- The source code of our implementation is publicly available [4].

The remainder of the paper is organized as follows: Sect. 2 describes the related work, Sect. 3 presents the general D2SJ algorithm, Sect. 4 describes the implementation details in Apache Spark, Sect. 5 reports the performance/scalability evaluation results, and Sect. 6 presents paths for future work.

2 Related Work

In the field of similarity-aware data processing, various types of similarity joins have been proposed. These include the distance range join, which identifies pairs with distances below a predefined threshold ε [1, 5–8], the k-Distance join that returns the k most-similar pairs [9], and the kNN-join which retrieves the k nearest neighbors in one dataset for each record in another dataset [10]. The distance range join has been extensively studied and is recognized as one of the most valuable similarity-aware operators. Because of this, it is usually referred to simply as similarity join.

In the realm of Big Data systems, Hadoop [11] and Spark [3] are two commonly used platforms. Hadoop, along with its programming framework MapReduce [12], facilitates two fundamental operations, namely *map* and *reduce*. The input data is divided into multiple map tasks that process the input data chunks in parallel. Each map call takes a pair $(k1, v1)$ and produces a list of $(k2, v2)$ pairs. The output of the map calls is subsequently transferred to reduce nodes, ensuring that all intermediate records with the same intermediate key $(k2)$ are routed to the same reducer node (shuffle phase). At each reducer node, the intermediate records corresponding to a given key $k2$, are grouped and processed in a single reduce call. Spark, a more recent framework and considered a successor of Hadoop, uses Resilient Distributed Datasets (RDDs) as its fundamental data structure and supports a broader range of operations that include various types of map, reduce, grouping, filtering, and set operations. Spark operations are primarily executed in a distributed fashion utilizing the main-memory resources of a computer cluster [3].

Multiple techniques have been proposed to implement the similarity join operator on big data frameworks such as Hadoop and Spark. Several of them were experimentally compared in [13, 14]. Some of these techniques such as the Ball Hashing, Subsequence, Splitting, Hamming Code, and Anchor Points approaches [15] support string/text data and the Hamming and Edit distance functions. Other techniques such as the MRSetJoin [16] and the V-Smart-Online Aggregation [17] were proposed for set-based data and applicable distance functions such as the Jaccard and Dice Similarity. More versatile techniques such as the MRSimJoin [8, 18] and MRThetaJoin [19] can be used with a wide range of data types and distance functions.

Several approaches have also been studied to diversify the output of common data analysis operators. Most of them consider the case of the range and k-NN search operations. Drosou and Pitoura proposed DisC diversity [20], an approach to identify the representatives of a set of tuples considering coverage and dissimilarity. Both properties are defined using a distance threshold r. In this approach, each record a in the original set is represented by a record d in the diverse set, i.e., $dist(a, d) \leq r$. Also, the objects in the diverse set should be dissimilar to each other, i.e., for every pair of records d_1 and d_2 in the diverse set, $dist(d_1, d_2) > r$. Vieira et al. proposed two approaches to diversify k-NN search queries [21]. These approaches are based on the use of a ranking framework that includes a component measuring the level of relevance (with respect to the query) of the selected k records and another one measuring the diversity (distance) among these records. The framework allows the user to set a parameter to specify the relative importance of each component. More recently, Ge and Chrysanthis proposed PrefDiv [22], a technique that aims at identifying a set of dissimilar records based on user-provided distance functions and diversity thresholds on specific attributes. For instance, given two records a and b, two specified attributes A_1 and A_2, their corresponding distance functions, f_1 and f_2, and thresholds, t_1 and t_2, a and b are considered diverse if $f_1(a.A_1, b.A_1) > t_1$ and $f_2(a.A_2, b.A_2) > t_2$. This approach also aims to maximize relevance using a utility function that measures the benefit of selecting a certain record. While these diversification approaches can be used in top-k search and range search operators, the authors did not explore how these techniques can be applied to the case of similarity join operators which link elements of two sets.

To the best of our knowledge, the only previous work directly addressing the problem of diversification in the context of similarity join was proposed by Santos et al. [2]. In this approach, given two sets of records R and S, for every record $r \in R$, the algorithm identifies all records $s \in S$ that are within ε from r. Then, all identified records s are processed one by one in order of distance from r. At this point, each record s is added to the diverse set of records in connection to *r*, denoted as *DivSet(r)*, if *s* does not belong to the area of influence (neighborhood) of any previously added record. This ensures that any added record *s* is sufficiently different than the previously added records. In our work, we use a similar notion of diversity but propose a fully distributed algorithm that is suitable to process big datasets.

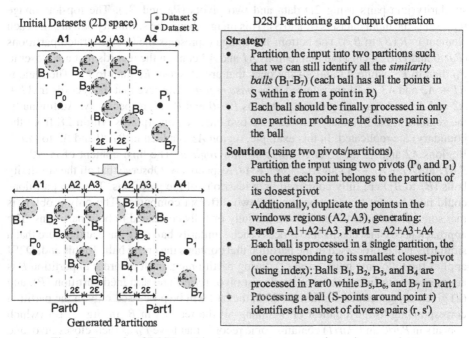

Fig. 1. Example of D2SJ partitioning and output generation using two pivots.

3 Distributed Diversity Similarity Join (D2SJ) Algorithm

The distributed diversity similarity join (D2SJ) algorithm presented in this section addresses the problem of generating a diversified subset of the similarity join output. D2SJ adopts a similar notion of diversity as the work in [2] but uses a fully distributed and parallelized approach that enables it to process very large datasets. D2SJ can be used with any data type and metric-space distance function, and is deterministic (multiple executions with the same input data and ε generate the same output).

Given two datasets that are joined (*R* and *S*), D2SJ identifies first the records in *S* that are located withing the distance threshold (ε) from the records in *R*. For every record *r* $\in$ *R*, the algorithm identifies the *similarity ball* around it (set of records *s* $\in$ *S* that are within ε from *r*, i.e., *dist(s, r)* $\leq$ ε). This process is performed in a distributed fashion but ensuring that each ball is eventually processed on a single node (to avoid duplicates in the output). In a second stage, the algorithm processes each ball (also in a parallel fashion) identifying the diverse set of records *s'* around each record *r* $\in$ *R*. For D2SJ to function across a cluster of computers to process vast amounts of data, the algorithm uses pivot-based partitioning to evenly distribute and parallelize the workload. Similar pivot-based partitioning was used in previous distributed algorithms, e.g., [7, 8, 23]. To make sure the algorithm identifies all the records in each similarity ball, in stage one, it duplicates some records from neighboring partitions.

Figure 1 shows an example of how D2SJ partitions and identifies the diversified similarity join pairs using 2D data and two pivots (P_0 and P_1). The top-left image represents the input datasets (R and S). This image shows the similarity ball around each element in R (B_1 to B_7). The bottom-left image represents the two generated partitions (*Part0* and *Part1*). Observe that regions *A1* and *A2* contain the records that are closer to P_0 than to P_1, while *A3* and *A4* the ones that are closer to P_1 than to P_0. The regions *A1* + *A2* and *A3* + *A4* are referred to as *base regions*. Observe that the records in *A1* + *A2* and *A3* + *A4* are assigned to partitions *Part0* and *Part1*, respectively. Additionally, the regions at the boundary between the two base regions (points within $2\mathcal{E}$ from the boundary) are replicated. In this example, region *A3* is added to *Part0* and *A2* to *Part1*. Regions *A2* and *A3* are referred to as *window regions*. The final content of *Part0* and *Part1* are *A1* + *A2* + *A3* and *A2* + *A3* + *A4*, respectively. Observe that all the similarity balls (B_1 to B_7) are fully contained in at least one of the partitions. The two partitions could now be sent to and processed by two different computers. The only problem is that some of the balls (B_2 to B_6) are partially or fully contained in both partitions. The approach needs a mechanism to process each similarity ball only once and ensure that a similarity ball is processed in the partition that contains the entire ball. To this end, D2SJ applies the following guidelines: (1) during partitioning, each record x in partition P is augmented with information of its closest pivot (*cPiv*) and assigned partition (P), and (2) given any generated similarity ball B, the ball will be processed only in the partition corresponding to the smallest *cPiv* among all the records in B. In Fig. 1, B4 (which appears in *Part0* and *Part1*) contains some records that have P_0 as their closes pivot and others that have P_1 as their closest one. Since the smallest one (based on index) is P_0, B_4 is processed in the partition linked to this pivot (*Part0*). Observe that while D2SJ requires replicating the records in the window regions, most useful queries involve a small $\mathcal{E}$ with a small effect on performance.

Algorithm 1 presents D2SJ's main algorithmic steps. Two sets of input data, R and S, are merged into one dataset (line 1), with pivots being selected from this combined set (line 2). After selecting the pivots, the algorithm partitions the data (lines 3–12), allowing for an even distribution of the data to be processed in each cluster node. Every input record *rec* is assigned to the partition of its closest pivot p_c (lines 5–6) and all the partitions of pivots p where *rec* belongs to the window regions between the partitions of p and p_c (lines 7–11). In general, the records in the window regions between two partitions (corresponding to pivots *p1* and *p2*) should be a superset of all the records within $2\mathcal{E}$ from the hyperplane that separates the partitions. However, this hyperplane does not always explicitly exist in a metric space. Instead, it is implicit and known as the generalized hyperplane. Since the distance of a record *rec* to the generalized hyperplane between two partitions for pivots *p1* and *p2* cannot always be computed exactly, a lower bound is used following [25] (line 8): *genHyperplaneDist(rec, p1, p2)* = (*dist(rec, p1)* − *dist(rec, p2)*) / 2. This expression can be replaced by the exact distance when this can be computed, e.g., for the Euclidean distance, *genHyperplaneDist* can be replaced by *euclideanHyperplaneDist*$(rec, p1, p2) = |(dist(rec, p1)^2 - dist(rec, p2)^2| / (2 \times dist(p1, p2)$. The partitioning phase also records the information of the closest pivot and assigned partition of each record (sequence of closest pivots and partitions if the execution requires multiple rounds). This information is used later in the process. The partitioning phase of D2SJ can be implemented using the map operations in Spark or Hadoop.

Algorithm 1: *DistDivSimJoin_Main*

Input: *input_R* (input dataset *R*), *input_S* (input dataset *S*), *eps* (radius),
  part_num (number of partitions), *mem_T* (memory threshold)

Output: diversified set of similaritry join pairs

```
1    input = input_R ∪ input_S
2    pivots = selectPivots(part_num, input)
3    //Partitioning - // rec: ⟨ID, dataset, value, assignedPartitionSeq, closestPivotSeq⟩
4    for each record rec in input do
5        pc = getClosestPivot(rec, pivots)
6        output ⟨pc, rec⟩ //intermediate output – base region
7        for each pivot p in pivots do
8            if ((distance(rec, p) - distance(rec, pc)) / 2 ≤ 2eps) then
9                output ⟨p, rec⟩ //intermediate output - window region
10           end if
11       end for
12   end for
13   //Shuffle: all the records sharing the same key will form a partition
14   //Similarity ball generation and diversification
15   for each partition Pᵢ do //each partition may be processeed in a different node
16       if (Pᵢ.memSize() > mem_T) do
17           store Pᵢ for processing in subsequent round
18       else
19           Bᵢ = IdentifySimBalls(Pᵢ, eps) //Bᵢ (ball set) format: {Bᵢ_k}
20               //Bᵢ_k: ⟨centerPoint, records, flags⟩, flags contains partitioning data
21           //Output Generation (preventing duplication)
22           for each similarity ball Bᵢ_k in Bᵢ do
23               generate minFlags for Ballᵢ_k //minFlags[q] = {index of first element in
24                                             //Ballᵢ_k.flags[q] equal to 1}
25               aPartitioningSeq = s.assignedPartitionSeq() //s is any record in Bᵢ_k
26               if (∀q, minFlags[q] = aPartitioningSeq[q]) then //if we are in the
27                                             //selected partition to process this similarity ball
28                   divBᵢ_k = Diversify(Bᵢ_k) //diversify this similarity ball
29                   output divBᵢ_k //final output
30               end if
31           end for
32       end if
33   end for
```

Alg. 1. Main D2SJ algorithm.

The intermediate records generated in the partitioning phase are grouped in the shuffle phase (line 13) such that all the records that belong to the same partition will form a single group. This task is implemented using the grouping operator in Spark and would be automatically performed in the shuffle phase of a Hadoop job. In the next phase, partitions are processed (1) identifying the similarity balls contained in each partition, (2) determining if a similarity ball should be processed on a given cluster node, and (3) diversifying and outputting the selected similarity ball (lines 14–33). Different partitions could be processed on different nodes. For a given partition, the algorithm first checks if the partition is small enough to be efficiently processed in a single node (line 16). If this is not the case, the partition is stored for further processing using the same D2SJ algorithm but applied to this single partition (line 17). This feature makes D2SJ a multi-round algorithm where at every round the small partitions are directly processed, and the large partitions are stored for processing in subsequent rounds. It is important to observe, however, that while D2SJ can be executed in multiple rounds, the best execution times in our experimental tests were obtained by increasing the number of pivots to generate a single round (with smaller partitions). When the partition is small enough to be processed in the current round, the algorithm identifies first the similarity balls contained in this partition (line 19). The details of this process are described later (Algorithm 2: *IdentifySimBalls*). Each similarity ball contains the records s within *eps* (ε) from a given record r used as a center point. The output of *IdentifySimBalls* is a set of similarity balls where each ball is composed of a center point (from R), the data records (from S), and information needed to ensure non-duplicated ball processing (*flags*). The *flags* component of a given ball B contains a sequence of flag arrays (one array per round that processed data that included this ball). This component is used to determine if the ball should be processed in the node processing the current partition or not (lines 23–30). For example, if four pivots are being used (p_0, p_1, p_2, p_3) and a single round is needed, $B.flags$ has the form $\{[f_0, f_1, f_2, f_3]\}$ and the content could be $\{[0, 0, 1, 1]\}$. A value of 1 at index i indicates that ball B contains at least one record whose closest pivot is p_i. In this example, B contains records with base region equal to p_2 and others with base region equal to p_3. A given ball of partition P_i (corresponding to pivot p_i) will be processed in the current node only if the minimum index of 1 in the flag array of this ball matches i. In the example, ball B will be processed only when this ball is detected in the partition of p_2 (because the smallest index with a value of 1 is 2). If a ball should be processed in the current node, the algorithm applies the diversification method (Algorithm 3: *Diversify*). This method processes a similarity ball and generates the subset of diversified similarity join pairs, where each pair is composed of the center point and one of the (S) records in the ball. The set of diverse SJ pairs is then added to the final output of D2SJ (line 29). The similarity ball generation and diversification phase can be performed using the reduce operations in Spark or Hadoop.

The details of *IdentifySimBalls* are presented in Algorithm 2. This algorithm identifies the similarity balls in an input partition. Each similarity ball is composed of a center point r (a record in R), a set of records (records in S within ε from r) and a set of flags which contain partitioning information. The algorithm separates first the records from R and S (lines 4–11). In our implementation, we use an initial algorithm to identify the similarity balls as this will be executed on a relatively small set of records and on a single node. Any other single-node similarity join algorithm could be integrated to identify the balls. In our case, a data structure for a ball is initialized in line 15. For each record r (from R) in the partition, the algorithm checks if the available records from S are within ε from r. All the qualifying records are added the ball of r (lines 16–21). The algorithm, then, generates the *flags* component of the ball using the closet-pivot information of the records in the ball (line 22). After this, the generated ball is added to the set of balls identified in the current partition (line 23).

Algorithm 2: *IdentifySimBalls*

Input: P_i (a partition), *eps* (radius)

Output: B_i (similarity balls in partition P_i, ball structure: ⟨*centerPoint, records, flags*⟩)

```
1   inputR = {}
2   inputS = {}
3   Bi = {}
4   for each record rec in Pi do
5       if (rec.dataset = 0) then //dataset
6                          //values: 0 (R), 1(S)
7           inputR.add(rec)
8       else
9           inputS.add(rec)
10      end if
11  end for
12  //Generation of similarity join balls
13  for each record r in inputR do //creates a
14      //ball around each records in R
15      b = {r, [], []}//r is the center point
16      for each record s in inputS do //find
17          //the records in S similar to r
18          if distance(r, s) ≤ eps then
19              b.records.add(s)
20          end if
21      end for
22      generateFlags(b) //updates b.flags
23      Bi.add(b) //adds the ball to the set
24  end for
25  return Bi
```

Alg. 2. Identification of Similarity Balls.

Algorithm 3: *Diversify*

Input: b (a similarity ball)

Output: b' (diversified similarity ball)

```
1   c = b.centerPoint
2   f = b.flags
3   sort(b.records, c) //sort the records in the
4       //ball in increasing distance from the
5       //center point c
6   b' = {c, [], f} //initializing the
7       //diversified ball
8   for every record s in b.records do
9       isDiverse = True
10      for each record d in b'.records do //this
11          //will be empty initially, but will
12          //get filled as diverse elements
13          //are discovered
14          if inInfluenceArea(s, d, c) then //if
15              //s is too similar to the diversified
16              //record d
17              isDiverse = False
18              break
19          end if
20      end for
21      if (isDiverse = True) then
22          b'.records.add(s)
23      end if
24  end for
25  return b'
```

Alg. 3. Diversification of Similarity Balls.

Algorithm 3 presents the details of the *Diversify* subroutine. The goal of this algorithm is to diversify the records of a similarity ball using a similar notion of diversity as in [2]. This algorithm receives a similarity ball (b) and returns a similarity ball (b') that contains a diverse subset of the records. The algorithm sorts first the records in the input ball as they will need to be processed in increasing distance from the center point (line 3). The structure for the diversified ball is initialized in line 6. The set of diverse records ($b'.records$) is initially empty. Then, the algorithm processes each record s from the input ball (lines 8–24). In each iteration, the algorithm verifies that s is diverse from every other record already in the diverse set (lines 10–20). If this is the case, s is added to the diverse set $b'.records$ (lines 21–23). Observe that if s fails the diversity test with an already added diverse record d in line 14, s is considered not diverse enough and the algorithm stops the process of checking with additional diverse records (lines 17–18). At the end, the method returns the diversity similarity ball b'.

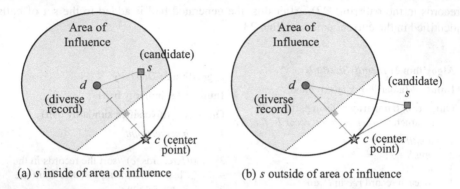

(a) s inside of area of influence   (b) s outside of area of influence

Fig. 2. Examples of different outcomes of *inInfluenceArea* with 2D data.

A key aspect of this algorithm is checking if a record s (that belongs to a ball centered in c) is contained in the area of influence of an already added diverse record d. This check is performed by *inInfluenceArea(s, d, c)*. Building on the work in [2], *inInfluenceArea(s, d, c)* returns True if $I(d,s) \geq I(d,c)$ and $I(d,s) \geq I(s,c)$, where I is the inverse of the distance function. Thus, *inInfluenceArea(s, d, c)* returns True if:

$$\left(\frac{1}{dist(d,\, s)} \geq \frac{1}{dist(d,\, c)}\right) \wedge \left(\frac{1}{dist(d,\, s)} \geq \frac{1}{dist(s,\, c)}\right),$$

or, equivalently, if $(dist(d,c) \geq dist(d,s) \wedge dist(s,c) \geq dist(d,s))$.

The intuition is that this check will return true if s (a record of a ball centered in c) belongs to the neighborhood of d. Figures 2.a and 2.b show examples of the two different outcomes for the case of 2D data and the Euclidean distance. Observe that in this case, the first condition $(dist(d,c) \geq dist(d,s))$ checks if s is contained in the circle centered in d with radius $\overline{dc}$, , and the second one $(dist(s,c) \geq dist(d,s))$ checks if s is closer to d than to c. The shaded area in both images is the area that would be considered the area of influence of record d. In Fig. 2.a, s belongs to this area and will be considered not diverse enough. In Fig. 2.b, s does not belong to the area of influence.

4   Implementation

Section 3 presented the algorithmic steps of D2SJ. This algorithm could be implemented on any MapReduce-based framework, e.g., Hadoop and Spark. Section 3 also indicated the main Spark and Hadoop operations for the core phases of the algorithm. As Spark is broadly considered a more efficient successor of Hadoop, we implemented the algorithm in Spark. The source code is available in [4].

In this section, we provide some additional implementation details. The implementation in Spark uses the RDD API. Spark's robust array of data processing operations enables a compact implementation. The *takeSample* operation is used to randomly select the pivots. Then, the *flatMapToPair* operation is used to implement the partitioning phase and the *partitionBy* operation to group the records that belong to the same partition (Shuffle phase). After this, *mapPartitionsToPair* is used in the implementation of *IdentifySimBalls*, which identified the similarity balls in a partition. Finally, *map* and *saveAsTextFile* are used to generate the final output.

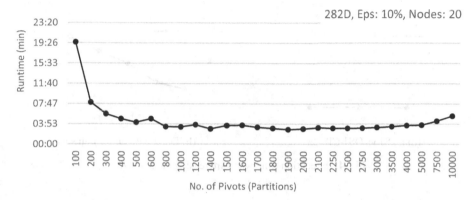

Fig. 3. Effect of the number of pivots (partitions) on execution time.

5 Performance Evaluation

5.1 Test Configuration

In this section we evaluate the performance and scalability properties of D2SJ. We also compare D2SJ with DSJ-CP, a direct Spark extension of the single-node algorithm presented in [2] (which uses a cartesian product to perform the similarity join). Both algorithms were implemented in Spark 3.0. Unless otherwise stated, all tests were executed using a cluster composed of 1 master and 20 worker nodes on the Google Cloud Platform. Each node used the Cloud Dataproc 2.0 image and had 4 virtual CPUs, 15 GB of memory, and 500 GB of disk space. The number of splits per Spark job was set to 2 × (# of worker nodes) × (# of vCPUs).

We used real data to perform our experiments. Specifically, we used the CoPhIR data collection [24], which is composed of visual descriptors extracted from 100 million images from Flickr. We used the following collections: Color Structure (CS, 64D), Scalable Color (SC, 64D), Edge Histogram (EH, 80D), Color Layout (CL, 12D), and Homogeneous Texture (HT, 62D). The datasets for different dimensionalities were generated as follows: 16D, 32D, and 64D: first 16, 32, and 64 attributes of CS, 128D: CS + SC, 208D: CS + SC + EH, and 282D: CS + SC + EH + CL + HT. The dataset for scale factor N (SFN) had $1,000,000 \times N$ records. These records were equally divided to form the R and S datasets. The value of $\mathcal{E}$ is expressed as the percentage of the maximum potential distance between two records.

Next, we compare how the various parameters affect D2SJ and DSJ-CP (except for varying the pivot count which is only applicable to D2SJ). Since DSJ-CP does not scale as well as D2SJ, we provide two graphs in each case. In the first, we scale down the experimental settings to maintain execution times under 10 h and avoid stack overflow errors in DSJ-CP's cartesian product. In the second, we evaluate D2SJ under larger workloads. The settings of each test appear in the top-right label of the figure.

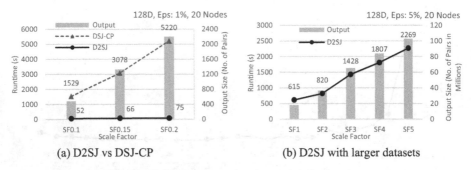

(a) D2SJ vs DSJ-CP   (b) D2SJ with larger datasets

Fig. 4. Execution time when increasing dataset size.

5.2 Performance Evaluation Results

Optimal Pivot Count. Figure 3 shows how D2SJ's execution time changes when the number of pivots increases (since a partition is generated for each pivot, this is equal to the number of partitions). This test uses the 282D SF5 dataset, a cluster with 20 worked nodes, and a large distance threshold ($\mathcal{E} = 10\%$). We observe that the execution time quickly decreases when the number of pivots increases initially. In general, larger numbers of pivots generate smaller execution times. However, exceeding 4,000 pivots leads to excessive replication in the window regions and increased execution times. The optimal pivot count is between 800 and 3500. These pivot counts solve the job in a single round. We use *numPivots* $= 400 \times$ SF in the remaining tests.

Increasing Scale Factor. Figure 4.a shows how the execution times of D2SJ and DSJ-CP (lines) and the output size (bars) vary when the scale factor (data size) increases. These tests used scaled-down parameters to enable the comparison ($\mathcal{E} = 1\%$, SF:[0.1–0.2]). We can observe that the execution times of D2SJ increase slowly as the scale factor increases. DSJ-CP's execution times, on the other hand, are significantly larger than those of D2SJ and grow rapidly. In fact, the execution time of DSJ-CP grows from being 29 times the execution time of D2SJ for SF 0.1 to 70 times for SF 0.2. Figure 4.b shows D2SJ's execution times with heavier settings ($\mathcal{E} = 5\%$ and SF:[1–5]). D2SJ's execution time grows gracefully following a semi-linear pattern. In this case, none of the DSJ-CP jobs were able to finish in under 10 h.

Increasing Scale Factor and Number of Cluster Nodes. A desired property in distributed algorithms is to have good scalability when the data size and number of nodes increase proportionally. While some overhead is expected with larger loads, a reduced overhead is desired. Figure 5.a shows the execution times of D2SJ and DSJ-CP as the data size and number of worker nodes increase from (4 nodes, SF 0.05) to (16 nodes, SF 0.2). The results with these scale-down settings show that D2SJ scales significantly better than DSJ-CP. The execution time of DSJ-CP with the largest SF is 11 times the one with the smallest SF. In the case of D2SJ, the increase is only 2.8 times. Figure 5.b presents D2SJ's execution time with larger workloads increasing from (4 nodes, SF 1) to (20 nodes, SF 5). We can observe that D2SJ scales well producing an execution time that is linear with a relatively small slope. D2SJ's execution time with SF5 (and 20 nodes) is only 2.2 times its execution time with SF1 (and 4 nodes).

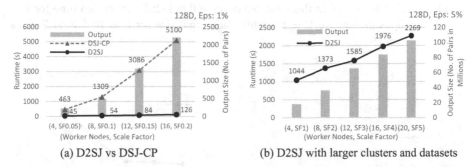

(a) D2SJ vs DSJ-CP (b) D2SJ with larger clusters and datasets

Fig. 5. Execution time when increasing dataset size and number of worker nodes.

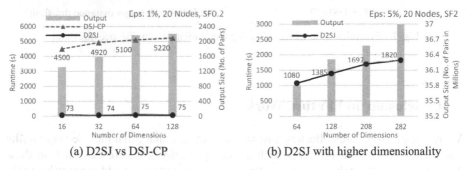

(a) D2SJ vs DSJ-CP (b) D2SJ with higher dimensionality

Fig. 6. Execution time when increasing the number of dimensions.

Increasing Number of Dimensions. To evaluate the effect of data dimensionality on execution time, we executed each algorithm with datasets of varying dimensionality while fixing the scale factor. Figure 6.a shows the execution time of both algorithms using SF 0.2 and 16D-128D datasets. In general, the execution time of both algorithms increases when dimensionality increases. D2SJ, however, has better scalability. While DSJ-CP's execution time with 128D represents an increase of 16% with respect to the 16D dataset, the increase is only of 2% for D2SJ. Figure 6.b shows the execution time of D2SJ with larger workloads (SF2 and 64D-282D). This figure shows that the execution time of D2SJ increases sublinearly in this dimensionality range.

Increasing Distance Threshold (ε). The use of diversification was motivated in part due to the large number of output records generated by traditional similarity join operations (with many output records being very similar to others). The work in [2] showed that the output of the diversified similarity join returns a very small fraction of the output of the standard similarity join. However, increasing the distance threshold (ε), still has a significant effect on the overall output size and execution time. In this experiment, we evaluate the execution time of both algorithms when ε increases. Figure 7.a shows the execution time of D2SJ and DSJ-CP when ε increases from 0.25% to 1%. We can observe that the execution time of D2SJ is significantly better than that of DSJ-CP. D2SJ's execution time grows from 67s ($\varepsilon = 0.25\%$) to 75s ($\varepsilon = 1\%$), while the growth for DSJ-CP is from 4680 s to 5220 s. Figure 7.b shows D2SJ's execution time with larger workloads (ε:[1%–10%]). In this case, we observe that D2SJ's execution time for $\varepsilon = 10\%$ is 4.2 times the one for $\varepsilon = 1\%$ while the output size for $\varepsilon = 10\%$ is 51.7 times the one for $\varepsilon = 1\%$.

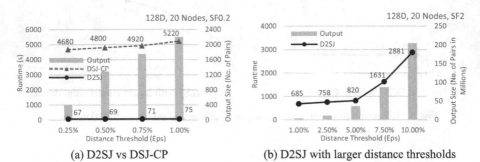

(a) D2SJ vs DSJ-CP   (b) D2SJ with larger distance thresholds

Fig. 7. Execution time when increasing the similarity join distance threshold.

6   Conclusion and Future Work

Many organizations are collecting vast amounts of data that often include very similar data items. When data operators such as the Similarity Join, are executed on these datasets, the results include many similar output pairs that do not add much value to the understanding of data patterns. To address this problem in the case of similarity joins, previous work explored the integration of a diversification step. This previous work, however, was proposed for small data on a single computer. In this paper, we present D2SJ, a distributed approach to solve the diversity similarity join problem with big data. D2SJ can be used with multiple data types and distance functions. We present a detailed description of D2SJ as well as implementation details in Apache Spark. Moreover, we also present experimental results with real datasets that show strong performance and scalability properties. Future areas of research building on the results of this work could include (1) the comparative study of additional ways to diversify the output of different types of similarity join for big data and (2) the development of efficient distributed algorithms supporting these new notions of diversity.

Acknowledgments. This project was supported by an award from the Google Cloud Research program. The authors would like to thank Steven Hu, Timothy Raymer, and Steven Anderson for their contributions in the preliminary stages of this project.

References

1. Böhm, C., Braunmüller, B., Krebs, F., Kriegel, H.-P.: Epsilon grid order: an algorithm for the similarity join on massive high-dimensional data. In: SIGMOD (2001)
2. Santos, L.F.D., Carvalho, L.O., Oliveira, W.D., Traina, A.J.M., Traina Jr., C.: Diversity in similarity joins. In: Amato, G., Connor, R., Falchi, F., Gennaro, C. (eds.) SISAP 2015. LNCS, vol. 9371, pp. 42–53. Springer, Cham (2015). https://doi.org/10.1007/978-3-319-25087-8_4
3. Apache. Spark. https://spark.apache.org/
4. SimCloud Research Team. D2SJ Source Code. https://ysilva.cs.luc.edu/SimCloud/downloads.html
5. Dohnal, V., Gennaro, C., Zezula, P.: Similarity join in metric spaces using ED-Index. In: Mařík, V., Retschitzegger, W., Štěpánková, O. (eds.) DEXA 2003. LNCS, vol. 2736, pp. 484–493. Springer, Heidelberg (2003). https://doi.org/10.1007/978-3-540-45227-0_48
6. Dohnal, V., Gennaro, C., Savino, P., Zezula, P.: Similarity join in metric spaces. In: Sebastiani, F. (ed.) ECIR 2003. LNCS, vol. 2633, pp. 452–467. Springer, Heidelberg (2003). https://doi.org/10.1007/3-540-36618-0_32
7. Jacox, E.H., Samet, H.: Metric space similarity joins. ACM Trans. Database Syst. **33**(2), 1–38 (2008). https://doi.org/10.1145/1366102.1366104
8. Silva, Y.N., Reed, J.M., Tsosie, L.M.: MapReduce-based similarity join for metric spaces. In: VLDB/Cloud-I (2012)
9. Hjaltason, G.R., Samet, H.: Incremental distance join algorithms for spatial databases. In: SIGMOD (1998)
10. Böhm, C., Krebs, F.: The k-nearest neighbour join: turbo charging the KDD process. KAIS **6**, 728–749 (2004)
11. Apache. Hadoop. https://hadoop.apache.org/
12. Dean, J., Ghemawat, S.: MapReduce: simplified data processing on large clusters. In: OSDI (2004)
13. Silva, Y.N., Reed, J., Brown, K., Wadsworth, A., Rong, C.: An experimental survey of MapReduce-based similarity joins. In: Amsaleg, L., Houle, M.E., Schubert, E. (eds.) SISAP 2016. LNCS, vol. 9939, pp. 181–195. Springer, Cham (2016). https://doi.org/10.1007/978-3-319-46759-7_14
14. Fier, F., Augsten, N., Bouros, P., Leser, U., Freytag, J.-C.: Set similarity joins on MapReduce: an experimental survey. Proc. VLDB Endow. **11**(10), 1110–1122 (2018). https://doi.org/10.14778/3231751.3231760
15. Afrati, F.N., Sarma, A.D., Menestrina, D., Parameswaran, A., Ullman, J.D.: Fuzzy joins using MapReduce. In: ICDE (2012)
16. Vernica, R., Carey, M.J., Li, C.: Efficient parallel set-similarity joins using MapReduce. In: SIGMOD (2010)
17. Metwally, A., Faloutsos, C.: V-SMART-join: a scalable MapReduce framework for all-pair similarity joins of multisets and vectors. Proc. VLDB Endow. **5**(8), 704–715 (2012). https://doi.org/10.14778/2212351.2212353
18. Silva, Y.N., Reed, J.M.: Exploiting MapReduce-based similarity joins. In: SIGMOD (2012)
19. Okcan, A., Riedewald, M.: Processing theta-joins using MapReduce. In: SIGMOD (2011)
20. Drosou, M., Pitoura, E.: DisC diversity: result diversification based on dissimilarity and coverage. In: CIKM (2010)

21. Vieira, M.R., et al.: On query result diversification. Inf. Syst. **42**, 57–77 (2014)
22. Ge, X., Chrysanthis, P.K.: PrefDiv: efficient algorithms for effective top-k result diversification. In: EDBT (2020)
23. Silva, Y.N., Sandoval, M., Prado, D., Wallace, X., Rong, C.: Similarity grouping in big data systems. In: Amato, G., Gennaro, C., Oria, V., Radovanović, M. (eds.) SISAP 2019. LNCS, vol. 11807, pp. 212–220. Springer, Cham (2019). https://doi.org/10.1007/978-3-030-32047-8_19
24. Bolettieri, P., et al.: CoPhIR: A Test Collection for Content-Based Image Retrieval. arXiv: 0905.4627 (2009)
25. Hjaltason, G.R., Samet, H.: Index-driven similarity search in metric spaces (survey article). TODS **28**, 517–580 (2003)

Indexing Challenge

Overview of the SISAP 2023 Indexing Challenge

Eric S. Tellez[1,2], Martin Aumüller[3], and Edgar Chavez[2(✉)]

[1] INFOTEC-IxM CONACyT, Aguascalientes, Aguascalientes, Mexico
eric.tellez@ieee.org
[2] CICESE, Ensenada, Baja California, Mexico
elchavez@cicese.mx
[3] ITU Copenhagen, Copenhagen, Denmark
maau@itu.dk

Abstract. This manuscript presents the premiere **SISAP 2023 Index-ing Challenge**, which seeks replicable and competitive solutions in the realm of approximate similarity search algorithms. Our aim is recall, all while optimizing build time, search time, and memory consumption. Using a subset of the deep features of a neural network model provided by the LAION-5B dataset, the challenge posed three tasks, each with its unique focus:

- **Task A**: Conduct classical approximate nearest neighbor search, ensuring an average recall of at least 0.9 for 10-NN queries.
- **Task B**: Find a succinct binary embedding of the original data that ensures high recall on the original data.
- **Task C**: Index and search binary representations from Task B.

Notably, an innovative and competitive binary mapping method emerged from the challenge. It also spotlighted graph methods as the preferred indexing technique for binary and real-valued high-dimensional vectors. However, these methods have little room for improvement. Enhancing memory efficiency, refining navigational strategies, and tackling the sec-ondary memory challenge are pivotal next steps.

Keywords: Approximate nearest neighbor search · Indexing and searching pipelines · Experimental comparison of search methods

1 Introduction

Similarity search algorithms are pivotal for efficiently retrieving similar items from vast datasets, underpinning tasks like information retrieval, multimedia indexing, and pattern recognition. As machine learning, dense retrieval, and computer vision become increasingly prevalent, similarity search methods must meet the quality and computational demands of both applications and the systems they run on.

The *curse of dimensionality* [3] dictates that all metric search algorithms falter when confronted with high-dimensional datasets. This matter necessitates

O. Pedreira and V. Estivill-Castro (Eds.): SISAP 2023, LNCS 14289, pp. 255–264, 2023.
https://doi.org/10.1007/978-3-031-46994-7_21

adopting approximate or probabilistic methods to balance speed against quality. Furthermore, there's an opportunity to trade-off between construction time and memory usage, leading to a variety of indexing solutions, each with its merits and drawbacks.

The SISAP Indexing Challenge[1] seeks to identify efficient similarity search algorithms that strike a balance between accuracy and practical constraints like build time, search time, and memory consumption. To facilitate this, we devised a test bed utilizing the LAION deep features English subset, segmented into 10M, 30M, and 100M benchmarks. Additionally, there are two query sets: public and private, each comprising 10k vectors. The public queries were made available during the call for papers, while the private ones were revealed post-submission and evaluation.

2   The Dataset

The LAION dataset, as detailed in [15], is an expansive public image collection comprising both images and textual descriptors. It has proven instrumental in training large visual and language deep-neural models, as cited in [4,13]. Every image within the collection is paired with a URL handle, simplifying the demonstration process. Moreover, the LAION consortium has made vector embeddings available using the Contrastive Language-Image Pre-Training (CLIP), specifically harnessing the OpenCLIP architecture [4]. These deep features manifest as 768-dimensional vectors, represented using 16-bit floating point numbers. The CLIP architecture was initially introduced in [12].

We employed three subsets from the English segment of the LAION collection (commonly referred to as LAION2B) as benchmarks. These subsets consist of 10, 30, and 100 million vectors, with vectors labeled as *Not Safe for Work* (NSFW) duly excluded. Further insights regarding the selection and packaging of these subsets can be found on the challenge's companion site.

3   Task Descriptions

The Indexing Challenge focuses on nearest neighbor queries, specifically on approximate k nearest neighbor queries. We have established three tasks that emulate various application scenarios, each catering to different needs in terms of quality, speed, and memory.

A key aspect of this challenge is reproducibility. Submissions were accepted in the form of Github repositories with operational Github Action (GHA) workflows.[2] Teams crafted their solutions by meticulously setting and benchmarking hyperparameters for each task and clearly detailing their choices in their GHA entry point.

---

[1] Official site of the challenge https://sisap-challenges.github.io/.

[2] Github Actions is a continuous integration platform that enables continuous testing of repositories within virtual machines.

We designed four benchmarks from the LAION2B dataset, each with a distinct number of vectors: 300K for development and 10M, 30M, and 100M workloads designated for the challenge. Furthermore, we designed two sets of public and private queries. Teams were tasked with designing their solutions and determining the hyperparameters based on the public query set. The private set was subsequently used to re-test and rank all solutions on our system. We computed gold standards for k nearest neighbor queries in public and private queries, which is the foundation for calculating the recall score in the final results. Public queries, along with their associated gold standards, were available from the commencement of the challenge, while private ones were unveiled post-validation. We expected teams to construct indexes that efficiently solve queries and excel under the specific conditions and metrics defined for each task. All tasks revolved around retrieving the approximate $k = 10$ nearest neighbors.[3]

During the challenge, Vladimir Míč (private communication) reported certain anomalies he detected in the public gold standard. He highlighted numerous distance value ties at k and $k + 1$ neighbors and instances of neighbors at distance 0, i.e., duplicates. Upon confirmation, we ascertained that these discrepancies were likely due to the prevalence of near duplicates in the LAION database [20]. For the private query set, we implemented measures to curb these issues, utilizing IEEE 754 floating point arithmetic to compute distance functions in the gold standard and excluding query objects where the k and $k + 1$ neighbors matched identically.[4] The subsequent segments of this section delve deeper into the intricacies of the tasks.

Task A: Searching the Original Embeddings. Task A focuses on high-throughput solutions with little loss in quality. The aim is to design the fastest search algorithm that hits a recall of at least 0.9 (on average over all queries). Teams adopted the strategy to build a single index and used a large collection of search parameters for each subset size. A small catch is that only the best-performing probe in the *private query set* gets to stay. Repositories should be ready to run right out of the box with all the settings in place. Note that teams likely used the public query set to fine-tune their settings.

Task B: Producing Binary Sketches. This task concerns the succinct representation of the original 768-dimensional real-valued vectors using fixed-length binary strings. These have a much smaller memory footprint and allow efficient distance calculations via SIMD instructions. The main goal of this task is to find embeddings that, using Hamming distance and a linear scan, produce a higher average recall than our baseline of 0.24. The baseline uses our current go-to method based on permutation binary sketches [18].

---

[3] Gold standards incorporate results up to $k = 1000$.
[4] This constraint was determined using the 100 million benchmark.

Task C: Indexing and Searching on Binary Sketches. For Task C, the challenge seeks solutions that let us first index and then search using these binary sketches while using the Hamming distance as our measuring stick. Participants can use the embeddings they came up with in Task B or go with the baseline embeddings. The benchmark? The fastest solutions with a recall close to ours, meaning they should achieve or surpass a recall of 0.216, i.e., 90% of our baseline.

4 Solutions Overview

This section describes the set of solutions to the SISAP Indexing Challenge. The solutions use diverse programming languages: C++, Rust, Java, and some use Python as a wrapper language. One baseline uses C++ with Python wrappers, and the rest uses the Julia programming language. The teams used different techniques to tackle the challenge: graph-based indexes, hashing-based indexes, linear projections, reranking, combinatorial and numerical optimization, among others.

4.1 Baselines

We included three baselines to compare with previous work: BL-SearchGraph, BL-FAISS-HSNW, and Bruteforce. Note that the first two are far from trivial solutions. The rest of this section describes our baselines and explains its construction and searching hyperparameters.

Bruteforce. This is a straightforward solution. It is implemented as an exhaustive search using the `SimilaritySearch.jl` package. However, as with the rest of our baselines, this approach takes advantage of the multithreading capabilities of our running infrastructure. Unsurprisingly, a well-implemented brute force algorithm can improve more sophisticated algorithms when the intrinsic dimension of the data is high.

BL-FAISS-HNSW. This baseline uses the HNSW index from FAISS.[5] The Hierarchical Navigable Small World (HNSW) index, see [8], is a graph-based index using a hierarchical structure to navigate the graph efficiently. It is created iteratively, adding one new object at a time. The ith element is inserted by adding edges from the ith element to a set of M approximate nearest neighbors using the graph containing the previous $i - 1$ objects; the hierarchy is maintained throughout the construction. The search algorithm consists of navigating the graph greedily using two priority queues. The first is the result set of size ef, and the second is a candidate list to prioritize the navigation. The search is conducted in rounds: The not-yet-visited current closest point to the query is inspected at each round. The search finishes when it is impossible to improve

---

[5] https://faiss.ai/.

the result set during the navigation. Due to its flexibility, the HNSW index is the *de-facto* standard in the industry; most vector databases also implement it. According to standard benchmarks [1], it is one of the faster metric indexes known. As a baseline, we set its parameters as follows. We set the $M = 32$ for all subsets and the *ef* parameter as 40 for construction. At the search stage, we probe with the following *ef* values: 32, 64, 128, 256, and 512.

BL-SearchGraph. This baseline uses the SearchGraph index from Julia's package SimilaritySearch.jl, see [16,17]. This index is a graph-based index similar to the HNSW, but instead of a hierarchy, it uses a small sample of disjoint neighbors to get fast navigation. The construction is also based on connecting the ith element with its neighbors, but it is simplified since there is no hierarchy. In contrast to HNSW, it uses variable-size neighborhoods using shrinking heuristics based on the Spatial Access Trees [10], with an upper bound defined as $M = O(\log i)$. It uses Beam Search (BS) as a search algorithm. The search stores candidates in a priority list of maximum size (beam size) and also limit what is considered to be inserted into the beam using a parameter $0 < \Delta < 2$; the result set is populated during the navigation, and the search finishes when the result set does not improve and the beam is empty. It supports single-pass automatic index optimization for a given quality score. It is a flexible alternative that supports automatic optimization and user-defined metrics, the latter due to Julia's just-in-time compiler. As a baseline, it was constructed with 0.9 as objective recall and a neighborhood size of $M = \log_{1.5} i$. During the search stage, we varied the optimized Δ parameter in the range $\Delta/1.05^2 \leq \Delta' < 2$ growing exponentially in a 1.05 factor.

4.2   Teams Solutions

Six teams submitted a candidate for evaluation; one team (HIOB) targeted all three tasks, one team (SWANN) focused on indexing binary sketches (Task C), the remaining teams (UTokyo, CRANBERRY, LMI, HSP) focused on efficient retrieval in the standard setting (Task A). Teams used their implementations and modifications or tuning of well-known approximate nearest neighbor search libraries.

UTokyo. This approach proposes a pipeline of dimensionality reduction, database subsampling, and entry point optimization to solve Task A. The pipeline is designed for graph-based indexes and optimizes the computational requirements in terms according to specified accuracy, runtime, and memory requirements. It employs black-box optimization for parameter tuning. In particular, the authors optimized a Navigating Spreading-out Graph (NSG) [7] with a neighborhood of 32 as index for their submission. More details are given in [11].

CRANBERRY. This approach combines several search techniques in a three-stage pipeline: data partitioning, candidate filtering, and reranking. The input database is divided into a Voronoi partition. The search algorithm locates the nearest partitions to the query to retrieve a list of potentially similar vectors. Then, 512-bit sketches and 24-dimensional prefixes of vectors are used to reduce the candidate list; an early termination strategy accompanies the filtering. The list of candidates is reranked using the original distance, that is, the 768-dimensional CLIP vectors and the cosine distance. CRANBERRY is designed to solve Task A. More details are found in [9].

SWANN. This approach uses a collection of tries together with the bit-sampling locality-sensitive hashing scheme [2]. Their solution targets Track C (indexing binary vectors). During index building, each binary vector is hashed $K \cdot L$ times, and each block of length-K bit strings is used to insert the vector into one out of L tries. The query vectors are hashed and looked up in the trie during the search. If too few candidates are found on the leaf level, the search is dynamically expanded to cover larger parts of the trie. More details are given in [14].

HIOB. This approach is based on creating binary sketches of a vector database explicitly designed for cosine similarity. The binarization is made through hyperplanes, i.e., encoding where the vector lies. The random sample consensus inspires the initialization of the set of encoding hyperplanes, RANSAC [5]. Then, the encoders are refined by maximizing bit independence and bit balance for binary sketches on the unit sphere. The iterative optimization process improves sketch quality through orthogonalization and is made in small batches, similar to stochastic gradient descent. In each iteration, a displacement vector is computed to update some hyperplanes. The bit assignments are recomputed after each batch. In the search stage, the bit-vectors under hamming distance are used to calculate a candidate list of size $n^\star$, this process is computed with a brute force procedure since the authors found no improvement on using HNSW or faiss [19]. The candidate set is reranked with the original database objects and cosine distance to get the k nearest neighbors.

We used the specified hyperparameters for each task and subset. For Task A and C, we used 256, 192, and 128 bit-vectors for 10M, 30M, and 100M subsets, respectively. The $n^\star$ parameter is increased with the dataset for Task A, going from more than a thousand to 60,000. Task B uses 1024-bit vectors with $n^\star = k$. Task C is similar to task A, but $n^\star = k$. More details are given in [19].

LMI. The Learned Metric Index uses an architecture of interconnected learned models; the LMI demonstrates notable performance characteristics, often surpassing traditional methods in terms of efficiency and effectiveness. Central to the LMI is a tree structure that harnesses machine learning, particularly neural networks, to shrink the search space, facilitating a sequential search among significantly fewer objects than the original dataset. This is followed by a more time-intensive bucket-level sequential search within identified data partitions. The distances between objects are ascertained through a trained neural network, resulting in a probability distribution matrix that captures object-category relationships. The approach is adaptable, with the procedure iterating over matrix columns based on similarity, treating the exact count as a parameter. The approach is more useful with the help of a GPU or TPU, which was not considered for this challenge.

Table 1. Task A results for all LAION subsets. Entries are sorted by best rank in the 10M subset. Query time is measured in seconds for the entire query set. OOM label means for *out of memory* and NR for *not run*.

Team	10M			30M			100M		
	Build time	Search time	Rank	Build time	Search time	Rank	Build time	Search time	Rank
HSP	1 h 21 m	0.34	1	4 h 16 m	0.49	1	17 h 15 m	0.51	1
UTokyo	38 m	0.49	2	2 h 35 m	0.71	2	OOM	–	–
BL-SearchGraph	13 m	0.61	3	53 m	1.09	4	5 h 55 m	1.67	2
BL-FaissHNSW	16 m	0.74	4	33 m	0.86	3	4 h 48 m	21.40	3
HIOB	7 m	35.89	5	8 m	89.97	5	13 m	247.01	4
CRANBERRY	1 h 57 m	107.05	6	5 h 49 m	192.02	6	17 h 29 m	589.76	5
LMI	7 h 4 m	450.25	7	NR	–	–	NR	–	–
Bruteforce	0 m	2,415.75	8	0 m	9,010.50	7	NR	–	–

HSP. The HSP team performed several modifications and tuning to the HNSW index, specifically on the *hnswlib*.[6] The authors reduced the memory requirements by hacking how the database is loaded and maintained in memory. This change allowed them to reduce the construction time by half. Another customization removes unnecessary functionality directed to vector databases and other search engines. The authors performed a broad ablation study and hyperparameter optimization to obtain a competitive setup for the challenge. Interestingly, one of the most critical parameters is the construction *ef*, which interchanges construction time by search quality. Note that construction was previously reduced, resulting in a net moderate increase in the building time. The parameters $M = 20$ and *ef*= 800 (construction) of the HSNW were determined to be the best choices for the provided workloads. The search state iterates on different values *ef* from 10 to 1000. The HSP team designed their solution for Task A. More details are given in the accompanying paper [6].

5   Results and Discussions

Evaluation Setup. Following the GHA setup, we prepared docker Linux images. The evaluation was conducted on 2× Intel(R) Xeon(R) CPU E5-2690 V4 CPUs

---

[6] Official site of the hnswlib project https://github.com/nmslib/hnswlib.

(28 cores, 56 hyperthreads) workstation with 512GiB of RAM. The original dataset resided on a 1TB SSD, but all solutions loaded data vectors and index data structures in memory. We encouraged participants to use multithreading or multiprocessing in the construction and searching stages to take advantage of the hardware—all participants except team SWANN employed multithreading. We enforced a time limit of 24 h for building the index and running the query workload.

Task A. Table 1 shows the results of Task A. As mentioned in Sect. 3, all teams built a unique index. We recorded the build time and, for each set of hyperparameters, the search time accumulated over 10k queries. From these timings, we present the shortest search time that exceeded the recall requirement.

All solutions worked on the 10M subset, and this performance is used to sort the table; five worked on 30M, and only three on 100. The HSP team presents the top-performing solution in all subsets. It achieved search times below a second, which put it in the tens of thousands of queries per second. On the downside, it has one of the most costly constructions. UTokyo performs the second best, having a better trade with construction time but having memory issues and being unable to run the 100M benchmark. Focusing on build time, team HIOB has an order of magnitude shorter build times but provides rather slow searches. Entries marked as NR were not run due to diverse causes, like very high computational resources or not given hyperparameters.

Table 2. Task B results. Recall values for 1024-dimensional bit-vectors.

Team	Recall		
	10M	30M	100M
HIOB	0.55	0.57	0.58
Baseline	0.24	0.24	0.25

Task B. Results of the second task are presented in Table 2. Here, the recall is used as the main performance score. Only the HIOB team participated, surpassing significantly our baseline by a factor of more than two. Interestingly, the 1024-dimensional bit-vectors are faster to compute and contain more metric information than approaches like PCA for the same given memory.

Task C. Table 3 shows the performance in Task C. Here, two teams were able to submit. The HIOB team uses the same configuration as in Task A but without reranking the results on the binary embedding using the original vectors. Since HIOB uses just 128-bit vectors for 100M, see Sect. 4, it achieves a lower recall behind the accepted threshold (AT). Using more bits improves their result quality at the cost of the search time. The SWANN team marked all entries beyond the AT, but its solution did not take advantage of the multicore architecture and was run in a single thread. Thus, they can improve their performance significantly if they solve queries in parallel.

Table 3. Task C results: Indexing binary vectors. The reference recall for bruteforce is around 0.24. Displayed are the top-performing parameters that surpass a recall of 0.216.

Team	10M			30M			100M		
	Build time	Search time	Rank	Build time	Search time	Rank	Build time	Search time	Rank
BL-SearchGraph	5 m	0.10	1	14 m	0.36	1	2 h 6 m	1.09	1
HIOB	7 m	36.56	2	8 m	90.35	2	–	–	–
Bruteforce	0 m	74.51	3	0 m	246.95	3	0 m	816.93	2
SWANN	3 m	159.82	4	12 m	717.54	4	1 h 3 m	3794.05	3

6   Conclusions

The SISAP 2023 Indexing Challenge's first edition was a successful event that brought together researchers worldwide to work on the problem of approximate similarity search. The challenge becomes a trigger for innovative methods; several indexing methods for vector spaces emerged, along with binary mappings, and indexes for the binary Hamming space, as well as insights into the strengths and weaknesses of different approaches.

One of the most notable findings of the challenge was the emergence of a new binary mapping method that is both competitive and efficient. The challenge highlighted the importance of graph-based indexing techniques for real-valued and binary high-dimensional vectors. While the methods developed in the SISAP 2023 Indexing Challenge represent a significant step forward, there is still room for improvement. Future work should enhance memory efficiency, refine navigational strategies, and tackle the secondary memory challenge.

The SISAP 2023 Indexing Challenge was a valuable opportunity to advance the state of the art in approximate similarity search. The challenge's findings will interest researchers and practitioners working in various fields, including information retrieval, machine learning, and computer vision.

References

1. Aumüller, M., Bernhardsson, E., Faithfull, A.J.: Ann-benchmarks: a benchmarking tool for approximate nearest neighbor algorithms. Inf. Syst. **87**, 101374 (2020)
2. Bawa, M., Condie, T., Ganesan, P.: LSH forest: self-tuning indexes for similarity search. In: WWW, pp. 651–660. ACM (2005)
3. Chávez, E., Navarro, G., Baeza-Yates, R., Marroquín, J.L.: Searching in metric spaces. ACM Comput. Surv. **33**(3), 273–321 (2001). https://doi.org/10.1145/502807.502808
4. Cherti, M., et al.: Reproducible scaling laws for contrastive language-image learning. In: Proceedings of the IEEE/CVF Conference on Computer Vision and Pattern Recognition, pp. 2818–2829 (2023)
5. Fischler, M.A., Bolles, R.C.: Random sample consensus: a paradigm for model fitting with applications to image analysis and automated cartography. Commun. ACM **24**(6), 381–395 (1981)

6. Foster, C., Kimia, B.: Computational enhancements of HNSW targeted to very large datasets. In: Similarity Search and Applications: 16th International Conference, SISAP 2023, A Coruña Spain, 9–11 October 2023, Proceedings. Springer, Heidelberg (2023)

7. Fu, C., Xiang, C., Wang, C., Cai, D.: Fast approximate nearest neighbor search with the navigating spreading-out graph. Proc. VLDB Endow. **12**(5), 461–474 (2019). https://doi.org/10.14778/3303753.3303754

8. Malkov, Y.A., Yashunin, D.A.: Efficient and robust approximate nearest neighbor search using hierarchical navigable small world graphs. IEEE Trans. Pattern Anal. Mach. Intell. **42**(4), 824–836 (2018). https://doi.org/10.1109/tpami.2018.2889473

9. Mic, V., Sedmidubsky, J., Zezula, P.: CRANBERRY: memory-effective search in 100M high-dimensional CLIP vectors. In: Similarity Search and Applications - 16th International Conference, SISAP 2023, Spain, Proceedings (2023)

10. Navarro, G.: Searching in metric spaces by spatial approximation. VLDB J. **11**, 28–46 (2002)

11. Oguri, Y., Matsui, Y.: General and practical tuning method for off-the-shelf graph-based index: Sisap indexing challenge report by team utokyo. In: Similarity Search and Applications: 16th International Conference, SISAP 2023, A Coruña Spain, 9–11 October 2023, Proceedings. Springer, Heidelberg (2023)

12. Radford, A., et al.: Learning transferable visual models from natural language supervision. In: International Conference on Machine Learning, pp. 8748–8763. PMLR (2021)

13. Rombach, R., Blattmann, A., Lorenz, D., Esser, P., Ommer, B.: High-resolution image synthesis with latent diffusion models. In: Proceedings of the IEEE/CVF Conference on Computer Vision and Pattern Recognition, pp. 10684–10695 (2022)

14. Romild, C.J., Schauser, T.H., Alexander Borup, J.: Enhancing approximate nearest neighbor search with binary-indexed lsh-tries, trie rebuilding & batch extraction. In: Proceedings of the 16th International Conference on Similarity Search and Applications (2023)

15. Schuhmann, C., et al.: Laion-5b: an open large-scale dataset for training next generation image-text models. Adv. Neural. Inf. Process. Syst. **35**, 25278–25294 (2022)

16. Tellez, E.S., Ruiz, G.: Similarity search on neighbor's graphs with automatic pareto optimal performance and minimum expected quality setups based on hyperparameter optimization. CoRR abs/2201.07917 (2022). https://arxiv.org/abs/2201.07917

17. Tellez, E.S., Ruiz, G.: Similaritysearch.jl: autotuned nearest neighbor indexes for Julia. J. Open Source Softw. **7**(75), 4442 (2022)

18. Tellez, E.S., Chavez, E.: On locality sensitive hashing in metric spaces. In: Proceedings of the Third International Conference on Similarity Search and Applications, pp. 67–74 (2010)

19. Thordsen, E., Schubert, E.: An alternating optimization scheme for binary sketches for cosine similarity search. In: Similarity Search and Applications: 16th International Conference, SISAP 2023, A Coruña Spain, 9–11 October 2023, Proceedings. Springer, Heidelberg (2023)

20. Webster, R., Rabin, J., Simon, L., Jurie, F.: On the de-duplication of laion-2b. arXiv preprint arXiv:2303.12733 (2023)

Enhancing Approximate Nearest Neighbor Search: Binary-Indexed LSH-Tries, Trie Rebuilding, and Batch Extraction

Christoffer J. W. Romild, Thomas H. Schauser[✉],
and Joachim Alexander Borup

IT University of Copenhagen, Copenhagen, Denmark
thomas.schauser@gmail.com

Abstract. Locality-Sensitive-Hashing (LSH) plays a crucial role in approximate nearest neighbour search and similarity-based queries. In this paper, we present a study on the performance of LSH for indexing and searching high-dimensional binary vectors under the hamming distance metric. The paper presents a simple implementation of LSH-tries, utilizing binary indexing to seamlessly traverse to neighbouring buckets. To speed up queries, two optimization techniques, Batch Extraction & Trie Rebuilding, are introduced and their impact on performance analyzed. For evaluation, we conducted benchmarks using 1024-bit binary sketches from the LAION dataset, showcasing the accuracy and efficiency of our solution w.r.t. the brute-force approach, optimization strategies, and dataset size. The results show that SWANN outperforms the brute-force approach, and exhibits sub-linear growth as the size of the dataset increases. Our results w.r.t. the aforementioned optimizations showcase the importance of the bucket distribution, and the impact of hash function configuration.

Keywords: Index Challenge · Approximate Nearest Neighbors · Locality-Sensitive Hashing · Binary-Indexed LSH-Tries · LSH-Forests

1 Introduction

The motivation for the paper has been to participate in the SISAP 2023 Implementation Challenge [8]. The challenge revolves around finding efficient ways to search for approximate nearest neighbors (ANN) in high-dimensional spaces. ANN search at a billion-scale (Big ANN) is a fundamental challenge in computer science. Notably, this challenge has applications in contemporary systems like image searches, recommendation engines and search systems [3].

Efficiently addressing this challenge, while considering memory utilization and performance optimization, is of paramount importance. Although tree-like approaches such as k-d trees exists, they suffer from scalability issues when

© The Author(s), under exclusive license to Springer Nature Switzerland AG 2023
O. Pedreira and V. Estivill-Castro (Eds.): SISAP 2023, LNCS 14289, pp. 265–272, 2023.
https://doi.org/10.1007/978-3-031-46994-7_22

dealing with a large number of dimensions – a conundrum often referred to as the "curse of dimensionality" [5]. To combat these challenges, Locality-Sensitive Hashing (LSH), a hash-based approximation technique, has emerged as a solution [6].

This paper builds on the findings of PUFFINN [2] by researching the performance of LSH under the Hamming distance metric on a dataset comprising 100 Million 1024-dimensional binary vectors.

The paper introduces two straightforward optimizations *Trie Rebuilding* (Subsect. 3.4) and *Batch Extraction* (Subsect. 3.5), and discuss their impact on performance. In section Sect. 3 we will present our version of the LSH-trie, which utilizes pre-built masks and binary indexing to facilitate seamless traversal between neighboring buckets.

1.1   (k, δ, q)-NN Problem Definition

Let $P = \{p_1, \ldots, p_N\}$ define a dataset of N points in the Hamming space and let each point be a $\mathcal{D}$-dimensional binary vector. Furthermore, let $\lambda(a, b)$ denote the Hamming distance metric between point a and b. The nearest neighbour of a point q can then be defined to be the point $p_i \in P$, such that $\lambda(q, p_i) \leq \lambda(q, p_n)$ for $n = \{1, \cdots, N\}$. Given P and $\lambda(a, b)$, the (k, δ, q)-NN problem is about finding k points in P, such that each is among the k closest points to q with a probability of at least $1 - \delta$.

2   Locality-Sensitive Hashing

The key to LSH's effectiveness lies in employing locality-sensitive hash families – special families of hash functions tailored to maintain pairwise similarity, reinforcing the property that nearby points are more likely to hash to the same value. The properties of LSH hash-families are formally defined in Definition 1.

Definition 1. A family of hash functions are locality-sensitive for points p and q in a metric space if and only if the following conditions hold.

$$\lambda(p, q) \leq R \rightarrow \Pr[\text{hash}(p) = \text{hash}(q)] \geq P_1 \tag{1}$$
$$\lambda(p, q) \geq cR \rightarrow \Pr[\text{hash}(p) = \text{hash}(q)] \leq P_2 \tag{2}$$

Where R is a threshold, c is an approximation factor, and $\Pr[E]$ is the probability of an event E occurring.

Single-Bit Hash Family. For SWANN we have utilized the single-bit hash family. Given a binary vector $p = \{0, 1\}^{\mathcal{D}}$, let the notation $p[i]$ denote the value of the i'th bit of p for $i \in \{1, \ldots, \mathcal{D}\}$.

$$\text{single\_bit\_hash}(x) = x[i] \tag{3}$$

Equation 3 depicts the single-bit hash function. The idea is select i at random, and return the value of that bit as the hash. A single-bit hash function does not impose a major optimization by itself, in clustering points together. However, multiple locality-sensitive hash functions can be chained to create a new LSH-function with improved properties.

Chaining LSH-functions naturally decreases P_2, the probability of distant points being hashed to the same cluster. By utilizing this chaining, the total number of clusters is increased to 2^n clusters.

3   Binary-Indexed LSH-Tries

The LSH-trie is a tree-like data structure, in which a chain of LSH-functions are employed to group closely related data points into clusters.

LSH-tries are organized using a binary-indexed approach, which entails structuring them in a manner that allows direct indexing of buckets based on the hash values of query points. See 3.2 for more details on this.

This indexing method facilitates seamless traversal not only within a bucket, but also to its neighboring buckets across the trie.

3.1   Building LSH-Tries

When constructing an LSH-trie, each of the hash functions is assigned to a particular depth within the trie. This assignment ensures that any point's traversal through the trie corresponds to a sequence of hash function applications. Consequently, a LSH-trie of depth d maps each data point to a binary string. The d letters of the string is determined by the outcome of the hash functions, such that the j'th character corresponds to the outcome of applying the j'th hash-function. As such, all combination of binary strings can be found within the numbers 0 to $2^d - 1$. Hence, the data structure can be implemented by storing the buckets in a 2^d-length array. To avoid the memory overhead of storing empty buckets, we use a hash-map to store buckets instead. For the indexing to work correctly the d hash functions used in the LSH-trie must be binary, i.e. yielding true or false, forming a perfectly balanced trie with 2^d leaf nodes. Figure 1 illustrates an LSH-trie made up of 4 hash functions, resulting in 4-bit binary string indexes.

3.2   Querying for Candidate Points

To determine the approximate nearest neighbors for a query point, the LSH-trie maps it to a specific bucket. If this bucket contains an adequate number of candidate points, those points are returned. Otherwise, neighboring buckets are examined to enhance candidate retrieval.

The LSH-trie provides a convenient mechanism to access adjacent buckets. By modifying the binary string representing the query point, one can navigate to nearby buckets. Specifically, flipping z bits in the binary string equates to moving to buckets that share the outcomes of d−z hash functions. This approach permits gradual adjustments in the number of flipped bits to efficiently explore buckets with decreasingly similar points.

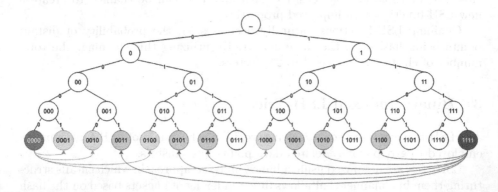

Fig. 1. An LSH-trie, with nodes of distance 1 and 2 from nodes 0000 and 1111 highlighted, respectively

Figure 1 illustrates the traversal from the bucket with the binary string "0000" to its four closest neighbors by flipping any one of its bits, equivalent to a distance of 1 in Hamming space. However, the number of neighboring buckets varies based on the distance from the query point. For instance, a Hamming distance of 1 results in four neighbors, while a distance of 2 results in six neighbors. To efficiently iterate through all buckets that share d − h hash functions, we utilize a pre-computed array of bitmasks indexed by the current depth h. The bitmasks for h is an ordered-set of all possible permutations containing h set bits for a binary string of length d.

3.3 LSH-Forests

The LSH-forest [3] is a data structure composed of multiple LSH-tries, forming a forest-like arrangement.

Constructing an LSH-forest involves inserting all of the data points into a set of LSH-tries. These tries share a consistent depth value, that determines the number of hash functions used and, in turn, the length of the binary strings.

When querying the k ANN of a specific data point, the LSH-forest extracts candidate points from each of the LSH-tries, and then returns the k points to the user.

The performance of the forest depends on the choice of depth d and number of tries in the forest L. The correct choice of d and L depends on the probabilities P_1 and P_2 for the underlying hash family used [2]. For a single-bit hash function, the depth can be calculated as:

$$d = \left\lceil \frac{\log N}{\log \frac{1}{P_2}} \right\rceil \quad (4)$$

$$L = \left\lceil P_1^{-d} \right\rceil \tag{5}$$

3.4 Rebuilding LSH-Tries

The LSH-trie rebuilding optimization occurs during the building of the index. The rebuilding optimization derives from the fact that different configurations of hash functions lead to different bucket distributions in the tries. How to rank a bucket distribution is difficult, but we have found γ, the size of the largest bucket in the trie, to be a good metric. This is due to a strong correlation between the size of γ and the worst-case performance of queries.

The optimization is simply to rebuild each trie $\tau > 1$ times and choosing the k LSH-tries that minimizes γ. By rebuilding the tries multiple time the likelihood of encountering good configurations must be improved, as we examine more possibilities.

SWANN deploys a multi-threaded version of the rebuilding process to utilize the cores available of the SISAP contest machine. To build k LSH-tries with τ optimization steps, a priority queue of LSH-tries ordered by γ is initialized. Now build k $\times$ τ LSH-tries with random hash-family configurations, and insert the tries into the queue. Once all the tries have been built, choose the k tries at the top of the priority queue.

3.5 Extracting in Batches

Another key optimization we've implemented is to extract candidate points from buckets in the LSH-tries in batches. The underlying idea is to minimize the number of distance computations for more points than necessary. In the worst-case w.o. the optimization, L $\times$ γ points would be included as candidate points before evaluating the exit criteria. Let β denote the batch size, the maximum number of candidate points extracted simultaneously from a single bucket. By partitioning each bucket into smaller batches, the optimization aims to limit the worst-case to L $\times$ β points. Equation 6 estimates β by multiplying some bucket factor α with the number of elements queried for, k.

$$\beta = \alpha + k, \quad \text{where } \alpha > 0 \tag{6}$$

In practice, this optimization improved our average running time by approximately three times.

4 Results

4.1 Methodology

To evaluate the performance and correctness of our solution, the benchmarking methodology is important [1]. Our benchmarks were conducted in Docker containers, to ensure reproducibility. We utilized the open-source C++ benchmark library by Google to execute parameterized benchmarks with multiple repetitions, enhancing result accuracy. Code to reproduce the benchmarks is available in on our GitHub repository.

Benchmarking involved a subset of the LAION-5B dataset [7], projected into 1024-bit binary sketches, along the Gold Standard list containing answers to the queries, both provided by the SISAP committee [4].

Our solution requires manual tuning of hyper-parameters P_1 and P_2.

4.2 Evaluation

In Fig. 2, our benchmarks cover 10-NN queries for up to 10 million 1024-bit points under the Hamming distance metric. Figure 2a shows that as dataset size increases, SWANN examines a smaller fraction, with an average number of candidate points close to $\beta \times L$.

Comparing Fig. 2b, SWANN outperforms our C++ brute-force approach for 10M points on the same hardware. The average query achieves at least a 90% recall in 0.024 s, about 48 times faster than brute-force.

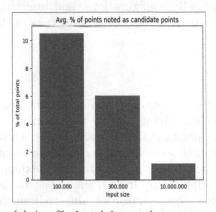 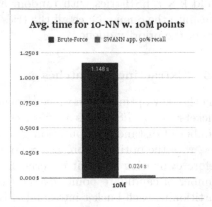

(a) Avg. % of total dataset that are candidate points for each query.

(b) Avg. time taken to query a dataset for 10 ANN w. atleast 90% recall.

Fig. 2. Benchmarks of ANN queries

Figure 3a illustrates the impact of the rebuilding optimization on the average queries per second for selected iterations between 1 to 8 for 300k points.

Figure 3b compares the build times for different rebuilding strategies when run on 20 cores. As seen, rebuilding the tries slows down the build time significantly but by multi-threading, the effect is diminished.

5   Implementation Refinement

The SISAP challenge evaluation is done using a 32-core Intel(R) Xeon(R) CPU E7-4809 workstation. As our solution uses a single thread for query evaluation,

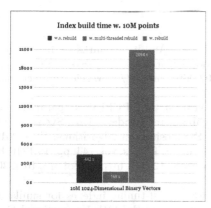
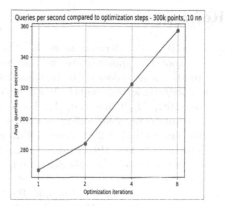

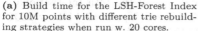

(a) Build time for the LSH-Forest Index for 10M points with different trie rebuilding strategies when run w. 20 cores.

(b) Avg. queries per second w. 90% recall for different number of trie rebuilding optimization steps.

Fig. 3. Benchmarks of trie rebuilding

an obvious improvement would be to segregate the work to multiple threads. Further, we could improve the performance of the Hamming distance computations by utilizing the hyper-optimized AVX-512 instruction set, as advised in the SISAP challenge.

6   Conclusion

In conclusion, this paper has addressed the challenge of efficient approximate nearest neighbor search in high-dimensional spaces using Locality-Sensitive Hashing.

We explored the application of binary-indexed LSH-tries under the Hamming distance metric on a large dataset of binary vectors and introduced optimization techniques, namely Batch Extraction and Trie Rebuilding, to enhance query efficiency.

SWANN's efficiency and accuracy were demonstrated through benchmarks with the LAION dataset, showcasing its superiority over brute-force approaches. By harnessing the power of binary indexed LSH-tries and the mentioned optimizations, SWANN achieves sub-linear growth in query times as dataset size increases.

We acknowledge that there remains potential for further improvement. In particular, refining the implementation could involve the integration of hardware-specific optimizations along with concurrent processing of queries through multi-threading.

References

1. Aumüller, M., Bernhardsson, E., Faithfull, A.: Ann-benchmarks: a benchmarking tool for approximate nearest neighbor algorithms. Inf. Syst. **87**, 101374 (2020) https://doi.org/10.1016/j.is.2019.02.006, https://www.sciencedirect.com/science/article/pii/S0306437918303685
2. Aumüller, M., Christiani, T., Pagh, R., Vesterli, M.: PUFFINN: parameterless and universally fast finding of nearest neighbors. In: Bender, M.A., Svensson, O., Herman, G. (eds.) 27th Annual European Symposium on Algorithms, ESA 2019, September 9–11, 2019, Munich/Garching, Germany. LIPIcs, vol. 144, pp. 10:1–10:16. Schloss Dagstuhl - Leibniz-Zentrum für Informatik (2019). https://doi.org/10.4230/LIPIcs.ESA.2019.10
3. Bawa, M., Condie, T., Ganesan, P.: LSH forest: self-tuning indexes for similarity search. In: Ellis, A., Hagino, T. (eds.) Proceedings of the 14th International Conference on World Wide Web, WWW 2005, Chiba, Japan, 10–14 May 2005, pp. 651–660. ACM (2005). https://doi.org/10.1145/1060745.1060840,
4. Chavez, E.L., Téllez, E.S., Aumüller, M.: Sisap 2023 datasets: 1024-bit binary sketches (hamming) (2023). https://sisap-challenges.github.io/datasets/#1024-bit_binary_sketches_hamming
5. Datar, M., Immorlica, N., Indyk, P., Mirrokni, V.S.: Locality-sensitive hashing scheme based on p-stable distributions (2004). https://www.cs.princeton.edu/courses/archive/spr05/cos598E/bib/p253-datar.pdf
6. Gionis, A., Indyk, P., Motwani, R.: Similarity search in high dimensions via hashing. In: Atkinson, M.P., Orlowska, M.E., Valduriez, P., Zdonik, S.B., Brodie, M.L. (eds.) VLDB 1999, Proceedings of 25th International Conference on Very Large Data Bases, 7–10 September 1999, Edinburgh, Scotland, UK, pp. 518–529. Morgan Kaufmann (1999). https://www.vldb.org/conf/1999/P49.pdf
7. Schuhmann, C., et al.: LAION-5B: An open large-scale dataset for training next generation image-text models (2022). arxiv.org/abs/2210.08402
8. Tellez, E.S., Aumüller, M., Chavez, E.: Overview of the SISAP 2023 indexing challenges. In: Similarity Search and Applications: 16th International Conference, SISAP 2023, A Coruña Spain, 9–11 October, Proceedings. Springer, Cham (2023)

General and Practical Tuning Method for Off-the-Shelf Graph-Based Index: SISAP Indexing Challenge Report by Team UTokyo

Yutaro Oguri(✉) and Yusuke Matsui

The University of Tokyo, 7-3-1 Hongō, Bunkyo-ku, Tokyo 113-8654, Japan
{oguri,matsui}@hal.t.u-tokyo.ac.jp

Abstract. Despite the efficacy of graph-based algorithms for Approximate Nearest Neighbor (ANN) searches, the optimal tuning of such systems remains unclear. This study introduces a method to tune the performance of off-the-shelf graph-based indexes, focusing on the dimension of vectors, database size, and entry points of graph traversal. We utilize a black-box optimization algorithm to perform integrated tuning to meet the required levels of recall and Queries Per Second (QPS). We applied our approach to Task A of the SISAP 2023 Indexing Challenge and got second place in the 10 M and 30 M tracks. It improves performance substantially compared to brute force methods. This research offers a universally applicable tuning method for graph-based indexes, extending beyond the specific conditions of the competition to broader uses.

Keywords: ANN search · Graph-based index · Black-box optimization

1 Introduction

The proliferation of deep learning has amplified the utility of Nearest Neighbor Search (NNS) in finding the closest vector within a set of embedding vectors for various documents. Particularly for million-scale data, the typical choice is Approximate Nearest Neighbor Search (ANNS). While different ANNS methods exist, graph-based techniques are superior in speed and accuracy, given that the data fits in RAM [9]. Renowned graph-based methods like NSG [4] and HNSW [8] are readily available through optimized libraries like `Faiss` [6].

While off-the-shelf graph indexes provide an efficient baseline, performance tuning becomes crucial to meet specific performance requirements. The evaluation of ANNS performance typically revolves around three metrics: accuracy (Recall@k), runtime (Queries Per Second; QPS), and memory usage. In practical scenarios, such as those presented in the SISAP competition, optimizing one metric often comes with constraints on accuracy, runtime, or memory. A performance tuning method for graph indexes under such constraints is non-trivial and remains an open area of investigation.

© The Author(s), under exclusive license to Springer Nature Switzerland AG 2023
O. Pedreira and V. Estivill-Castro (Eds.): SISAP 2023, LNCS 14289, pp. 273–281, 2023.
https://doi.org/10.1007/978-3-031-46994-7_23

Our work proposes a practical approach for performance tuning off-the-shelf
state-of-the-art graph-based method (e.g., NSG [4]) according to specified accuracy, runtime, and memory requirements. We focus on three key factors: vector
dimensionality reduction, database subsampling, and entry point optimization
for graph traversal. We employ black-box optimization for parameter tuning. Our
method is flexible and adaptable to various datasets and performance demands.

We participated in Task A in SISAP Indexing Challenge [12] and got second
place in the final score. In Task A, we use the LAION2B dataset [10] to perform
k-nearest neighbor search ($k = 10$). The dataset consists of 16-bit float vectors
with 768 dimensions. Under the condition of exceeding a recall of 0.9, the faster
the search speed, the higher the score you will receive. We use several subsets
(300K, 10 M, and 30 M size) for evaluation. The submitted code is available at
https://github.com/mti-lab/UTokyo-sisap23-challenge-submission.

We specifically apply our method to optimize the runtime of NSG index [4]
within constraints on accuracy and memory usage. NSG index is a graph-based
index approximating MRNG (Monotinic Relative Neighborhood Graph) [4]
structure. The time complexity of the search is close to logarithmetic time.

This work makes two key contributions.

1. Through exhaustive experiments, we demonstrate that dimensions, database
   size, and the entry point of the graph traversal serve as valuable parameters
   for performance tuning.
2. We introduce a practical and universal method for performing constrained
   optimization in ANN, considering accuracy, runtime, and memory metrics,
   utilizing black-box optimization.

2   Preliminary Study and Findings

2.1   Preliminary Study

We first evaluate representative types of indexes with subsets of LAION5B [10]
provided in the competition to choose the baseline. The subset size is 300K,
and the query set is 10K public queries provided in the competition. [12]. The
evaluated indexes include the brute force approach, graph-based, PQ-based, and
IVF-based index. Evaluation metrics are Recall@k, QPS, and memory usage.
Let ground truth k-nearest neighbors be R and approximate nearest neighbors
be $\hat{R}$. Recall@k is defined by $\frac{|R \cap \hat{R}|}{k}$. QPS is the average number of processed
queries per second. Memory usage represents the index's memory footprint.

All implementations utilize indexes provided by Faiss [6], a well-optimized
ANNS library with C++ and Python bindings. We run preliminary experiments
on an Intel(R) Xeon(R) Platinum 8259CL CPU @ 2.50 GHz with 512 GB RAM.

From the outcomes illustrated in Fig. 1, we found NSG is promising for Task
A. We also demonstrate that a graph-based index is the best choice when a
memory capacity is sufficient, as often suggested [9]. Among records whose recall
is more than 0.9, NSG index runs 22.2 times faster than the brute force method.
In addition, despite the memory efficiency and better QPS of the PQ-based

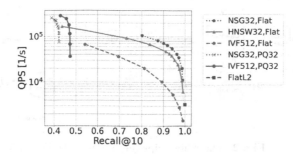

Fig. 1. A preliminary experiment comparing various indexes: The FlatL2 means brute-force approach. Other indexes have a common format consisting of two parts separated by a comma. The former means the index name. "NSG32" means NSG [4] index whose number of links per vertex is 32. (2) "HNSW32" means HNSW [8] index whose number of links per vertex is 32. (3) "IVF512" means an inverted file index that divides the dataset into 512 clusters. The latter means the precision of data. "Flat" means original database vectors, and "PQ32" means quantized vectors of 32-byte PQ [7] code. Note that we did not re-rank the quantized vectors.

approach, we cannot employ it due to its low accuracy. Thus, we select NSG index as a baseline.

In addition, we conducted performance profiling and found that the bottle-neck of NSG index is the computation of L2 distances. It occupies a significant fraction (more than 90%) of the whole computational cost during the search phase. We used `perf` as a profiling tool.

2.2 Findings

Based on these results, we propose three key tuning parameters to improve the search speed: the dimensionality of database vectors, the size of the database, and the entry point for graph traversal. Reducing the dimensionality of vectors and subsampling the database directly reduces the cost of computing L2 distances. In addition, we can change where to start the graph traversal. It is another parameter to be tuned. We aim to optimize these three parameters to improve QPS without compromising Recall@10.

Efficiently selecting these parameters in ANN is not straightforward since Recall@10 and QPS are trade-offs. In addition, the increase in speed due to these parameters is not monotonic due to the various complex factors involved. Moreover, conducting a simple grid search is inefficient. Therefore, this paper proposes a practical framework for tuning an off-the-shelf graph-based index, specifically the NSG index. This framework is a generic method that applies to other types of constraints or different targets.

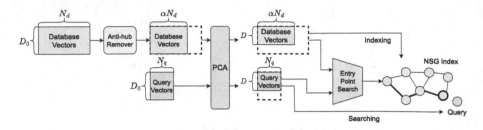

Fig. 2. The whole pipeline of our method.

3   Method

Figure 2 shows the whole pipeline of our method. It subsamples the database and reduces the dimensionality of vectors. In the search phase, it selects the entry point where the graph traversal begins. We explain the details of each reduction method in Sect. 3.1 and how to effectively optimize them in Sect. 3.2. Note that we do not modify the graph index itself. Thus, this approach is independent of the implementation of NSG graph index.

3.1   Components in Pipeline

Dimensionality Reduction. We employ Principal Component Analysis (PCA). It is a linear dimensionality reduction algorithm that projects data to a lower dimensional subspace. It reduces the dimension from D_0 to $D(\leq D_0)$ (Fig. 2). It can directly reduce the computational cost of L2 distance calculation. The reduced dimension D is an indexing parameter to be tuned.

Database Subsampling. We employ AntiHub Removal [11] to subsample the database effectively. This reduction method is based on hubness in data. It reduce the size of database from N_d to $\alpha N_d (0 \leq \alpha \leq 1)$ (Fig. 2). It can improve accuracy at a given memory consumption level while maintaining the same QPS. This approach is a compelling tuning candidate because we can apply it with dimensionality reduction methods. The ratio α is also a parameter for indexing.

Optimizing Entry Point. If we have multiple entry point candidates, starting with the one closest to the query dramatically speeds up the search [2,5]. We propose a novel and straightforward entry point selection method utilizing k-means clustering. It first divides the entire dataset into k clusters and computes a centroid of each cluster (i.e., a centroid is the nearest vector to the mean vector of the cluster). Given a query, we select the closest centroid to the query as an entry point. This approach enables to start traversal from a near point to the query. It prevents excessively long search paths. The number of clusters k is a parameter for building the entry point searcher.

Our approach works well in parallel, even when queries arrive in a batch. Faiss is good at a parallel search at the query level within a batch. However, because our approach requires each query in a batch to have a different optimal entry point, batch processing can become inefficient (Algorithm 1). To address this issue, we propose a gather-style parallel-friendly approach (Algorithm 2). It divides queries into multiple subsets based on optimal entry points and performs batch processing separately for each subset. This approach achieves the same result as Algorithm 1, but with more room for parallel execution (L1 and L6).

Algorithm 1. An implementation with naive approach

```
1  for query_id , query in enumerate(queries):
2        ep = search_entrypoint(query)
3        set_entrypoint(index , ep)
4        # single query
5        results[query_id] = index.search(query , k)
```

Algorithm 2. An implementation for query batch

```
1  epts = search_entrypoints(queries) # runs in batch
2  for ep in np.unique(epts):
3        query_ids = (epts == ep)
4        query_batch = queries[query_ids , :]
5        set_entrypoint(index , ep)
6        # runs in batch
7        results[query_ids] = index.search(query_batch , k)
```

3.2 Parameter Tuning with Black-Box Optimization

We apply a black-box optimization technique to tune parameters D, α, and k to maximize QPS under memory usage constraints and Recall@10, as specified in Task A. As we cannot compute the gradient of QPS with respect to the tunable parameters, we employ black-box optimization. It is an optimization method that does not need derivatives. We use Optuna [1], a framework for black-box optimization, to implement it. Optuna offers various efficient optimization algorithms. We explore two different strategies under constraints: 1) single-objective optimization with constraint and 2) multi-objective optimization.

Single-Objective Optimization with Constraint. Single-objective optimization with constraint is formulated as shown in Eqs. (1) and (2). Optuna

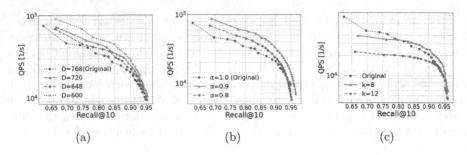

Fig. 3. Ablation Study for Each Components (30 M subset): (a) PCA + NSG [4], (b) Antihub Removal [11] + PCA and (c) entry point Search with k-means + NSG.

has a sampler that narrows the parameter search space considering optimization history. TPE (Tree-structured Parzen Estimator) sampler [3] supports this type of optimization. It is important to note that it does not guarantee that the obtained solution will always satisfy the constraints; we can only treat them as soft constraints.

$$\text{maximize QPS} \tag{1}$$

$$\text{subject to Recall@}k \geq 0.9. \tag{2}$$

Multi-objective Optimization. Multi-objective optimization is formulated as shown in Eq. 3. It can include multiple objective functions, and each of them is desired to be maximized or minimized. TPE sampler [3] also supports it. The result is a Pareto frontier, a set of parameter points that achieve the best trade-offs. Since QPS and Recall@k are competing objectives, we can apply multi-objective optimization for Task A.

$$\text{maximize QPS, Recall@}k. \tag{3}$$

4   Experiment

We evaluate the impact of each of these components on QPS and Recall@k. Then, we conduct an experiment to tune everything integratively with Optuna. We used the same dataset and query set as Sect. 2.1. The tested subset size is 300K, 10 M, and 30 M. The whole experiments are conducted on the same environment as Sect. 2.1.

4.1   Ablation Study

Dimensionality Reduction + NSG (ours) vs Vanilla NSG. In addition to the vanilla NSG, we apply PCA for dimensionality reduction. We varied the reduced dimension D and measured QPS and Recall@k. The results shown in

Fig. 3 (a) demonstrates that applying PCA can increase QPS without compromising accuracy. The best configuration with the condition of Recall@$k \geq 0.9$ is $D = 600$. Its QPS is 1.53 times greater than the best records in the vanilla NSG [4].

Subsampling + NSG (ours) vs Vanilla NSG. We apply Antihub Removal to reduce the size of the database. Figure 3 (b) shows the performance with various subsampling ratios α. It demonstrates that applying subsampling to the database improves efficiency while maintaining accuracy. The best configuration among them is $\alpha = 0.9$, which exhibits 1.61 times greater QPS than the vanilla NSG [4].

Entry Point Optimization + NSG (ours) vs Vanilla NSG. We compare performance among various entry point candidates with k-means. Figure 3 (c) demonstrates that optimizing the entry point with k-means can potentially increase the QPS in the high accuracy regime. The best configuration shows 1.30 times greater QPS than the vanilla one, while its Recall@10 is 0.9 or greater.

4.2   Parameter Tuning

We conducted parameter tuning with black-box optimization. Our ablation study demonstrates that all three aspects can improve performance, and the trends are consistent across different subset sizes for all tuning components. Therefore, we conducted the tuning using a 300K subset for efficiency.

The result demonstrates that multi-objective optimization outperforms single-objective optimization with constraints. When compared over the same tuning time (about 3.5 h), the best configuration with the former method is 1.85 times faster than that with the latter.

Table 1 shows the best results for each subset. We apply tuned parameters for the subset 300K. We choose the best setting among some records for other subsets. It demonstrates that performances for all subsets significantly increased compared to vanilla NSG [4] and brute-force method.

Table 1. The best results for each subset size (Recall@$k \geq 0.9$).

Size	Recall@10(↑) Ours	QPS [1/s] (↑) Ours	Vanilla NSG [4]	Brute-force
300K	0.9208	$\mathbf{1.104 \times 10^5}$ (×**34.16**)	7.186×10^4 (×22.23)	3.232×10^3 (×1.0)
10M	0.9082	$\mathbf{3.822 \times 10^4}$ (×**1078**)	2.881×10^4 (×812.5)	35.46 (×1.0)
30M	0.9030	$\mathbf{3.010 \times 10^4}$ (×**1188**)	1.860×10^4 (×734.6)	25.32 (×1.0)

5   Discussion

5.1   Applicability to General Settings

Our framework is practical in other general ANN problems. If there are more complex constraints than the ones in this work, our method may not be suitable, requiring a more complex approach. However, many real-world ANN tunings are oriented towards improving the three axes - Recall, QPS, and Memory - in a straightforward manner. Thus, our approach is applicable in other settings.

In addition, we need to investigate whether the methods proposed here are adequate for graph indexes other than NSG. Since the search for the entry point and the reduction of dimensionality and database are not techniques bound by the specific circumstances of NSG, we can expect their applicability.

5.2   Comparison to a Previous Work

SimilaritySearch.jl [14] also introduces an autotuning method for graph-based indices, leveraging a beam search algorithm for parameter tuning [13]. Like our methodology, it models the problem as a black-box optimization to optimize recall and efficiency. SimilaritySearch.jl uses the count of distance computations as its efficiency metric. In contrast, our approach models efficiency using an average QPS measured ten times. A shared limitation for both methods is the presumption of consistent query distributions during tuning and search. If the assumption is invalid, it might lead to suboptimal outcomes or drastic performance drops.

5.3   Limitation and Future Work

Our method cannot satisfy the memory constraint with a 100 M subset. It requires further dimensionality and data size reduction, but the problem is that it takes far more time to tune it.

We select conservative parameters to satisfy accuracy for unknown queries in our 10M and 30M submissions. We recognize the need for using more diverse query sets other than public queries for tuning to ensure robust performance.

Lastly, we only used the 300K subset for tuning in these experiments, as the impact of the three parameters we tuned showed consistent trends across all subset sizes. Although it would be ideal to perform tuning on larger subsets, it is exceedingly time-consuming when working with larger subsets. This is because we have to rebuild the index every time the parameters D and α change with each trial. We need to explore practical strategies to reduce the duration.

6   Conclusion

In conclusion, this study proposes a successful tuning method for an off-the-shelf graph-based ANN index. By adjusting vector dimension, database size, and graph traversal entry points and utilizing a black-box optimization, we significantly improve Recall@k and QPS performance. We applied our approach to the SISAP Indexing Challenge and significantly outperformed brute force methods. It is also applicable under general conditions.

Acknowledgments. This work was supported by JST AIP Acceleration Research JPMJCR23U2, Japan.

References

1. Akiba, T., Sano, S., Yanase, T., Ohta, T., Koyama, M.: Optuna: a next-generation hyperparameter optimization framework. In: Proceedings SIGKDD (2019)
2. Arai, Y., Amagata, D., Fujita, S., Hara, T.: LGTM: a fast and accurate kNN search algorithm in high-dimensional spaces. In: Strauss, C., Kotsis, G., Tjoa, A.M., Khalil, I. (eds.) DEXA 2021. LNCS, vol. 12924, pp. 220–231. Springer, Cham (2021). https://doi.org/10.1007/978-3-030-86475-0_22
3. Bergstra, J., Bardenet, R., Bengio, Y., Kégl, B.: Algorithms for hyper-parameter optimization. In: Proceedings NIPS2021, pp. 2546–2554. Curran Associates Inc. (2011)
4. Fu, C., Xiang, C., Wang, C., Cai, D.: Fast approximate nearest neighbor search with the navigating spreading-out graph. Proc. VLDB Endow. **12**, 461–474 (2017)
5. Iwasaki, M., Miyazaki, D.: Optimization of indexing based on k-nearest neighbor graph for proximity search in high-dimensional data. preprint: arXiv:1810.07355 (2018)
6. Johnson, J., Douze, M., Jégou, H.: Billion-scale similarity search with GPUs. IEEE Trans. Big Data **7**(3), 535–547 (2019)
7. Jègou, H., Douze, M., Schmid, C.: Product quantization for nearest neighbor search. IEEE Trans. PatternAnal. Mach. Intell. **33**(1), 117–128 (2011)
8. Malkov, Y.A., Yashunin, D.A.: Efficient and robust approximate nearest neighbor search using hierarchical navigable small world graphs. IEEE Trans. Pattern Anal. Mach. Intell. **42**, 824–836 (2016)
9. Matsui, Y., Aumüller, M., Xiao, H.: CVPR 2023 tutorial on neural search in action (2023)
10. Schuhmann, C., et al.: LAION-5b: an open large-scale dataset for training next generation image-text models. In: NeurIPS 2022 Datasets and Benchmarks Track (2022)
11. Tanaka, K., Matsui, Y., Satoh, S.: Efficient nearest neighbor search by removing anti-hub. In: Proceedings ICMR2021, pp. 285–293. ACM (2021)
12. Tellez, E.S., Aumüller, M., Chavez, E.: Overview of the SISAP 2023 indexing challenges. In: Similarity Search and Applications: 16th International Conference, SISAP 2023, A Coruña Spain, 9–11 October, Proceedings. Springer, Cham (2023)
13. Tellez, E.S., Ruiz, G.: Similarity search on neighbor's graphs with automatic pareto optimal performance and minimum expected quality setups based on hyperparameter optimization (2022)
14. Tellez, E.S., Ruiz, G.: SimilaritySearch.jl autotuned nearest neighbor indexes for Julia. J. Open Source Softw. **7**(75), 4442 (2022)

SISAP 2023 Indexing Challenge –
Learned Metric Index

Terèzia Slanináková$^{(\boxtimes)}$ [iD], David Procházka [iD], Matej Antol [iD], Jaroslav Olha [iD],
and Vlastislav Dohnal [iD]

Faculty of Informatics and Institute of Computer Science, Masaryk University,
Botanická 68a and Šumavská 15, 602 00 Brno, Czech Republic
{xslanin,davidprochazka,olha,dohnal}@mail.muni.cz, antol@muni.cz

Abstract. This submission into the SISAP Indexing Challenge examines the experimental setup and performance of the Learned Metric Index, which uses an architecture of interconnected learned models to answer similarity queries. An inherent part of this design is a great deal of flexibility in the implementation, such as the choice of particular machine learning models, or their arrangement in the overall architecture of the index. Therefore, for the sake of transparency and reproducibility, this report thoroughly describes the details of the specific Learned Metric Index implementation used to tackle the challenge.

Keywords: sisap indexing challenge · learned metric index · similarity search · machine learning for indexing · performance benchmarking

1 Introduction

The Learned Metric Index (LMI) [1] functions as a hierarchical tree structure composed of nodes housing machine learning models. These models are trained to classify query objects, simulating the behavior of traditional index nodes. However, instead of ascertaining object positions based on distances, queries are resolved by executing a series of predictions. This departure from the usual index creation and query assessment approach results in markedly different performance characteristics, frequently outperforming conventional similarity search methods in terms of both efficiency and effectiveness.

The building process to form LMI's tree structure prepares a machine learning model for each internal node. Such a model is trained on a sub-section of data assigned to it by its parent node. We applied K-Means to obtain the data partitions in this paper. Leaf nodes constitute storage of data objects.

The publication of this paper and the follow-up research was supported by the Czech Science Foundation project No. GF23-07040K (all authors but V. Dohnal) and by the ERDF "CyberSecurity, CyberCrime and Critical Information Infrastructures Center of Excellence" (No.CZ.02.1.01/0.0/0.0/16_019/0000822) – V. Dohnal. Computational resources were supplied by the project "-Infrastruktura CZ" (e-INFRA CZ LM2018140) and ELIXIR CZ Research Infrastructure (ID LM2018131) supported by the Ministry of Education, Youth and Sports of the Czech Republic.

O. Pedreira and V. Estivill-Castro (Eds.): SISAP 2023, LNCS 14289, pp. 282–290, 2023.
https://doi.org/10.1007/978-3-031-46994-7_24

The search algorithm traverses the tree by making inferences for the query object. The higher the inference value, the higher the priority of accessing the branch. Leaf nodes are gathered until the specified number of objects have been accessed, which serves as a stop condition. The objects stored within the leaf nodes make up the candidate objects, which are then ranked based on their distance from the query object. This way the final query answer is formed.

In the following sections, we provide an insight into the implementation and configuration used for the SISAP 2023 Indexing Challenge, and report the results of our experimental evaluation.

2   Related Work

The idea of a learned index as a series of increasingly specific machine learning models was originated by Kraska et al. [9], and has since seen several applications on simple structured data, such as approximate data-aware index structures [4] or a learned database system [8]. An analogous concept was later proposed for similarity searching in complex data [1], and then extended into the data-driven version [11,14] used in this submission.

Other applications of learned models in the domain of similarity searching in metric data have also been proposed, such as ANN-tree [7], the FLANN library [10], Neural Locality-Sensitive Hashing (Neural LSH) [3]. Furthermore, a learned model that approximates bounds on k nearest neighbor distances and computes reverse nearest neighbors has been introduced in [2]. Finally, Hünemörder et al. [5] explored the application of various predictive models to train an index for approximate nearest-neighbor queries.

3   Implementation

The design and implementation decisions of the LMI are geared towards optimizing a fast approximate nearest neighbor (ANN) search for ten nearest objects while achieving a recall of at least ninety percent. We detail the construction and searching stages of LMI, as well as the approaches considered and adopted within our submission, in the following sections.

3.1   Construction

The index construction is composed of two phases – a partitioning phase and a learning phase. In the **partitioning** phase, a clustering algorithm produces categories of similar data, which serve as inputs to the learning phase. A neural network is then employed to learn the associations between individual data objects and their categories, thereby forming an index with one internal (root) node and n leaf nodes (buckets) corresponding to the number of categories. Utilizing the neural network on top of the clustering allows for faster probability-based navigation in query answering.

Partitioning algorithms on large data volumes oftentimes run into scalability issues caused by their big space and time complexity. To avoid these problems, K-Means is used in favor of K-Medoids and DBSCAN[1]. In our experience, the latter two fail to cluster volumes larger than 100K, whereas K-Means successfully clusters the entire 10M dataset in a matter of seconds (up to a few minutes). K-Means operates in Euclidean space, however, the dataset uses the cosine distance. To resolve this inconsistency, the data is L_2-normalized, where it holds that the negative squared Euclidean distance is proportional to the cosine distance [17].

The **learning** phase involves solving a supervised multi-class classification problem, which can be addressed with an arbitrary classifier [1,14]. In this challenge, LMI uses a *single* fully-connected neural network, in particular, a multi-layer perceptron (MLP).[2] The reason is that, as is explained in the following section, the search procedure for this challenge is optimized toward considering queries in parallel, and a big hierarchical structure slows down this kind of processing by producing more categories that need to be accessed sequentially. The learning phase is the more time-intensive of the two, usually taking up to several hours depending on the number of epochs the training runs for (but always less than twelve hours). The entirety of the dataset is used for the training.

3.2 Searching

The search process is composed of navigation and a bucket-level sequential search. The main role of navigation, powered by the inference operation of the neural network, is to vastly reduce the search space, resulting in a sequential search in far fewer than the original 10M objects. Subsequently, the bucket-level sequential search is done in the identified data partitions (buckets). This step is more time-demanding as it computes the pair-wise distances between the remaining data objects and the queries in the original (768 dimensional) space. The process is visualized and described in Fig. 1.

To further reduce the time necessary for the retrieval, LMI adopts the following optimizations: **(1) parallelism of query processing** and **(2) object filtering**. Firstly, since all queries are known beforehand and there are many more queries than categories, LMI leverages the fact that many queries share one category and need to compute distances to the same objects. Therefore, all queries are processed in parallel during navigation.[3] Secondly, the most similar categories (buckets) are searched and used to populate the final answer first. Any subsequent bucket accesses can therefore capitalize on the recorded value of the k-th distance in the final answer and disregard objects with a distance greater than this value. This results in a much smaller matrix of mutual distances to be sorted, a particularly expensive operation. Irrespective of the data

---

[1] K-Means is adopted from the FAISS library [6], K-Medoids' implementation with FasterPAM [13] is used, and DBSCAN [12] is taken from the scikit-learn library, v. 0.24.2.

[2] The neural network is implemented in the PyTorch library (v 1.1.0), using the ReLU activation function and Adam optimization algorithm.

[3] Python's NumPy library [16] (v. 1.19.5) is used to achieve this optimization.

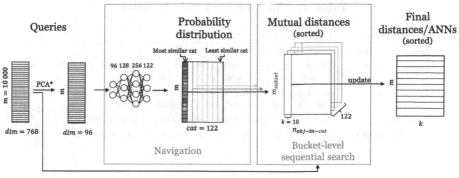

* Provided by the organizers of the challenge.

Fig. 1. Schema of the search process. Navigation starts by passing all 10 000 queries ($dim = 96$) through the trained neural network, resulting in a probability distribution matrix, which captures the relationship of every query object to every category. To gather input for bucket-level sequential search, the procedure iterates over several columns of this matrix, starting from the most similar. The exact number is treated as a hyper-parameter (number of buckets). Each column is broken down into individual categories (colored cells), and processed separately. The bucket-level sequential search starts by evaluating and sorting the distances between the data objects from a given category and the relevant queries (both with $dim = 768$), forming a matrix of mutual distances. The first k columns are then used to continuously update the final output – two matrices of k closest distances and ANNs.

dimensionality employed during index construction or navigation, LMI utilizes the original data descriptors during bucket-level sequential search. Based on our observations, only the unreduced version of the data consistently managed to effectively locate the k-NNs among candidate objects using sequential search.

Observing the LMI configuration with the fastest search time, the most expensive operation in terms of search time is computing the matrices of mutual distances (72.5% of overall search time) to objects in the buckets, followed by sorting these matrices to extract the k nearest distances (16%). Identifying the objects within a given category accounts for 9.7% and constructing the probability distribution matrix (navigation) only for 0.016%.

4   Experimental Evaluation

The search performance of LMI is evaluated in more than 2k experiments with varying hyper-parameters. The three provided dataset versions are used in the building and navigation process: *clip768*, *pca96*, and *pca32*, while, as noted in Sect. 3, the bucket-level sequential search operates exclusively with *clip768*. Apart from the main goal of the challenge (recall above 90% on the ground truth set), we optimized for reasonable RAM consumption (below 40 GB) in conjunction with training time (under 12 h).

Table 1. Explored hyper-parameter values.

Hyper-parameter	Explored values
Dataset version	pca32, pca96, clip768
No. categories	10, 20, 30, 50, 80, 90, 95, 100, 105, 110, 112, 114, 115, 116, 118, 120, 121, 122, 123, 124, 125, 130, 150, 200
No. epochs	1, 10, 20, 30, 40, 45, 50, 55, 60, 80, 100, 120, 160, 180, 190, 195, 200, 205, 208, 209, 210, 211, 212, 220, 250, 300
Learning rate	0.001, 0.005, 0.007, 0.008, 0.009, 0.01, 0.05, 0.09, 0.1, 0.11
No. visited buckets	1, 2, 3, 4, 5, 6, 7, 8, 9, 10, 12, 15, 20, 24, 25, 40, 50, 100

4.1 Hyper-parameter Selection

LMI's efficiency is primarily affected by three kinds of hyper-parameters: the size of the structure, the size of the candidate set, and the model parameters. Given that the instance of LMI is only one level deep, its size is determined by the number of categories, i.e., leaf nodes or buckets. Navigation quality is closely linked to the hyper-parameters of the single inner node, i.e., MLP, where the learning rate and number of epochs are carefully considered. The learning rate mainly affects the quality of the navigation, whereas the number of epochs influences the training time. In all experimental scenarios, an MLP architecture comprising one hidden layer containing 128 neurons is utilized. This architecture provides good performance for the task at hand, and we leave fine-tuning of the architecture as future work. Finally, the stop condition is defined in the number of buckets limiting the size of the candidate set. The full enumeration of the explored hyper-parameter values is listed in Table 1.

The hyperparameter search was performed in two steps. First, a coarse set of hyperparameters was evaluated, indicating their general trends, followed by a restricted hyperparameter search around the observed optima.

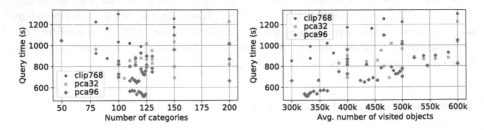

Fig. 2. (Left) Relationship between overall search time and the number of categories. (Right) Relationship between overall search time and the average number of visited objects. Only experiments achieving recall ≥ 90% are shown.

Table 2. Building time and query performance of the best-performing setup for each dataset version with the hyper-parameter values.

Dataset	Epochs	Learning rate	Categ.	Build time (h)	Stop condition	Query time (s)	Recall (%)
pca32	190	0.050	105	7.2	5	689.58	90.19
pca96	**205**	**0.009**	**122**	**8.2**	**4**	**514.92**	**90.88**
clip768	180	0.010	100	11.5	4	839.35	92.88

4.2 Results

The evaluation of numerous hyper-parameter values reveals the following trends. The overall execution time is most significantly influenced by the quantity of categories and the number of visited buckets, which directly corresponds to the count of objects processed during bucket-level sequential search, see Fig. 2. Furthermore, when these parameters are fixed, the fastest configurations remain relatively stable in their learning rate, and their build time is proportional to the number of epochs.

Table 2 presents the setups that performed the best on each dataset and highlights the differences between them. LMI constructed on *pca96* provided the fastest response to 10 000 queries, with a training time of over eight hours. The index inspected four buckets while surpassing the required recall in 514.92 s, i.e., 51.5 ms per query.

The construction times and, to a certain extent, the search times indicate that it is beneficial to use datasets produced by dimensionality reduction techniques. Using *pca96*, LMI finishes the building three hours faster than when using *clip768*. The smaller size descriptors of *pca32* further reduce the construction cost but do not show superior query performance. LMI successfully uses the additional principal components of *pca96* to reduce query processing time by more than 174 s compared with *pca32*. However, setups based on *pca32* have to visit one additional bucket in order to rank among the best-performing setups, costing additional time. This result might indicate a navigation impairment, which could be attributed to insufficient dimensionality. Compared to *clip768*, LMI with *pca32* holds an advantage in terms of epoch numbers and dataset division granularity. Dimensionality reduction should therefore be considered in situations where LMI processes high-dimensional data; however, the optimal dimensionality after the reduction is to be further investigated.

The observed optimal number of categories, the first high-impact hyper-parameter, is between 100 and 125 across all dataset versions. Furthermore, tuning this hyper-parameter with finer-grained steps within the interval is crucial for query time and further time reduction. It decreases the query time by more than 180 s, implying a strong impact of the number of categories on query processing time. As seen in Fig. 2, the number of categories also roughly correlates with the value of the stop condition.

Table 3. Instructions to reproduce the results

Step	Step-by-step instructions
1	Ensure that `pip`, `conda`, and `git` are installed

2	Clone the repository at the commit SHA of the submitted version
	`$ git clone https://github.com/TerkaSlan/sisap23-laion-challenge-learned-index`
	`$ git reset --hard bc9b76b`

3	Create a virtual environment and install all the dependencies
	`$ conda create -n env python=3.8 && conda activate env`
	`$ conda install matplotlib pandas scikit-learn`
	`$ pip install h5py flake8 setuptools tqdm faiss-cpu`
	`$ pip install torch --index-url https://download.pytorch.org/whl/cpu`
	`$ pip install --editable .`

4	Run the script with the chosen hyper-parameters `$ python search/search.py --n-categories=122 --epochs=205 --lr=0.009 -bp=4`

5	Evaluate
	`$ python eval/eval.py`
	`$ python eval/plot.py res.csv`
	`$ cat res.csv`

The optimal value of the second high-impact hyper-parameter, the stop condition, proves to be sufficient up to ten. LMI can visit only a fraction of the total buckets, typically four or five out of more than a hundred, to exceed the required recall. Navigation can thus efficiently and effectively pinpoint the relevant buckets and present them to the bucket-level sequential search.

4.3 Reproducing the Results

The software accompanying this paper is made available on GitHub[4] and is designed in a way to support ease of use, reproducibility, and seamless integration with the challenge's evaluation pipelines. Table 3 guides the reader through the installation, running, and evaluation of our submission.

All of our experiments were carried out on a single-core AMD EPYC 7532 CPU with 40 GB RAM and on Linux Debian 5.10 operating system. The code was written in Python 3.8. Since *search time* is hardware-dependent, we expect it to be weakly reproducible, while we expect *recall* to be strongly reproducible [15].

5   Summary and Conclusions

With the constraint of solving the approximate nearest neighbor search with recall above ninety percent, our method shows fast navigation times and introduces several optimizations to reduce bucket-level sequential search times.

In future work, we aim to investigate the trade-off relationship between accuracy and time in regard to the architecture of the index (the number of inner nodes) and the architecture of the neural networks (the number of hidden layers and neurons), as well as scalability to larger data volumes.

---

[4] https://github.com/TerkaSlan/sisap23-laion-challenge-learned-index.

References

1. Antol, M., Oĺha, J., Slanináková, T., Dohnal, V.: Learned metric index - proposition of learned indexing for unstructured data. Inf. Syst. **100** (2021)
2. Berrendorf, M., Borutta, F., Kröger, P.: k-Distance approximation for memory-efficient RkNN retrieval. In: Amato, G., Gennaro, C., Oria, V., Radovanović, M. (eds.) SISAP 2019. LNCS, vol. 11807, pp. 57–71. Springer, Cham (2019). https://doi.org/10.1007/978-3-030-32047-8_6
3. Dong, Y., Indyk, P., Razenshteyn, I.P., Wagner, T.: Learning space partitions for nearest neighbor search. In: 8th International Conference on Learning Representations, ICLR, Addis Ababa, Ethiopia, 26–30 April 2020 (2020)
4. Galakatos, A., Markovitch, M., Binnig, C., Fonseca, R., Kraska, T.: FITing-tree: a data-aware index structure. In: Proceedings of the International Conference on Management of Data (SIGMOD), pp. 1189–1206. ACM (2019)
5. Hünemörder, M., Kröger, P., Renz, M.: Towards a learned index structure for approximate nearest neighbor search query processing. In: Reyes, N., et al. (eds.) SISAP 2021. LNCS, vol. 13058, pp. 95–103. Springer, Cham (2021). https://doi.org/10.1007/978-3-030-89657-7_8
6. Johnson, J., Douze, M., Jégou, H.: Billion-scale similarity search with GPUs. IEEE Trans. Big Data **7**(3), 535–547 (2019)
7. Lin, K.-I., Yang, C.: The Ann-tree: an index for efficient approximate nearest neighbor search. In: Proceedings Seventh International Conference on Database Systems for Advanced Applications. DASFAA 2001, pp. 174–181, April 2001
8. Kraska, T., et al.: SageDB: a learned database system. In: CIDR 2019, 9th Biennial Conference on Innovative Data Systems Research, Asilomar, CA, USA, 13–16 January 2019, Online Proceedings (2019). www.cidrdb.org
9. Kraska, T., Beutel, A., Chi, E.H., Dean, J., Polyzotis, N.: The case for learned index structures. In: Proceedings of the 2018 International Conference on Management of Data. SIGMOD '18, pp. 489–504. Association for Computing Machinery (2018)
10. Muja, M., Lowe, D.G.: Fast approximate nearest neighbors with automatic algorithm configuration. In: International Conference on Computer Vision Theory and Applications (VISAPP), pp. 331–340 (2009)
11. Olha, J., Slanináková, T., Gendiar, M., Antol, M., Dohnal, V.: Learned indexing in proteins: substituting complex distance calculations with embedding and clustering techniques. In: Skopal, T., Falchi, F., Lokoč, J., Sapino, M.L., Bartolini, I., Patella, M. (eds.) SISAP 2022. LNCS, vol. 13590, pp. 274–282. Springer, Cham (2022). https://doi.org/10.1007/978-3-031-17849-8_22
12. Sander, J., Ester, M., Kriegel, H.P., Xu, X.: Density-based clustering in spatial databases: the algorithm GDBScan and its applications. Data Min. Knowl. Disc. **2**(2), 169–194 (1998)
13. Schubert, E., Rousseeuw, P.J.: Faster k-medoids clustering: improving the PAM, CLARA, and CLARANS algorithms. In: Amato, G., Gennaro, C., Oria, V., Radovanović, M. (eds.) SISAP 2019. LNCS, vol. 11807, pp. 171–187. Springer, Cham (2019). https://doi.org/10.1007/978-3-030-32047-8_16
14. Slanináková, T., Antol, M., Oĺha, J., Kaňa, V., Dohnal, V.: Data-driven learned metric index: an unsupervised approach. In: Reyes, N., et al. (eds.) SISAP 2021. LNCS, vol. 13058, pp. 81–94. Springer, Cham (2021). https://doi.org/10.1007/978-3-030-89657-7_7
15. Slanináková, T., Antol, M., Olíha, J., Dohnal, V., Ladra, S., Martínez-Prieto, M.A.: Reproducible experiments with learned metric index framework. Inf. Syst. 102255 (2023). https://doi.org/10.1016/j.is.2023.102255

16. Van Der Walt, S., Colbert, S.C., Varoquaux, G.: The numpy array: a structure for efficient numerical computation. Comput. Sci. Eng. **13**(2) (2011)
17. Zhang, C., Koishida, K., Hansen, J.H.: Text-independent speaker verification based on triplet convolutional neural network embeddings. IEEE/ACM Trans. Audio Speech Lang. Process. **26**(9), 1633–1644 (2018)

Computational Enhancements of HNSW Targeted to Very Large Datasets

Cole Foster[(✉)] and Benjamin Kimia

Brown University, Providence 02912, USA
{cole_foster,benjamin_kimia}@brown.edu

Abstract. The Hierarchical Navigable Small World (HNSW) Graph is
a graph-based approximate similarity search algorithm that achieves fast
and accurate search through a hierarchical structure providing long-range
and short-range links. The HNSW remains as a state-of-the-art method,
as shown by this submission to the SISAP 2023 Indexing Challenge. This
submission introduces a modification to the implementation of HNSW
that avoids the cost of a batched construction and drastically reduces
disk-space of the saved index when working with large datasets. Through
the lens of this competition, this work provides a careful analysis of several important factors for high-performance applications, including the
removal of unnecessary functionality, the use of SIMD vectorization for
distance computations, and optimal utilization of cache through spatial
locality and cache prefetching.

Keywords: similarity search · large datasets · memory efficiency

1 Introduction

The navigation between two distant locations by car requires the traversal of the
road network, a scalable system of roads that enables both long-range and short-
range travel. This type of network has the "small world" property [7], *i.e.*, the
path between two nodes scales logarithmically with network size. Small world
networks are found many places in nature, often in relation to navigation and
social networks [2].

The small world property has proven useful in a variety of domains, including
the field of similarity search, where the goal is to search a dataset to return
objects similar to a given query object. Graph-based methods of similarity search
first construct a graph on the elements of the dataset and traverse the edges of
this graph to approach the local neighborhood of the query. The Navigable Small
World (NSW) [10] Graph, introduced in 2014, is an approximate graph-based
method that provides the small-world property by having both long-range and
short range-links. This provides excellent navigability, but the presence of both
types of links increases the average degree of the graph and reduces its efficiency
at search time.

A few years later, the Hierarchical Navigable Small World (HNSW) [9] Graph
was proposed as an improvement to the NSW by leveraging a hierarchical

O. Pedreira and V. Estivill-Castro (Eds.): SISAP 2023, LNCS 14289, pp. 291–299, 2023.
https://doi.org/10.1007/978-3-031-46994-7_25

Fig. 1. A conceptual understanding of the hierarchical of the HNSW [9] through the analogy of a road trip. The upper levels of the HNSW contain long-range links, analogous to highways or freeways, while the bottom level has short-range links similar to local streets. These satellite images of Providence, RI, USA are captured by Google Earth [5].

organization of the graph. The key intuition of HNSW is to use the layers to separate links based on length: returning to the road network analogy, the upper layers of the hierarchy contain a sparse sampling of the dataset which provide long-range links, analogous to high-volume roads like highways or freeways, and is used for long-distance traversal on the graph. Conversely, the bottom-layer of the hierarchy contains short-range links, similar to local streets, since it is dense with all members of the dataset, Fig. 1. This separation of links by length is intuitive: when traveling a long distance, one starts on high-speed roads like major highways, and once closer to the destination, more fine-grained roads like local streets can be used.

SISAP Indexing Challenge: The Similarity Search and Applications (SISAP) 2023 Indexing Challenge [13] features a competition to index 100 million 768D vector embeddings of the LAION-5B dataset [12]. The goal of this competition is to have the fastest time to perform kNN search (k=10) on 10,000 queries with over 90% recall. This challenge further specifies the constraints of a 512GB RAM and 24 h wall time limit.

Overview: Despite the introduction of several new algorithms [4,14] in the past few years, the HNSW remains a state-of-the-art algorithm in similarity search as evident through this submission to the SISAP Indexing Challenge. Our submission features modifications to the *hnswlib*[1] implementation of HNSW, namely, modification of internal storage of the dataset for improved memory efficiency with large datasets and the removal of unnecessary functionality for optimal efficiency. There are several public implementations of the HNSW: the one we use, *hnswlib*, is a lightweight, header-only library written in C++ while *faiss*[2] is a part of Facebook's collection of different indexing methods. It has been

---

[1] https://github.com/nmslib/hnswlib.git.
[2] https://github.com/facebookresearch/faiss.git.

observed that *hnswlib* is faster than *faiss* implementation[3] [1,8] and this is also highlighted here. This work also provides a careful analysis of several important concepts for high-performance applications, including the use of vectorization in high-dimensional distance computations and the importance of optimizing cache performance. Finally, we provide an analysis of the the impact of hyper-parameters on the performance of HNSW. The results of the competition show that our submission[4] achieved the highest throughput (queries per second) with over 90% recall on all subsets (10M, 30M, and 100M) of the LAION dataset.

2 Memory Efficiency for Large Datasets

The 768D embeddings of the 100 million text-image pairs of the LAION dataset [12], produced by CLIP [11], result in a large, high-dimensional dataset for this challenge. Although these embeddings are stored as half-precision (16-bit) floating point numbers, most existing algorithms rely on single-precision (32-bit), thus requiring 3KB of memory to store each embedding. In total, this dataset requires 143.05GB of memory in its half-precision form and 286.10GB in its single-precision form. Thus, proper memory use is an important point of concern in this competition, as it is when working with any large dataset.

The *hnswlib* library of HNSW is designed with cache efficiency in mind. Throughout the search procedure, the HNSW algorithm will simultaneously tra verse the edges of the graph and compute distances to nodes in the graph. The *hnswlib* implementation improves *spatial locality* [3] by storing the vector representation of each point alongside its neighborhood information in one large, contiguous block of memory. This memory organization improves the number of "cache hits" which in turn provides better memory transfer efficiency.

On the other hand, the *faiss* implementation stores the vector representations separately from the graph structure to provide compatibility with the rest of the library, but this comes at the cost of more "cache misses". It is important to note that both implementations store their own copy of the dataset. For this reason, users must be careful, particularly with large datasets, to avoid the duplication of the entire dataset: two copies of the LAION dataset, in single precision, would take up 572.20GB of RAM which surpasses the 512GB limit of the competition.

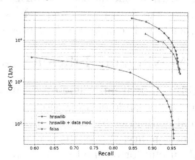

Fig. 2. Recall vs. throughput (queries per second) on LAION 100M subset, up and to the right is better. Our submission features a memory-efficient modification to *hnswlib* that results in a loss of search speed.

---

[3] https://ann-benchmarks.com/.
[4] https://github.com/cole-foster/sisap-2023.git.

To avoid this duplication, one approach is to construct the index in batches. Specifically, the original dataset can remain in disk memory and one batch at a time can be loaded to the RAM for its incremental addition to the index. However, this batched approach, requiring costly disk to RAM transfers, increases the overall construction using *hnswlib* took 13.85 h while the batched approach required 15.46 h using batches of 100,000. Note that all results of this paper, unless otherwise specified, use the HNSW hyperparameters of $M = 20$ and *efConstruction*=800 and are evaluated on a workstation with a 32-Core Intel(R) Xeon(R) Gold 6242 CPU@2.80GHz and 756GB of RAM.

Rather than incurring the additional construction time costs of the batched approach, our submission opts to load the full dataset into RAM and modify the *hnswlib* implementation to use a pointer to this memory location. By using the original memory location, our submission avoids the increased construction cost of the batched approach and further avoids copying each vector representation throughout the incremental construction. With these changes, our submitted approach required 13.18 h for construction, slightly less than the *hnswlib* implementation in both forms. However, this benefit comes at the detriment of losing spatial locality, which is evident by a drop in search speed, Fig. 2. Note that this modification to *hnswlib* is still much faster than the *faiss* implementation, suggesting spatial locality alone does not explain the major difference in performance between *hnswlib* and *faiss*.

Another significant benefit of this memory modification is that our submitted index can be serialized *without* the dataset, drastically reducing the disk memory usage. Specifically, the *faiss* and *hnswlib* implementations must save the entire, single-precision dataset (286.10GB) within their index, requiring 309.00GB and 309.40GB, respectively. On the other hand, our submitted approach stores the index without the dataset, requiring only 17.16GB of disk space. Of course, this assumes the dataset is separately stored on the disk, and in the case of this competition, it can be the half-precision version that only requires 143.05GB.

3 Maximizing Efficiency at Online Search Time

In this competition, where every microsecond matters, it is imperative to investigate all possible optimizations of the HNSW algorithm. Often times, publicly available libraries focus on compatibility, ease of use, and offering a wide range of functionality. While this is beneficial to the larger community, it is not necessarily useful for optimal performance.

Speed Improvement by Discarding Unnecessary Functionality: The first, most simple improvement that can be made is to remove unnecessary functionality from the implementation, analogous to "trimming the fat". Building on the memory modification, our submission removes the search-time functionality to (*i*) avoid deleted nodes, (*ii*) filter out specific nodes, and (*iii*) record statistical information. The cost of this functionality is small in comparison to the cost of distance computations and memory transfer, and thus provides a small, but visible impact on efficiency, Fig. 3(a).

Importance of Vectorization in Distance Computations: Distance computations, particularly for high-dimensional vectors, contain a large number of repetitive, non-sequential instructions that can be performed in parallel. While CPU-based systems only have a few cores, and in the case of HNSW they are all handling different queries, it is possible to use vectorization on a single core through Single Instruction/Multiple Data (SIMD) instructions [6]. Figure 3(b) shows the comparative performance of different types of SIMD instructions used for distance computations during the search procedure. Note that *hnswlib* contains functionality to use AVX512 instructions when available, while the *faiss* implementation only supports AVX2 instructions, which may partially explain the significant performance difference between the two libraries [8].

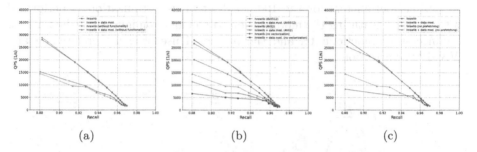

(a)   (b)   (c)

Fig. 3. Observing the comparative impact of (a) unnecessary functionality, (b) different types of SIMD vectorization for distance computations, and (c) cache prefetching on the 100M subset. The differences between "hnswlib" and "hnswlib + data modification" serve as an observation on the importance of spatial locality.

Critical Nature of Caching: The memory configuration of the *hnswlib* implementation provides efficient memory access by increasing "cache hits" with spatial locality. However, our modification the memory configuration separates the vector representations from the graph information, increasing the likelihood of "cache misses". The comparative performance of *hnswlib* with and without this memory modification in Fig. 3 emphasizes impact of *spatial locality* on search speed. Another way to improve cache performance is through *cache prefetching* [3], a technique where developers hint towards memory locations that will be used in the near future, allowing the memory to be transferred into the cache before it is used. This concept is fully utilized by *hnswlib*, which further optimizes the cache utilization of the approach. To understand the benefit of cache prefetching, observe in Fig. 3(c) that disabling cache prefetching has a significant impact on performance.

Admission of Error: While analyzing the final results of the competition, it has come to our attention that our submission inadvertently had cache prefetching disabled. Thus, the performance of our submission is even further improved by correctly enabling the cache prefetching functionality, as suggested by Fig. 3(c).

4   Hyperparameter Selection for HNSW

The HNSW algorithm has two hyper-parameters to select: M, which controls the maximum degree of the graph, and *efConstruction*, which determines the size of the candidate list used to choose neighbors during construction. With only two hyperparameters, we perform a simple grid search on the 10M subset to choose the optimal parameters for the 100M subset. This optimization is performed on the *hnswlib* featuring the modification to memory configuration, the removal of unnecessary functionality, and the preserved use of cache prefetching and AVX512 vectorization.

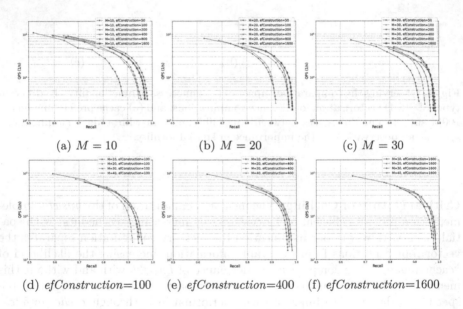

    (a) $M = 10$ (b) $M = 20$ (c) $M = 30$

(d) *efConstruction*=100   (e) *efConstruction*=400   (f) *efConstruction*=1600

Fig. 4. The impact of the hyperparameters M and *efConstruction* on the performance of our submission on the 10M subset.

To visualize the impact of M and *efConstruction* on performance, Fig. 4 fixes either of the two hyper-parameters and observes the impact of varying the other. Table 1 shows the grid search optimization based on the target of the competition, which is the maximum queries-per-second (QPS) with 90% recall. It is clear that *efConstruction* provides the greatest impact on performance yet heavily influences the construction time, Table 2. Thus, *efConstruction* should be chosen as the largest time afforded by the time limit. Surprisingly, the parameter M has a low impact on performance, provided that *efConstruction* is greater than 200 and M is greater than 5.

Table 1. The impact of hyperparameters on the maximum queries-per-second with at least 90% recall. Best viewed in color.

efConstruction	M=5	M=10	M=15	M=20	M=25	M=30	M=35	M=40
50	N/A	N/A	4,124.80	8,653.46	5,428.77	7,384.09	11,732.78	4,189.05
100	N/A	17,988.60	25,071.56	10,704.81	21,138.30	21,341.14	26,023.84	19,719.42
200	12,801.05	17,526.32	18,700.81	21,916.05	24,860.16	31,302.62	28,288.30	29,526.87
400	12,004.72	28,581.44	29,196.84	23,602.70	29,992.69	24,881.28	27,959.25	25,870.50
800	14,026.79	28,208.72	26,152.74	30,832.04	28,293.23	21,933.03	32,422.35	33,836.27
1600	14,529.17	27,290.25	24,844.08	31,178.00	26,026.81	27,882.65	26,929.86	31,024.27

Table 2. The impact of hyperparameters on construction time (minutes) for the 10M subset. Best viewed in color.

efConstruction	M=5	M=10	M=15	M=20	M=25	M=30	M=35	M=40
50	3.43	4.40	5.50	5.46	5.73	6.07	5.79	6.05
100	12.72	8.12	10.29	10.80	11.26	11.23	12.72	12.10
200	9.84	14.94	17.32	23.37	21.49	25.95	28.25	25.02
400	17.76	31.16	33.81	38.41	43.22	49.39	48.04	48.76
800	28.66	48.56	64.16	78.82	88.11	98.26	95.77	90.99
1600	53.73	61.44	147.09	145.01	161.95	157.92	219.37	182.25

5   Results of the Competition

The final results of the SISAP 2023 Indexing Challenge are shown in Fig. 5, which features the Recall vs. Queries-Per-Second (QPS) trade-off for each submission on the three subsets (10M, 30M, 100M). All submissions were evaluated on the same workstation featuring a 28-Core Intel(R) Xeon(R) CPU E5-2690 V4 with 512GB of RAM. Our submission featured the *hnswlib* implementation with the memory modification and the removal of unnecessary functionality. The inner product was used as a similarity measure, the selected hyperparameters were $M = 20$ and *efConstruction*=800, and the search parameter *ef* was varied from 10 to 1000 to obtain the recall vs. queries-per-second curve. Our submission performed the best in all three subsets by achieving the highest throughput with over 90% recall, besting the *faiss* implementation (faissHNSW), the Search Graph [14] (SearchGraph), the Navigable Spreading-Out Graph [4] (NSG32), and several others.

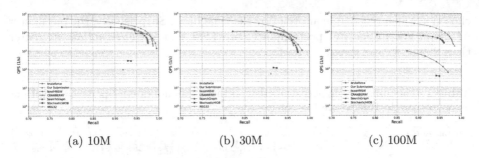

(a) 10M   (b) 30M   (c) 100M

Fig. 5. The final results of the competition, up and to the right is better. Our submission, in orange, performed the best in each subset by achieving the greatest queries-per-second with a recall over 90%. (Color figure online)

6   Conclusion

This submission to the SISAP 2023 Indexing Challenge [13] proves that the HNSW [9] is still a state-of-the-art method for approximate similarity search, especially with large datasets. This approach modified the memory configuration of the *hnswlib* implementation of HNSW to avoid the increased time of a batched construction and drastically reduce disk space upon serialization. Our work demonstrates that the *hnswlib* implementation of HNSW performs much better than the *faiss* implementation, which is made clear by its order-of-magnitude improvement on the 100M subset. Finally, this work highlights the importance several factors for fast searching, including the removal of unnecessary functionality, the utilization of SIMD vectorization in high-dimensional distance computations, and the optimization of cache efficiency.

Acknowledgement. We gratefully acknowledge the support of NIH Award 1S10OD025181 and NSF award 1910530.

References

1. Aumüller, M., Bernhardsson, E., Faithfull, A.: ANN-benchmarks: a benchmarking tool for approximate nearest neighbor algorithms. Inf. Syst. **87**, 101374 (2020)
2. Boguna, M., Krioukov, D., Claffy, K.C.: Navigability of complex networks. Nat. Phys. **5**(1), 74–80 (2009)
3. Ericson, C.: Memory Optimization. Santa Monica, Sony Computer Entertainment (2003)
4. Fu, C., Xiang, C., Wang, C., Cai, D.: Fast approximate nearest neighbor search with the navigating spreading-out graph. Proc. VLDB Endowment **12**(5), 461–474 (2019)
5. Google: Google Earth (2023) Brown University 41°39'36"N, 71°24'08"W (2023)
6. Intel: Intel® 64 and IA-32 Architectures Software Developer's Manual. www.intel.com/content/www/us/en/developer/articles/technical/intel-sdm.html (2023)
7. Kleinberg, J.M.: Navigation in a small world. Nature **406**(6798), 845–845 (2000)

8. Liu, T.Z.: Kids! Use hnswlib for HNSW. https://terencezl.github.io/blog/2022/09/28/kids-use-hnswlib/ (2022). Accessed 25 Aug 2023
9. Malkov, Y.A., Yashunin, D.A.: Efficient and robust approximate nearest neighbor search using hierarchical navigable small world graphs. IEEE Trans. Pattern Anal. Mach. Intell. **42**(4), 824–836 (2018)
10. Malkov, Y., Ponomarenko, A., Logvinov, A., Krylov, V.: Approximate nearest neighbor algorithm based on navigable small world graphs. Inf. Syst. **45**, 61–68 (2014)
11. Radford, A., et al.: Learning transferable visual models from natural language supervision. In: International Conference on Machine Learning, pp. 8748–8763. PMLR (2021)
12. Schuhmann, C., et al.: LAION-5B: an open large-scale dataset for training next generation image-text models. Adv. Neural. Inf. Process. Syst. **35**, 25278–25294 (2022)
13. Tellez, E.S., Aumüller, M., Chavez, E.: Overview of the SISAP 2023 indexing challenges. In: Similarity Search and Applications: 16th International Conference, SISAP 2023, A Coruña Spain, October 9–11, Proceedings. Springer (2023)
14. Tellez, E.S., Ruiz, G., Chavez, E., Graff, M.: A scalable solution to the nearest neighbor search problem through local-search methods on neighbor graphs. Pattern Anal. Appl. **24**(2), 763–777 (2021)

CRANBERRY: Memory-Effective Search in 100M High-Dimensional CLIP Vectors

Vladimir Mic[1(✉)] ⓘ, Jan Sedmidubsky[2] ⓘ, and Pavel Zezula[2] ⓘ

[1] Department of Computer Science, Aarhus University, Aarhus, Denmark
v.mic@cs.au.dk

[2] Faculty of Informatics, Masaryk University, Brno, Czech Republic

Abstract. Recent advances in cross-modal multimedia data analysis necessarily require efficient similarity search on the scales of hundreds of millions of high-dimensional vectors. We address this task by proposing the CRANBERRY algorithm that specifically combines and tunes several existing similarity search strategies. In particular, the algorithm: (1) employs the Voronoi partitioning to obtain a query-relevant candidate set in constant time, (2) applies filtering techniques to prune the obtained candidates significantly, and (3) re-rank the retained candidate vectors with respect to the query vector. Applied to the dataset of 100 million 768-dimensional vectors, the algorithm evaluates 10NN queries with 90 % recall and query latency of 1.2 s on average, all with a throughput of 15 queries per second on a server with 56 core-CPU, and 4.7 q/sec. on a PC.

Keywords: approximate similarity searching · high-dimensional data · indexing · filtering · LAION dataset

1 Introduction

The recent boom in cross-modal multimedia analysis, such as text-to-image and text-to-video retrieval, has attracted much attention from the IR community. For example, the LAION-5B dataset[1] offers several billions of images and associated text annotations, both of the modalities represented by high-dimensional vectors. We target the SISAP 2023 Indexing Challenge[2] [8] (shortly Challenge) that serves as a competition benchmark for evaluating k-nearest neighbour (kNN) similarity queries over 100M 768-dimensional CLIP vectors extracted from a subset of the LAION dataset. Formally, we assume domain D of the searched vectors and the cosine distance function $d : D \times D \mapsto \mathbf{R}_0^+$ that quantifies the dissimilarity of two vectors. We focus on an efficient evaluation of kNN(q) queries where $q \in D$ is a query vector and $k \in \mathbf{N}$: having a dataset $X \subseteq D$ and $q \in D$, kNN(q) query searches for the k most similar vectors $o \in X$ to q. The Challenge addresses the evaluation of 10NN queries in the collections of 10M, 30M, and 100M vectors.

This work was supported by a research grant (VIL50110) from VILLUM FONDEN.

[1] https://laion.ai/blog/laion-5b/.
[2] https://sisap-challenges.github.io/.

Disregarding a target hardware infrastructure, we generally expect that dataset X is maintained in the secondary storage (e.g., SSD or HDD) since very large volumes of vectors can hardly fit into the main memory. Therefore, our main objective is to minimize secondary-storage accesses, which we consider the main bottleneck. Based on these assumptions, we aim at utilising main memory to keep only IDs and compact representations of all dataset vectors, accompanied by additional space characteristics. To achieve this, we propose the CRANBERRY algorithm (similarity searChing using voRonoi pArtitioNing, Binary skEtches and Relational similaRitY), which constitutes a unique combination of retrieval techniques that have been treated in the past only separately.

2   The CRANBERRY Algorithm

The CRANBERRY algorithm consists of the following steps that combine several searching techniques specifically connected into a single pipeline.

- **Data partitioning.** We adopt the Voronoi partitioning to decompose dataset X of 768D vectors into non-overlapping partitions (Voronoi cells), each of them consisting of a set of similar vectors. In the retrieval phase, the nearest partitions to the query vector q are quickly identified to return a candidate set of IDs of vectors that are potentially similar to the query vector.
- **Filtering candidates.** We adopt two filtering techniques to further prune the set of candidates obtained from the previous step. In particular, we learn and extract two types of representations from original vectors: 512bit sketches and 24D prefixes of vectors shortened by the PCA to 256D. In the retrieval phase, the sketch and PCA filters accompanied by an early termination strategy are applied to return a very small set of candidates' IDs.
- **Refining candidates.** The remaining IDs from the previous step are used to read original 768D representations from data storage. Such original vectors are compared against the 768D query vector q using the cosine distance to obtain a final ranking.

If the dataset X is stored in the secondary storage, the refinement is the only step that reads data from this storage. Information needed for the data partitioning and filtering is stored in the main memory. The whole process is schematically illustrated in Fig. 1 and discussed in the following sections in more detail. The implementation for reproducibility of the results, along with other implementation details, is available at the GitHub repository: https://github.com/xsedmid/sisap23-laion-challenge-CRANBERRY.

2.1   Data Partitioning

The only data partitioning we use is the one-layer Voronoi partitioning [10] with the set $P \subseteq D$ of 20,000 reference vectors called *pivots*, selected at random. The Voronoi partitioning assigns each vector $o \in X$ to the closest partition cl_{pmin} given by $pmin \in P$ such that $d(o, pmin) \leq d(o, p)$ for each $p \in P$.

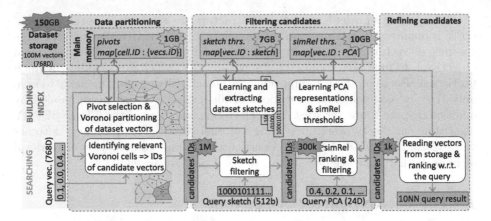

Fig. 1. A schematic illustration of indexing and searching phases of the CRANBERRY on 100M dataset of 768-dimensional CLIP vectors

When a query vector q comes, all 20,000 distances $d(q, p), p \in P$ are evaluated to identify the nearest pivots to q. The first-level candidate set is made of IDs of vectors from the partitions with the closest pivots to q, and the number of partitions is given to create the candidate set with at most $\mathcal{V}$ IDs. In other words, adding IDs from the next partition would increase the number of candidates over threshold $\mathcal{V}$. Threshold $\mathcal{V}$ is set automatically using empirical knowledge: We observed suitable $\mathcal{V}$ values 100,000; 200,000; 400,000; and 1 million for the datasets of size 300k, 10M, 30M, and 100M, respectively. Datasets of different sizes use $\mathcal{V}$ given by linear interpolation of these known thresholds $\mathcal{V}$. Identification of candidate vectors with the Voronoi partitioning is very efficient since it uses only the main-memory map associating each partition ID with assigned IDs of vectors.

2.2   Filtering Candidates

CRANBERRY algorithm filters candidates using a sketch filter, a PCA-based filter, and an early termination strategy. Applied filters require storing just the IDs and compact representations of the original vectors in the main memory. This leads to an order of magnitude less space requirement compared to the original 768D representations. We describe these filters in more detail in the following.

Learning Binary Sketches. We transform dataset X to sketches $sk(o), o \in X$ by technique *GHP_50_512* [3,4] learned before the index build. *GHP_50_512* produces sketches with 512 bits using 512 instances of a *generalised hyperplane partitioning* (GHP) [10]. Each bit i of sketches $sk(o), o \in X$ is defined by pivots p_{i0}, p_{i1} and denotes which of them is closer to o according to $d(o, p_{i0}), d(o, p_{i1})$. All pivots come from 20,000 pivots used by the Voronoi partitioning and define sketches with approximately balanced and low-correlated bits [4].

Sketch Filtering. To further prune the candidate set, we employ the so-called *Secondary filtering with sketches* [5]. This filtering is applicable just together with the sketching technique that we use, regardless length of sketches.

The technique of the Secondary filtering with sketches defines a mapping of the Hamming distances $h(sk(o_1), sk(o_2))$ to the cosine distances $d(o_1, o_2)$. Having the searching radius $x = d(q, o_k)$ given by the distance to the kth nearest neighbour found so far, the mapping is utilised to define the search radius in the space of sketches, i.e., the threshold on the Hamming distance. This threshold defines vectors o to be filtered out due to a high Hamming distance $h(sk(q), sk(o))$. We emphasise that (1) this filtering is dynamic, i.e., the threshold on the Hamming distance decreases with decreasing search radius $x = d(q, o_k)$ during the query execution, and (2) it takes into account differences between particular query vectors q, i.e., their local intrinsic dimensionality [1].

The CRANBERRY algorithm sorts the candidates o identified by the Voronoi partitioning according to the Hamming distances $h(sk(q), sk(o))$. This sorting is done in parallel due to a possibly high number of candidates. Having the candidates sorted according to the Hamming distances of sketches, the mapping of distances defines a direct stop condition for the candidate set filtering.

Relational Similarity (simRel) Filtering. If candidate $o \in X$ goes through the sketch filter, it is checked by the *relational similarity function simRel* [6,7]. The simRel compares triplets of vectors with the following semantics:

$$simRel(q, o_1, o_2) = \begin{cases} 1 & q, o_1 \text{ are more similar to each other than } q, o_2 \\ 2 & q, o_1 \text{ are less similar to each other than } q, o_2 \\ 0 & \text{both similarities are the same or the difference in the} \\ & \text{similarities is as small as its proper investigation does} \\ & \text{not pay off, and they can be treated arbitrarily} \end{cases}$$

The CRANBERRY uses the *simRel* implementation from [6] which is applicable to the Euclidean and Cosine spaces. It shrinks vectors $o \in X$ by the Principal Component Analysis (PCA) to vectors $o^{PCA(L)}$ of length $L = 256$ and tries to estimate which of $o_1^{PCA(L)}, o_2^{PCA(L)}, o_1, o_2 \in X$ is more similar to $q^{PCA(L)}$ using the first 24 dimensions of shrunk vectors. Since the variance of values in coordinates of $o^{PCA(L)}$ decreases with increasing index of coordinate, the prefixes of shrunk vectors are often sufficient to correctly estimate the $simRel(q, o_1, o_2)$ result.

The *simRel* filtering proposed in [6] splits the filtered set into three parts: (1) highly similar vectors to q, (2) vectors with unknown relation to q, and (3) dissimilar vectors to q. If a given vector has an unknown relation to q, it is immediately refined. The set of similar vectors to q according to the *simRel* are being improved during the *simRel* filtering, so they are refined just occasionally to shrink the search radius $x = d(q, o_k)$ and thus possibly shrink the threshold on the Hamming distance $h(sk(q), sk(o))$, i.e., increase the power of the Secondary filtering with sketches for the remaining candidates.

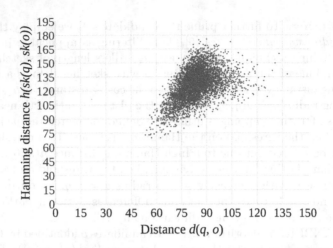

Fig. 2. Example of $d(q,o)$ and corresponding Hamming distances $h(sk(q), sk(o))$ for a given $q \in D$. Adopted illustration [2] made for 256bit sketches

Early Termination Strategy. The early termination of the query execution is essential for the CRANBERRY's efficiency. Figure 2 illustrates[3] that for a given $q \in D$, Hamming distances $h(sk(q), sk(o))$, $o \in X$ generally grow with distances $d(q,o)$ and the variance of the Hamming distances decreases with decreasing distance $d(q,o)$ – notice the almost triangular shape of the plotted area. This is important for the early termination strategy.

The CRANBERRY algorithm limits the number of candidates refined during the query execution. If the threshold is achieved, the filtering with sketches is interrupted and just the remaining vectors defined by the *simRel* as similar to q are refined. Therefore, the threshold on the refined vectors is soft and we use value 800 in the case of all datasets.

2.3 Refining Candidates

IDs of vectors that remain after the filtering are used to read the original 768D vectors from data storage. If the dataset is kept in secondary storage, this can be one of the most expensive operations. The cosine distance $d(q,o)$ between the query q and each read vector o is calculated to return the 10 nearest neighbours. The resulting query answer is approximate in a general case.

3 Experiments

We evaluate 10NN queries with 10,000 query vectors on 3 subsets of the LAION2B dataset. The 10M, 30M and 100M datasets contain 10,109,960 vectors;

---

[3] This figure is adopted from [2] and reports sketches of length 256 bits made for a different dataset. Illustrated properties are independent of specific data.

30,369,256 vectors; and 102,041,055 vectors, respectively[4]. The query vectors are not in the datasets. Experiments ran on 2 computers. Computer A is a PC with the Intel Core i9-9900K processor, NVMe SSD 512 GB, and 96 GB DDR4 RAM. Server B has the Intel Xeon E5-2690 v4 processor with 56 cores, and 512 GB DDR4 RAM.

Table 1. CRANBERRY: Averages of 10NN search with 10,000 query vectors

Dataset	Refined vectors	Recall	Computer A – NVMe SSD			Server B – RAM		
			Latency	Throughput	Σ time	Latency	Throughput	Σ time
100M	821	0.901	1.07 s	4.7 q/s	2,139 s	1.20 s	15.0 q/s	669 s
30M	855	0.902	0.40 s	12.6 q/s	796 s	0.37 s	48.7 q/s	205 s
10M	876	0.909	0.16 s	30.9 q/s	324 s	0.20 s	88.0 q/s	114 s

Table 2. Decreasing recall (the last column) with the increasing number of refined vectors (the first column), search in the 100M dataset

# refined $o \in X$ out of 100M	# q	percentage of queries with the recall							Avg. recall	Avg. dist of q to true 10NN
		1	0.9	0.8	0.7	0.6 – 0.2	0.1	0		
10	126	78 *					2*	21*	0.779*	0.001
11 – 50	158	87	9	2	1	1			0.978	0.019
51 – 200	152	80	14	4	1			1	0.967	0.032
201 – 800	226	81	17		1	< 0.5			0.976	0.046
803	4,097	64	23	7	3	3	< 0.5	< 0.5	0.937	0.123
804 – 901	456	67	21	6	3	3			0.943	0.148
902	609	62	23	7	4	4	< 0.5		0.929	0.153
903 – 910	1,795	57	21	9	6	7	< 0.5	< 0.5	0.905	0.182
911 – 930	1,332	43	22	13	8	14	< 0.5	< 0.5	0.855	0.230
931 – 1,000	909	29	19	17	9	25	1	< 0.5	0.774	0.296
1,001 1,100	125	20	13	9	8	46	2	2	0.642	0.359
1,101 – 1,390	15	13	7	7	20	53			0.647	0.335
Average: 821	10,000	57	21	8.5	4.8	7.8	0.2	0.4	0.901	0.165

3.1   Results of the CRANBERRY

Table 1 summarises the average search statistics. The CRANBERRY filters out 99.9992 % of the 100M dataset so just 821 vectors out of 102,041,055 refined on average. These 821 vectors contain at least 9 out of 10 NN per average query – the average recall is 0.901. The filtering power enables efficient search even with the dataset stored in the secondary storage. The query throughput is 4.7 queries per second on Computer A where the 100M dataset is on the NVMe SSD and

---

[4] https://sisap-challenges.github.io/datasets/.

15 queries per second on Server B where the dataset is in the main memory. The query throughput on the smaller datasets grows up to 88 queries per second in the case of Server B and the 10M dataset.

Table 2 illustrates varying numbers of refined vectors when querying the 100M dataset. The first and last columns illustrate decreasing filtering power when a distance between q and its 10th NN increases. The CRANBERRY thus adapts to the difficulty of queries, often formalised by the local intrinsic dimensionality [1]. Most of the queries with the 10th NN in a very small distance need to refine just 10 $o \in X$ thanks to the Secondary Filtering with Sketches [5] – see the first row in Table 2. If a query is evaluated efficiently, users should expect an answer of above-average quality, otherwise, the engine should have spent more resources on the evaluation. The first row of Table 2 shows that 78 %, i.e., 98 out of 126 queries that need to refine just 10 $o \in X$ are evaluated precisely, 2 %, i.e., 2 out of 126 searches find just 1 out of 10 NN, and 21 %, i.e., 26 out of 126 find none of 10 NN. However, we evaluate the recall as the size of the intersection of the returned answer with the precise answer (of size 10) divided by 10. Therefore, if the true 10th NN of q is in the same distance as the 11th NN, the recall can decrease by chance. Some of our query vectors have thousands of NNs in the same distance [9]. Luckily, these nearest neighbours are usually within very small distances to q which increases the filtering power of the Secondary Filtering by Sketches up to the maximum, i.e., the CRANBERRY refines just 10 vectors during the query execution, and the recall is 0 even though the error on distance [10] is 0 as well. We confirm that all 156 queries that need to refine less than 15 vectors $o \in X$ have zero error on distance.

The rest of Table 2 illustrates that the recall generally decreases when the number of refined vectors increases. For instance, 15 queries with the most refined vectors, i.e., from 1,100 to 1,390, have an average recall of just 0.647. The last row clarifies that 57 % of queries are evaluated precisely, the average number of the refined vectors is 821, and the average recall is 0.901. The parallel brute force evaluation of all 10,000 queries with just one reading of the dataset from an SSD takes approximately 155,443 s i.e., $\approx$ 43 h using Computer A.

Table 3. Contributions of the CRANBERRY's steps to the filtering power and filtering error, search in the 100M dataset

	Voronoi P.		Voronoi P. + Sketches		Voronoi P. + Sketches + simRel		all + early term. = CRANBERRY	
	Refined o	Recall	Refined o	Recall	Refined o	Recall	Refined o	Recall
Min	841,743	0	10	0	10	0	10	0
1st quartile	987,139	1.0	35,128	1.0	26,407	0.9	803	0.9
Median	993,611	1.0	192,890	1.0	75,631	1.0	836	1.0
3rd quartile	997,394	1.0	616,322	1.0	145,882	1.0	909	1.0
Max	1,000,000	1.0	1,000,000	1.0	627,375	1.0	1,390	1.0
Average	989,724	0.979	336,621	0.966	98,892	0.951	821	0.901

3.2   Contribution of Particular Techniques

Table 3 illustrates contributions of CRANBERRY's steps to the filtering power and decrease of the recall – averages are evaluated over 10,000 query vectors on the 100M dataset. The Voronoi partitioning identifies almost 1M vectors. Just 336,621 of them remain after the secondary filtering with sketches, and 98,892 remain after the filtering with the relational similarity. The early termination of query executions leads to the recall 0.901 and 821 refined vectors. The recall decreases from 0.979 after the Voronoi partitioning to 0.966 after the sketch filtering, and to 0.951 after the *simRel* filtering. Experiments also confirmed that the number of the *simRel* computations grows over-linearly with respect to the filtered set size [7]. This clarifies why the *simRel* is used as the last filter in the CRANBERRY. We have also evaluated an ablation study by varying selected hyper-parameters. The detailed description and evaluation are available online at https://github.com/xsedmid/sisap23-laion-challenge-CRANBERRY.

4   Conclusions

The proposed CRANBERRY algorithm applied to the 100M dataset of 768-dimensional vectors is capable of extreme space pruning by accessing only 0.0008 % of the original vectors while reaching the recall higher than 90 %. This is achieved by a convenient pipeline of the Voronoi partitioning, filtering techniques based on learned bit-sketches plus PCA representations, early termination strategy, and final refinement. The adopted techniques could not have achieved such results when treated separately, even with additional tuning of parameters. The great advantage of the proposed approach is also its main-memory space complexity requiring roughly an order of magnitude less space compared to the original dataset representation. This enables the indexing of datasets bigger by an order of magnitude in comparison with the approaches keeping the dataset in the main memory such as FAISS[5]. In the future, we would like to target larger datasets by utilising more powerful space partitioning instead of the Voronoi partitioning, as the follow-up sketch filtering consumes a significant portion of time due to computations and sorting of many Hamming distances. We would also like to investigate batch execution of queries as the current implementation supports only one-by-one query processing.

References

1. Amsaleg, L., et al.: Estimating local intrinsic dimensionality. In: Proceedings of the International Conference on Knowledge Discovery and Data Mining, pp. 29–38. KDD 2015, ACM, USA (2015)
2. Mic, V.: Binary Sketches for Similarity Search. Dissertation thesis, Masaryk University Brno (2020). https://theses.cz/id/c7kstr/. supervisor: Pavel Zezula

---

[5] https://github.com/facebookresearch/faiss.

3. Mic, V., Novak, D., Zezula, P.: Improving sketches for similarity search. In: Tenth Doctoral Workshop on Mathematical and Engineering Methods in Computer Science (MEMICS 2015), pp. 45–57 (2015)
4. Mic, V., Novak, D., Zezula, P.: Designing sketches for similarity filtering. In: 2016 IEEE 16th International Conference on Data Mining Workshops (ICDMW), pp. 655–662 (2016)
5. Mic, V., Novak, D., Zezula, P.: Binary sketches for secondary filtering. ACM Trans. Inf. Syst. **37**(1), 1:1–1:28 (2018)
6. Mic, V., Zezula, P.: Concept of relational similarity search. In: Skopal, T., Falchi, F., Lokoc, J., Sapino, M.L., Bartolini, I., Patella, M. (eds.) Similarity Search and Applications. SISAP 2022. LNCS, vol. 13590. Springer, Cham (2022). https://doi.org/10.1007/978-3-031-17849-8_8
7. Mic, V., Zezula, P.: Filtering with relational similarity (2023)
8. Tellez, E.S., Aumüller, M., Chavez, E.: Overview of the SISAP 2023 indexing challenges. In: Similarity Search and Applications: 16th International Conference, SISAP 2023, A Coruña Spain, October 9–11, Proceedings. Springer (2023)
9. Webster, R., Rabin, J., Simon, L., Jurie, F.: On the de-duplication of LAION-2B (2023)
10. Zezula, P., Amato, G., Dohnal, V., Batko, M.: Similarity Search - The Metric Space Approach, vol. 32. Springer (2006). https://doi.org/10.1007/0-387-29151-2

Author Index

O. Pedreira and V. Estivill-Castro (Eds.): SISAP 2023, LNCS 14289, pp. 309–310, 2023.
https://doi.org/10.1007/978-3-031-46994-7

Printed in the United States
by Baker & Taylor Publisher Services